Lecture Notes in Bioinformatics 4210

Edited by S. Istrail, P. Pevzner, and M. Waterman

Editorial Board: A. Apostolico   S. Brunak   M. Gelfand
T. Lengauer   S. Miyano   G. Myers   M.-F. Sagot   D. Sankoff
R. Shamir   T. Speed   M. Vingron   W. Wong

Subseries of Lecture Notes in Computer Science

T0216770

Corrado Priami (Ed.)

Computational Methods in Systems Biology

International Conference, CMSB 2006
Trento, Italy, October 18-19, 2006
Proceedings

 Springer

Series Editors

Sorin Istrail, Brown University, Providence, RI, USA
Pavel Pevzner, University of California, San Diego, CA, USA
Michael Waterman, University of Southern California, Los Angeles, CA, USA

Volume Editor

Corrado Priami
University of Trento
ICT, Dept. for Information and Communication Technology
Via Sommarive 14, 38050 Povo (TN), Italy
E-mail: priami@dit.unitn.it

Library of Congress Control Number: 2006933640

CR Subject Classification (1998): I.6, D.2.4, J.3, H.2.8, F.1.1

LNCS Sublibrary: SL 8 – Bioinformatics

ISSN   0302-9743
ISBN-10   3-540-46166-3 Springer Berlin Heidelberg New York
ISBN-13   978-3-540-46166-1 Springer Berlin Heidelberg New York

Springer is a part of Springer Science+Business Media

springer.com

© Springer-Verlag Berlin Heidelberg 2006
Printed in Germany

Typesetting: Camera-ready by author, data conversion by Scientific Publishing Services, Chennai, India
Printed on acid-free paper   SPIN: 11885191   06/3142   5 4 3 2 1 0

Preface

The CMSB (Computational Methods in Systems Biology) conference series was established in 2003 to help catalyze the convergence of modellers, physicists, mathematicians, and theoretical computer scientists from fields such as language design, concurrency theory, program verification, and molecular biologists, physicians, and neuroscientists interested in a systems-level understanding of cellular physiology and pathology.

The community of scientists becoming interested in this new field is growing rapidly as witnessed by the increasing number of submissions. This year we received 68 papers of which we accepted 22 for publication in this volume.

Luca Cardelli and David Harel gave two invited talks at the conference showing the computer science perspective in the emerging field of dynamical modelling and simulation of biological systems. Orkun Soyer gave two invited talks on the systems biology perspective.

Finally, we organized a poster session to favor discussion and cross-fertilization of different fields as we feel it essential to making interdisciplinary research grow.

July 2006   Corrado Priami

Organization

Programme Committee of CMSB 2006

Charles Auffray, CNRS (France)
Muffy Calder, University of Glasgow (UK)
Luca Cardelli, Microsoft Research Cambridge (UK)
Diego Di Bernardo, Telethon Institute of Genetics and Medicine (Italy)
David Harel, Weizmann Institute (Israel)
Monika Heiner, University of Cottbus (Germany)
Ela Hunt, University of Zurich (Switzerland)
Franois Kepes, CNRS / Epigenomics Program, Evry (France)
Marta Kwiatkowska, University of Birmingham (UK)
Cosimo Laneve, University of Bologna (Italy)
Eduardo Mendoza, LMU (Germany) and University of the Philippines-Diliman
    (Philippines)
Bud Mishra, New York University (USA)
Satoru Miyano, University of Tokyo (Japan)
Christos Ouzounis, European Bioinformatics Institute (UK)
Gordon Plotkin, University of Edinburgh (UK)
Corrado Priami, Chair, The Microsoft Research - University of Trento Centre
    for Computational and Systems Biology, Italy
Alessandro Quattrone, University of Florence (Italy)
Magali Roux-Rouqui, CNRS-UPMC (France)
David Searls, Senior Vice-President, Worldwide Bioinformatics - GlaxoSmithK-
    line (USA)
Adelinde Uhrmacher, University of Rostock (Germany)
Alfonso Valencia, Centro Nacional de Biotecnologia-CSIC (Spain)

Local Organizing Committee

Matteo Cavaliere and Elisabetta Nones - The Microsoft Research University of
Trento Centre for Computational and Systems Biology (Italy), and the University of Trento Events and Meetings Office.

List of Referees

H. Adorna, P. Adritsos, P. Amar, A. Ambesi-Impiombato, Y. Atir, P. Baldan, M.
Bansal, E. Blanzieri, L. Brodo, N. Busi, A. Casagrande, M. Cavaliere,
D. Chu, F. Ciocchetta, J.-P. Comet, R. del Rosario, G. Dellagatta, L. Demattè,
P. Degano, M.L. Guerriero, J. Hillston, A. Kaban, V. Khare, C. Kuttler,

I. Lanese, P. Lecca, G. Norman, R. Mardare, M. Miculan, P. Milazzo, V. Mysore, G. Nuel, C. Pakleza, T. Pankowski ,D. Parker, C. Piazza, A. Policriti, S. Pradalier, D. Prandi, P. Quaglia, A. Romanel, A. Sadot, S. Sedwards, Y. Setty, K. Sriram, O. Tymchyshyn, H. Wiklicky, G. Zavattaro.

Acknowledgement

The workshop was sponsored and partially supported by the Microsoft Research - University of Trento Centre for Computational and Systems Biology.

Table of Contents

Modal Logics for Brane Calculus 1
    M. Miculan, G. Bacci

Deciding Behavioural Properties in Brane Calculi 17
    N. Busi

Probabilistic Model Checking of Complex Biological Pathways 32
    J. Heath, M. Kwiatkowska, G. Norman, D. Parker,
    O. Tymchyshyn

Type Inference in Systems Biology 48
    F. Fages, S. Soliman

Stronger Computational Modelling of Signalling Pathways Using Both
Continuous and Discrete-State Methods 63
    M. Calder, A. Duguid, S. Gilmore, J. Hillston

A Formal Approach to Molecular Docking............................ 78
    D. Prandi

Feedbacks and Oscillations in the Virtual Cell VICE 93
    D. Chiarugi, M. Chinellato, P. Degano, G. Lo Brutto,
    R. Marangoni

Modelling Cellular Processes Using Membrane Systems with Peripheral
and Integral Proteins 108
    M. Cavaliere, S. Sedwards

Modelling and Analysing Genetic Networks: From Boolean Networks
to Petri Nets 127
    L.J. Steggles, R. Banks, A. Wipat

Regulatory Network Reconstruction Using Stochastic Logical
Networks 142
    B. Wilczyński, J. Tiuryn

Identifying Submodules of Cellular Regulatory Networks 155
    G. Sanguinetti, M. Rattray, N.D. Lawrence

Incorporating Time Delays into the Logical Analysis of Gene Regulatory
Networks 169
    H. Siebert, A. Bockmayr

A Computational Model for Eukaryotic Directional Sensing 184
    A. Gamba, A. de Candia, F. Cavalli, S. Di Talia, A. Coniglio,
    F. Bussolino, G. Serini

Modeling Evolutionary Dynamics of HIV Infection 196
    L. Sguanci, P. Liò, F. Bagnoli

Compositional Reachability Analysis of Genetic Networks 212
    G. Gössler

Randomization and Feedback Properties of Directed Graphs Inspired
by Gene Networks 227
    M. Cosentino Lagomarsino, P. Jona, B. Bassetti

Computational Model of a Central Pattern Generator 242
    E. Cataldo, J.H. Byrne, D.A. Baxter

Rewriting Game Theory as a Foundation for State-Based Models of
Gene Regulation 257
    C. Chettaoui, F. Delaplace, P. Lescanne, M. Vestergaard,
    R. Vestergaard

Condition Transition Analysis Reveals TF Activity Related to
Nutrient-Limitation-Specific Effects of Oxygen Presence in Yeast 271
    T.A. Knijnenburg, L.F.A. Wessels, M.J.T. Reinders

An In Silico Analogue of In Vitro Systems Used to Study Epithelial
Cell Morphogenesis... 285
    M.R. Grant, C.A. Hunt

A Numerical Aggregation Algorithm for the Enzyme-Catalyzed
Substrate Conversion 298
    H. Busch, W. Sandmann, V. Wolf

Possibilistic Approach to Biclustering: An Application to
Oligonucleotide Microarray Data Analysis........................... 312
    M. Filippone, F. Masulli, S. Rovetta, S. Mitra, H. Banka

Author Index 323

Modal Logics for Brane Calculus

Marino Miculan and Giorgio Bacci

Dept. of Mathematics and Computer Science
University of Udine, Italy
mm@uniud.it

Abstract. The Brane Calculus is a calculus of mobile processes, intended to model the transport machinery of a cell system. In this paper, we introduce the *Brane Logic*, a modal logic for expressing formally properties about systems in Brane Calculus. Similarly to previous logics for mobile ambients, Brane Logic has specific spatial and temporal modalities. Moreover, since in Brane Calculus the activity resides on membrane surfaces and not inside membranes, we need to add a specific logic (akin Hennessy-Milner's) for reasoning about membrane activity.

We present also a proof system for deriving valid sequents in Brane Logic. Finally, we present a model checker for a decidable fragment of this logic.

1 Introduction

In [4], Cardelli has proposed a schematic model of biological systems as three different and interacting abstract machines. Following the approach pioneered in [13], these abstract machines are modelled using methodologies borrowed from the theory of concurrent systems.

The most abstract of these three machines is the *membrane machine*, which focuses on the dynamics of biological membranes. At this level of abstraction, a biological system is seen as a hierarchy of compartments, which can interact by changing their position. In order to model this machinery, Cardelli has introduced the *Brane Calculus* [3], a calculus of mobile nested processes where the computational activity takes place *on* membranes, not inside them. A process of this represents a system of nested membranes; the evolution of a process corresponds to membrane interactions (phagocytosis, endo/exocytosis, ...).

Having such a formal representation of the membrane machine, a natural question is how to express formally also the *biological properties*, that is, the "statements" about a given system. Some examples are the following:

"If a macrophage is exposed to target cells that have been evenly coated with antibody, it ingests the coated cells." [1, Chap.6, p.335]
"The [...] Rous sarcoma virus [...] can transform a cell into a cancer cell." [1, Chap.8, p.417]
"The virus escapes from the endosome" [1, Chap.8, p.469]

In our opinion, it is highly desirable to be able to express formally (i.e., in a well-specified logical formalism) this kind of properties. First, this would avoid the intrinsic ambiguity of natural language, ruling out any misinterpretation of

C. Priami (Ed.): CMSB 2006, LNBI 4210, pp. 1–16, 2006.
© Springer-Verlag Berlin Heidelberg 2006

the meaning of a statement. Secondly, such a logical formalism can be used for defining *specifications of systems*, i.e. requirements that a system must satisfy. These specifications can be used in *(semi)automatic verification* of existing systems (using model-checking or static analysis techniques), or in *(semi)automatic synthesis* of new systems (meeting the given specification). Finally, the logical formalism yields naturally a formal notion of *system equivalence:* two systems are equivalent if they satisfy precisely the same properties. Often this equivalence implies observational equivalence (depending on the expressive power of the logical formalism), so a subsystem can be replaced with a logically equivalent one (possibly synthetic) without altering the behaviour of the whole system.

The aim of this work is to take a step in this direction. We introduce the *Brane Logic*, a modal logic specifically designed for expressing properties about systems described using the Brane Calculus. Modal logics are commonly used in concurrency theory for describing behaviour of concurrent systems. In particular, we take inspiration from Ambient Logic, the logic for Ambient calculus [5]. Like Ambient Logic, our logic features *spatial* and *temporal* modalities, which are specific logical operators for expressing properties about the topology and the dynamic behaviour of nested systems. However, differently from Ambient Logic, we need to define also a specific logic for expressing properties of membranes themselves. Each membrane can be seen as a flat surface where different agents can interact, but without nestings. Thus membranes are more similar to CCS than to Ambients; as a consequence, the logic for membranes is similar to Hennessy-Milner's logic [8], extended with spatial connectives as in [2].

After having defined Brane Logic and its formal interpretation over the Brane Calculus (Section 3), in Section 4 we consider *sequents*, and introduce a set of valid *inference rules* (with many derivable corollaries). Several examples throughout the paper will illustrate the expressive power of the logic. Finally, in Section 5, we single out a fragment of the calculus and of the logic for which the satisfiability problem is decidable and for which we give a model checker algorithm. Conclusions, final remarks and directions for future work are in Section 6.

2 Summary of Brane Calculus

In this paper we focus on the basic version of Brane Calculus without communication primitives and molecular complexes. For a description of the intuitive meaning of the language and the reduction rules, we refer the reader to [3].

Syntax of (Basic) Brane Calculus

Systems Π :	$P, Q ::= \diamond \mid \sigma \langle\!\langle P \rangle\!\rangle \mid P \circ Q \mid !P$
Membranes Σ :	$\sigma, \tau ::= \mathbf{0} \mid \sigma\vert\tau \mid a.\sigma \mid !\sigma$
Actions Ξ :	$a, b ::= \eth_n \mid \eth_n^\perp(\sigma) \mid \eth_n \mid \eth_n^\perp \mid \circledcirc(\sigma)$

---

where n is taken from a countable set Λ of *names*. We will write a, $\langle\!\langle P \rangle\!\rangle$ and $\sigma\langle\!\langle\rangle\!\rangle$, instead of $a.\mathbf{0}$, $\mathbf{0}\langle\!\langle P \rangle\!\rangle$ and $\sigma\langle\!\langle\diamond\rangle\!\rangle$, respectively.

The set of free names of a system P, of a membrane σ and of an action a, denoted by $\mathsf{FN}(P)$, $\mathsf{FN}(\sigma)$, $\mathsf{FN}(a)$ respectively, are defined as usual; notice that in this syntax there are no binders.

As in many process calculi, terms of the Brane Calculus can be rearranged according to a structural congruence relation ($\equiv$). For a formal definition see [3].

The dynamic behaviour of Brane Calculus is specified by means of a reduction relation ("reaction") between systems $P \Rightarrow Q$, whose rules are the following:

Operational Semantics

$$\eth_n^\perp(\rho).\tau|\tau_0(\!|Q|\!) \circ \eth_n.\sigma|\sigma_0(\!|P|\!) \Rightarrow \tau|\tau_0(\!|\rho(\!|\sigma|\sigma_0(\!|P|\!)|\!) \circ Q|\!) \qquad \text{(React phago)}$$

$$\eth_n^\perp.\tau|\tau_0(\!|\eth_n.\sigma|\sigma_0(\!|P|\!) \circ Q|\!) \Rightarrow \sigma|\sigma_0|\tau|\tau_0(\!|Q|\!) \circ P \qquad \text{(React exo)}$$

$$\circledcirc(\rho).\sigma|\sigma_0(\!|P|\!) \Rightarrow \sigma|\sigma_0(\!|\rho(\!|\diamond|\!) \circ P|\!) \qquad \text{(React pino)}$$

$$\frac{P \Rightarrow Q}{\sigma(\!|P|\!) \Rightarrow \sigma(\!|Q|\!)} \qquad \frac{P \Rightarrow Q}{P \circ R \Rightarrow Q \circ R} \qquad \text{(React loc, React comp)}$$

$$\frac{P \equiv P' \quad P' \Rightarrow Q' \quad Q' \equiv Q}{P \Rightarrow Q} \qquad \text{(React equiv)}$$

We denote by $\Rightarrow^*$ the usual reflexive and transitive closure of $\Rightarrow$.

As in [3], the Mate-Bud-Drip calculus is easily encoded, as follows:

Derived membrane constructors and reaction

Mate : $\mathsf{mate}_n.\sigma \triangleq \eth_n.\eth_{n'}.\sigma \qquad \mathsf{mate}_n^\perp.\tau \triangleq \eth_n^\perp(\eth_{n'}^\perp.\eth_{n''}).\eth_{n''}^\perp.\tau$

$\mathsf{mate}_n.\sigma|\sigma_0(\!|P|\!) \circ \mathsf{mate}_n^\perp.\tau|\tau_0(\!|Q|\!) \Rightarrow^* \sigma|\sigma_0|\tau|\tau_0(\!|P \circ Q|\!)$

Bud : $\mathsf{bud}_n.\sigma \triangleq \eth_n.\sigma \qquad \mathsf{bud}_n^\perp(\rho).\tau \triangleq \circledcirc(\eth_n^\perp(\rho).\eth_{n'}).\eth_{n'}^\perp.\tau$

$\mathsf{bud}_n^\perp(\rho).\tau|\tau_0(\!|\mathsf{bud}_n.\sigma|\sigma_0(\!|P|\!) \circ Q|\!) \Rightarrow^* \rho(\!|\sigma|\sigma_0(\!|P|\!)|\!) \circ \tau|\tau_0(\!|Q|\!)$

Drip : $\mathsf{drip}_n.(\rho).\sigma \triangleq \circledcirc(\circledcirc(\rho).\eth_n).\eth_n^\perp.\sigma$

$\mathsf{drip}_n(\rho).\sigma|\sigma_0(\!|P|\!) \Rightarrow^* \rho(\!|\!|\!) \circ \sigma|\sigma_0(\!|P|\!)$

---

3 The Brane Logic

In this section we introduce a logic for expressing properties of systems of the Brane Calculus, called *Brane Logic*. Like similar temporal-spatial logics, such as Ambient Logic [5] and Separation Logic [14], Brane Logic features special modal connectives for expressing spatial properties (i.e., about relative positions) and behavioural properties. The main difference between its closest ancestor (Ambient Logic), is that Brane Logic can express properties about the actions which can take place *on membranes*, not only in systems. Thus, there are actually two spatial logics, interacting each other: one for reasoning about membranes (called *membrane logic*) and one for reasoning about systems (the *system logic*).

Syntax The syntax of the Brane Logic is the following:

Syntax of Brane Logic

System formulas Φ

$\mathcal{A}, \mathcal{B} ::= \mathbf{T} \mid \neg\mathcal{A} \mid \mathcal{A} \vee \mathcal{B}$ (classical propositional fragment)

$\qquad \diamond$ (void system)

$\qquad \mathcal{M}(\!|\mathcal{A}|\!) \mid \mathcal{A}@\mathcal{M}$ (compartment, compartment adjoint)

$\qquad \mathcal{A} \circ \mathcal{B} \mid \mathcal{A} \triangleright \mathcal{B}$ (spatial composition, composition adjoint)

$\qquad \Diamond\mathcal{A} \mid \Diamond\!\!\!\!\!\diamond\, \mathcal{A}$ (eventually modality, somewhere modality)

$\qquad \forall x.\mathcal{A}$ (quantification over names)

Membrane formulas Ω

$$\mathcal{M}, \mathcal{N} ::= \mathbf{T} \mid \neg\mathcal{M} \mid \mathcal{M} \vee \mathcal{N} \quad \text{(classical propositional fragment)}$$

$$\mathbf{0} \qquad\qquad\qquad \text{(void membrane)}$$

$$\mathcal{M}|\mathcal{N} \mid \mathcal{M} \blacktriangleright \mathcal{N} \quad \text{(spatial composition, composition adjoint)}$$

$$\langle\alpha\rangle\mathcal{M} \qquad\qquad \text{(action modality)}$$

Action formulas Θ

$$\alpha, \beta ::= \eth_\eta \mid \eth_\eta^\perp(\mathcal{M}) \quad \text{(phago, co-phago)}$$

$$\eth_\eta \mid \eth_\eta^\perp \qquad\qquad \text{(exo, co-exo)}$$

$$\circledcirc(\mathcal{M}) \qquad\qquad \text{(pino)}$$

$$\eta ::= n \mid x \qquad\qquad \text{(terms)}$$

Given a formula $\mathcal{A}$, its free names $\mathsf{FN}(\mathcal{A})$ are easily defined, since there are no binders for names. Similarly, we can define the set of free variables $\mathsf{FV}(\mathcal{A})$, noticing that the only binder for variables is the universal quantifier. As usual, a formula $\mathcal{A}$ is *closed* if $\mathsf{FV}(\mathcal{A}) = \varnothing$.

For sake of simplicity, we will use the shorthands $\mathcal{M}\langle\!\langle \mathcal{D}$ and $\langle\alpha\rangle$ in place of $\mathcal{M}\langle\!\langle\diamond\mathcal{D}$ and $\langle\alpha\rangle\mathbf{0}$ respectively.

We give next an intuitive explanation of the most unusual constructors.
- P satisfies $\mathcal{M}\langle\!\langle\mathcal{A}\mathcal{D}$ if $P \equiv \sigma\langle\!\langle Q\mathcal{D}$, where σ and Q satisfy $\mathcal{M}$ and $\mathcal{A}$ respectively.
- @ e $\triangleright$ are very useful for expressing security and safety properties. A system P satisfies $\mathcal{A}@\mathcal{M}$ if, when P is enclosed in a membrane satisfying $\mathcal{M}$, the resulting system satisfies $\mathcal{A}$. Similarly, a system P satisfies $\mathcal{A} \triangleright \mathcal{B}$ if, when P is put aside a system enjoying $\mathcal{B}$, the whole system satisfies $\mathcal{A}$.
- A membrane σ satisfies $\langle\alpha\rangle\mathcal{M}$ if σ can perform an action satisfying α, yielding a residual satisfying $\mathcal{M}$.
- $\mathcal{M}|\mathcal{N}$ and its adjoint $\mathcal{M} \blacktriangleright \mathcal{N}$ are analogous to $\mathcal{A} \circ \mathcal{B}$ and $\mathcal{A} \triangleright \mathcal{B}$ respectively.

Satisfaction Formally, the meaning of a formula is defined by means of a family of *satisfaction* relations, one for each syntactic sort of logical formulas[1]

$$\vDash\,\subseteq \Pi \times \Phi \qquad \vDash\,\subseteq \Sigma \times \Omega \qquad \vDash\,\subseteq \Xi \times \Theta$$

These relations are defined by induction on the syntax of the formulas. Let us start with satisfaction of systems. First, we have to introduce the *subsystem* relation $P \downarrow Q$ (read "Q is an immediate subsystem of P"), defined as

$$P \downarrow Q \triangleq \exists P' : \Pi, \sigma : \Sigma. P \equiv \sigma\langle\!\langle Q\mathcal{D}|P'$$

We denote by $\downarrow^*$ the reflexive-transitive closure of $\downarrow$.

Then, we can define the satisfaction of system formulas.

Satisfaction of System Formulas

$$\forall P : \Pi \qquad\qquad P \vDash \mathbf{T}$$

$$\forall P : \Pi, \mathcal{A} : \Phi \qquad P \vDash \neg\mathcal{A} \quad \triangleq P \nvDash \mathcal{A}$$

$$\forall P : \Pi, \mathcal{A}, \mathcal{B} : \Phi \quad P \vDash \mathcal{A} \vee \mathcal{B} \triangleq P \vDash \mathcal{A} \vee P \vDash \mathcal{B}$$

$$\forall P : \Pi \qquad\qquad P \vDash \diamond \qquad \triangleq P \equiv \diamond$$

$$\forall P : \Pi, \mathcal{A} : \Phi, \mathcal{M} : \Omega \; P \vDash \mathcal{M}\langle\!\langle\mathcal{A}\mathcal{D} \triangleq \exists P' : \Pi, \sigma : \Sigma. P \equiv \sigma\langle\!\langle P'\mathcal{D} \wedge P' \vDash \mathcal{A} \wedge \sigma \vDash \mathcal{M}$$

---

[1] We will use the same symbol $\vDash$ for the three relations, since they are easily distinguishable from the context.

$$\forall P : \Pi, \mathcal{A}, \mathcal{B} : \Phi \qquad P \vDash \mathcal{A} \circ \mathcal{B} \triangleq \exists P', P'' : \Pi.P \equiv P' \circ P'' \wedge P' \vDash \mathcal{A} \wedge P'' \vDash \mathcal{B}$$

$$\forall P : \Pi, \mathcal{A} : \Phi, x : \vartheta \qquad P \vDash \forall x.\mathcal{A} \triangleq \forall m : \Lambda.P \vDash \mathcal{A}\{x \leftarrow m\}$$

$$\forall P : \Pi, \mathcal{A} : \Phi \qquad P \vDash \Diamond\mathcal{A} \triangleq \exists P' : \Pi.P \Longrightarrow^* P' \wedge P' \vDash \mathcal{A}$$

$$\forall P : \Pi, \mathcal{A} : \Phi \qquad P \vDash \diamondsuit\mathcal{A} \triangleq \exists P' : \Pi.P \downarrow^* P' \wedge P' \vDash \mathcal{A}$$

$$\forall P : \Pi, \mathcal{A} : \Phi, \mathcal{M} : \Omega \quad P \vDash \mathcal{A}@\mathcal{M} \triangleq \forall \sigma : \Sigma.\sigma \vDash \mathcal{M} \Rightarrow \sigma\langle\!|P|\!\rangle \vDash \mathcal{A}$$

$$\forall P : \Pi, \mathcal{A}, \mathcal{B} : \Phi \qquad P \vDash \mathcal{A} \triangleright \mathcal{B} \triangleq \forall P' : \Pi.P' \vDash \mathcal{A} \Rightarrow P \circ P' \vDash \mathcal{B}$$

This definition relies on the satisfaction of membrane formulas, which we define next. To this end, we need to introduce a notion of *membrane observation*, by means of a *labelled transition system* (LTS) $\sigma \xrightarrow{l} \tau$ for membranes. A crucial point is how to define correctly the labels (i.e., the observations) l of this LTS.

The evident similarity between membranes and Milner's CCS [12] could suggest to define observations simply as *actions;* e.g., we could take $a.\sigma \xrightarrow{a} \sigma$. However, an important difference between membranes and CCS is that in latter case, the labels are τ and communications over channels, i.e. names (possibly together with terms, which are separated from processes in any case). On the other hand, actions in membranes form a whole language, which incorporates also the membranes themselves. Thus, observing actions over the membranes would mean to observe explicitly (also) membranes instead of some abstract logical property. For instance, in the transition $\eth(\sigma).\tau \xrightarrow{\eth(\sigma)} \tau$ we have a specific membrane σ in the label. This kind of observation is too "fine-grained" and intensional with respect to the rest of the logic, which never deals with specific membranes but only with their properties.

Therefore, we choose to take as labels the *action formulas*, instead of actions. Thus the LTS is a relation $\sigma \xrightarrow{\alpha} \tau$, which reads as "$\sigma$ performs an action satisfying α, and reduces to τ". This LTS is defined by the following rules:

Labelled Transition System for Membranes

$$\frac{a \vDash \alpha}{a.\sigma \xrightarrow{\alpha} \sigma}(\text{prefix}) \qquad \frac{\sigma \xrightarrow{\alpha} \sigma'}{\sigma|\tau \xrightarrow{\alpha} \sigma'|\tau}(\text{par}) \qquad \frac{\sigma \equiv \sigma' \quad \sigma' \xrightarrow{\alpha} \tau' \quad \tau' \equiv \tau}{\sigma \xrightarrow{\alpha} \tau}(\text{equiv})$$

Notice that in the (prefix) rule we use the satisfaction relation for actions:

Satisfaction of action formulas

$$\forall a : \Gamma, n : \Lambda \qquad a \vDash \eth_n \triangleq a = \eth_n$$

$$\forall a : \Gamma, n : \Lambda, \mathcal{M} : \Omega \quad a \vDash \eth_n^\perp(\mathcal{M}) \triangleq \exists \sigma : \Sigma.a = \eth_n^\perp(\sigma) \wedge \sigma \vDash \mathcal{M}$$

$$\forall a : \Gamma, n : \Lambda \qquad a \vDash \eth_n \triangleq a = \eth_n$$

$$\forall a : \Gamma, n : \Lambda \qquad a \vDash \eth_n^\perp \triangleq a = \eth_n^\perp$$

$$\forall a : \Gamma, \mathcal{M} : \Omega \qquad a \vDash \circledcirc(\mathcal{M}) \triangleq \exists \sigma : \Sigma.a = \circledcirc(\sigma) \wedge \sigma \vDash \mathcal{M}$$

This relation is defined in terms of the satisfaction of membrane formulas:

Satisfaction of membrane formulas

$$\forall \sigma : \Sigma \qquad \sigma \vDash \mathbf{T}$$

$$\forall \sigma : \Sigma, \mathcal{M} : \Omega \qquad \sigma \vDash \neg\mathcal{M} \triangleq \sigma \nvDash \mathcal{M}$$

$$\forall \sigma : \Sigma, \mathcal{M}, \mathcal{N} : \Omega \qquad \sigma \vDash \mathcal{M} \vee \mathcal{N} \triangleq \sigma \vDash \mathcal{M} \vee \sigma \vDash \mathcal{N}$$

$$\forall \sigma : \Sigma \qquad\qquad \sigma \models \mathbf{0} \qquad \triangleq \sigma \equiv \mathbf{0}$$
$$\forall \sigma : \Sigma, \mathcal{N}, \mathcal{M} : \Omega \qquad \sigma \models \mathcal{M}|\mathcal{N} \quad \triangleq \exists \sigma', \sigma'' : \Sigma.\sigma \equiv \sigma'|\sigma'' \wedge \sigma' \models \mathcal{M} \wedge \sigma'' \models \mathcal{N}$$
$$\forall \sigma : \Sigma, \alpha : \Theta \qquad \sigma \models \langle \alpha \rangle \mathcal{M} \quad \triangleq \exists \sigma' : \Sigma.\sigma \xrightarrow{\alpha} \sigma' \wedge \sigma' \models \mathcal{M}$$
$$\forall \sigma : \Sigma, \mathcal{M}, \mathcal{N} : \Omega \qquad \sigma \models \mathcal{M} \blacktriangleright \mathcal{N} \quad \triangleq \forall \sigma' : \Sigma.\sigma' \models \mathcal{M} \Rightarrow \sigma|\sigma' \models \mathcal{N}$$

Notice that the truth of $\langle \alpha \rangle \mathcal{M}$ is defined using the LTS we defined before. Thus, the LTS, the satisfaction of action formulas, and the satisfaction of membrane formulas are three mutually defined judgments.

Derived connectives In the following table, we introduce several useful derived connectives which can be defined as shorthands of longer formulas, together with an intuitive description of their meaning. This description can be easily checked by unfolding the formal meaning, using the satisfaction relations above.

Some derived connectives

$\mathcal{A} \,\phi\, \mathcal{B} \triangleq \neg(\neg\mathcal{A} \circ \neg\mathcal{B})$	system decomposition
$\mathcal{A}^\forall \triangleq \mathcal{A} \,\phi\, \mathbf{F}$	every subsystem (also non proper) satisfies $\mathcal{A}$
$\mathcal{A}^\exists \triangleq \mathcal{A} \circ \mathbf{T}$	some subsystem satisfies $\mathcal{A}$
$\mathcal{A} \propto \mathcal{B} \triangleq \neg(\mathcal{B} \triangleright \neg\mathcal{A})$	system fusion
$\mathcal{A} \Rrightarrow \mathcal{B} \triangleq \neg(\mathcal{A} \circ \neg\mathcal{B})$	fusion adjoint
$\mathcal{M} \parallel \mathcal{N} \triangleq \neg(\neg\mathcal{M}	\neg\mathcal{N})$	membrane decomposition
$\mathcal{M}^\forall \triangleq \mathcal{M} \parallel \mathbf{F}$	every part of the membrane satisfies $\mathcal{M}$
$\mathcal{M}^\exists \triangleq \mathcal{M}	\mathbf{T}$	some part of the membrane satisfies $\mathcal{M}$
$\mathcal{M} \ltimes \mathcal{N} \triangleq \neg(\mathcal{N} \blacktriangleright \neg\mathcal{M})$	membrane fusion
$\mathcal{M} \Mapsto \mathcal{N} \triangleq \neg(\mathcal{M}	\neg\mathcal{N})$	fusion adjoint

---

Derived connectives for Mate-Bud-Drip

$\langle \text{mate}_\eta \rangle \mathcal{M} \triangleq \langle \circlearrowleft_\eta \rangle \langle \circlearrowleft_{\eta'} \rangle \mathcal{M}$	mate
$\langle \text{mate}_\eta^\perp \rangle \mathcal{N} \triangleq \langle \circlearrowleft_\eta^\perp (\langle \circlearrowleft_{\eta'}^\perp \rangle \langle \circlearrowleft_{\eta''} \rangle) \rangle \langle \circlearrowleft_{\eta''}^\perp \rangle \mathcal{N}$	co-mate
$\langle \text{bud}_\eta \rangle \mathcal{M} \triangleq \langle \circlearrowleft_\eta \rangle \mathcal{M}$	bud
$\langle \text{bud}_\eta^\perp (\mathcal{K}) \rangle \mathcal{N} \triangleq \langle \circledcirc (\langle \circlearrowleft_\eta^\perp (\mathcal{K}) \rangle \langle \circlearrowleft_{\eta'} \rangle) \rangle \langle \circlearrowleft_{\eta'}^\perp \rangle \mathcal{N}$	co-bud
$\langle \text{drip}_\eta (\mathcal{N}) \rangle \mathcal{M} \triangleq \langle \circledcirc (\langle \circledcirc (\mathcal{N}) \rangle \langle \circlearrowleft_\eta \rangle) \rangle \langle \circlearrowleft_\eta^\perp \rangle \mathcal{M}$	drip

---

Let us describe shortly the meaning of the most important derived connectives; not surprisingly, these are close to similar ones in the Ambient Logic.

System decomposition is the dual of composition, and it is useful to describe invariant properties of systems. A system satisfies $\mathcal{A} \,\phi\, \mathcal{B}$ if, for any decomposition of the system in two parts, a part satisfies $\mathcal{A}$ or the other $\mathcal{B}$. As a consequence, the formula $\mathcal{A}^\forall$ means that any decomposition satisfies $\mathcal{A}$, or satisfies $\mathbf{F}$. Since $\mathbf{F}$ is never satisfied, this means that in every possible decomposition, a part satisfies $\mathcal{A}$; hence, every immediate subsystem satisfies $\mathcal{A}$. Thus, the formula

$(\mathcal{M}(\mathbf{T}\mathbb{D} \Rightarrow \mathcal{M}(\mathcal{N}(\mathbf{T}\mathbb{D}\mathbb{D})^{\forall}$ means "every membrane satisfying $\mathcal{M}$ in the system, must contain just a membrane satisfying $\mathcal{N}$".

Dually, $\mathcal{A}^{\exists}$ means that there exists a decomposition of the system where a component satisfies $\mathcal{A}$. Thus, the formula $\mathcal{M}(\mathcal{N}(\mathbf{T}\mathbb{D}^{\exists}\mathbb{D}$ states that the system is composed by a membrane satisfying $\mathcal{M}$, which contains at least another membrane satisfying $\mathcal{N}$.

Other interesting applications of derived constructors are, e.g., $\Box\mathcal{M}(\mathbf{T}\mathbb{D}$ ("the system will be always composed by a single membrane, satisfying $\mathcal{M}$), and $\boxtimes\neg(\mathcal{M}(\mathbf{T}\mathbb{D}^{\exists})$ ("nowhere there is a membrane satisfying $\mathcal{M}$"). This last formula expresses a *purity* condition (like, e.g., "nowhere there exists a bacterium/virus identified by $\mathcal{M}$", i.e., "the system is free from infections of type $\mathcal{M}$").

The fusion $\mathcal{A} \propto \mathcal{B}$ means that there exists a system satisfying $\mathcal{B}$ such that, when put together with the actual system, the whole system satisfies $\mathcal{A}$. Dually, $\mathcal{A} \Rightarrow \mathcal{B}$ means that in any decomposition of the system, whenever a part satisfies $\mathcal{A}$ then the other satisfies $\mathcal{B}$.

We end this section with a basic property of satisfaction relations, that is, that satisfaction is preserved by structural congruence.

Proposition 1 (Satisfaction is up to $\equiv$)
   1. $(\sigma \vDash \mathcal{M} \wedge \sigma \equiv \tau) \Rightarrow \tau \vDash \mathcal{M}$ *2.* $(P \vDash \mathcal{A} \wedge P \equiv Q) \Rightarrow Q \vDash \mathcal{A}$

4 Validity and Proof System

In this section, we investigate *validity* of formulas or, more generally of *sequents* and *inference rules*. Validity is defined in terms of satisfaction; more precisely, a closed system/membrane/action formula is *valid* if it is satisfied by every system/membrane/action.

4.1 Interpretation of Sequents and Rules

For sequents and rules we will adopt a notation similar to that of Ambient Logic [5]. A sequent will have exactly one premise and one conclusion, denoted as $\mathcal{A} \vdash \mathcal{B}$; in this way we do not have to decide any (somewhat arbitrary) intrepretation of commas in sequents.

Formally, validity of formulas, sequents and rules is as follows:

─────────── **Validity of formulas, sequents and rules** ───────────

$$\mathbf{vld}(\mathcal{A}) \triangleq \forall P : \Pi.P \vDash \mathcal{A} \quad \mathcal{A} \text{ (closed) is valid}$$
$$\mathcal{A} \vdash \mathcal{B} \triangleq \mathbf{vld}(\mathcal{A} \Rightarrow \mathcal{B}) \quad \text{Sequent}$$
$$\mathcal{A} \dashv\vdash \mathcal{B} \triangleq \mathcal{A} \vdash \mathcal{B} \wedge \mathcal{B} \vdash \mathcal{A} \quad \text{Double sequent}$$

$$\frac{\mathcal{A}_1 \vdash \mathcal{B}_1 \cdots \mathcal{A}_n \vdash \mathcal{B}_n}{\mathcal{A}_0 \vdash \mathcal{B}_0} \triangleq \mathcal{A}_1 \vdash \mathcal{B}_1 \wedge \cdots \wedge \mathcal{A}_n \vdash \mathcal{B}_n \Rightarrow \mathcal{A}_0 \vdash \mathcal{B}_0 \quad \begin{array}{l}\text{Inference} \qquad \text{rule} \\ (n \geq 0)\end{array}$$

$$\frac{\mathcal{A}_1 \vdash \mathcal{B}_1 \cdots \mathcal{A}_n \vdash \mathcal{B}_n}{\mathcal{A}_0 \dashv\vdash \mathcal{B}_0} \triangleq \mathcal{A}_1 \vdash \mathcal{B}_1 \wedge \cdots \wedge \mathcal{A}_n \vdash \mathcal{B}_n \Rightarrow \mathcal{A}_0 \dashv\vdash \mathcal{B}_0 \quad \text{Double conclusion}$$

$$\frac{\mathcal{A}_1 \vdash \mathcal{B}_1}{\mathcal{A}_2 \vdash \mathcal{B}_2} \triangleq \frac{\mathcal{A}_1 \vdash \mathcal{B}_1}{\mathcal{A}_2 \vdash \mathcal{B}_2} \wedge \frac{\mathcal{A}_2 \vdash \mathcal{B}_2}{\mathcal{A}_1 \vdash \mathcal{B}_1} \qquad \text{Double rule}$$

4.2 Logical Rules

In this section we collect several valid sequents and rules for the Brane Logic. We distinguish between "inference rules", which can be seen as proper theorems validated by the interpretation above, and "derived rules", that is corollaries derived by solely applying the inference rules. We omit the rules for propositional calculus which are the same of Ambient Logic [5].

Composition The spatial nature of Brane Logic leads to important rules for reasoning about composition and decomposition of systems and membranes.

─────── **Rules for composition of systems and membranes** ───────

$(\circ\diamond)\quad \dfrac{}{\mathcal{A} \circ \diamond \dashv\vdash \mathcal{A}}$

$(A\circ)\quad \dfrac{}{\mathcal{A} \circ (\mathcal{B} \circ \mathcal{C}) \dashv\vdash (\mathcal{A} \circ \mathcal{B}) \circ \mathcal{C}}$

$(\circ\vee)\quad \dfrac{}{(\mathcal{A} \vee \mathcal{B}) \circ \mathcal{C} \vdash \mathcal{A} \circ \mathcal{C} \vee \mathcal{B} \circ \mathcal{C}}$

$(\circ\varphi)\quad \dfrac{}{\mathcal{A}' \circ \mathcal{A}'' \vdash (\mathcal{A}' \circ \mathcal{B}'') \vee (\mathcal{B}' \circ \mathcal{A}'') \vee (\neg\mathcal{B}' \circ \neg\mathcal{B}'')}$

$(|0)\quad \dfrac{}{\mathcal{M}|0 \dashv\vdash \mathcal{M}}$

$(A|)\quad \dfrac{}{\mathcal{M}|(\mathcal{N}|\mathcal{K}) \dashv\vdash (\mathcal{M}|\mathcal{N})|\mathcal{K}}$

$(|\vee)\quad \dfrac{}{(\mathcal{M} \vee \mathcal{N})|\mathcal{K} \vdash \mathcal{M}|\mathcal{K} \vee \mathcal{N}|\mathcal{K}}$

$(|\;\|)\quad \dfrac{}{\mathcal{M}'|\mathcal{M}'' \vdash (\mathcal{M}'|\mathcal{N}'') \vee (\mathcal{N}'|\mathcal{M}'') \vee (\neg\mathcal{N}'|\neg\mathcal{N}'')}$

$(\circ\neg\diamond)\quad \dfrac{}{\mathcal{A} \circ \neg\diamond \vdash \neg\diamond}$

$(X\circ)\quad \dfrac{}{\mathcal{A} \circ \mathcal{B} \vdash \mathcal{B} \circ \mathcal{A}}$

$(\circ\vdash)\quad \dfrac{\mathcal{A}' \vdash \mathcal{B}' \quad \mathcal{A}'' \vdash \mathcal{B}''}{\mathcal{A}' \circ \mathcal{A}'' \vdash \mathcal{B}' \circ \mathcal{B}''}$

$(\circ\triangleright)\quad \dfrac{\mathcal{A} \circ \mathcal{C} \vdash \mathcal{B}}{\mathcal{A} \vdash \mathcal{C} \triangleright \mathcal{B}}$

$(|\neg0)\quad \dfrac{}{\mathcal{M}|\neg 0 \vdash \neg 0}$

$(X|)\quad \dfrac{}{\mathcal{M}|\mathcal{N} \vdash \mathcal{N}|\mathcal{M}}$

$(|\vdash)\quad \dfrac{\mathcal{M}' \vdash \mathcal{N}' \quad \mathcal{M}'' \vdash \mathcal{N}''}{\mathcal{M}'|\mathcal{M}'' \vdash \mathcal{N}'|\mathcal{N}''}$

$(|\blacktriangleright)\quad \dfrac{\mathcal{M}|\mathcal{K} \vdash \mathcal{N}}{\mathcal{M} \vdash \mathcal{K} \blacktriangleright \mathcal{N}}$

Most of these rules have a direct and intuitive meaning. For instance, $\circ\diamond$ and $\circ\neg\diamond$ state that $\diamond$ is part of any system, and if a part of a system is not void then the whole system is not void. Notice that rule $(\circ\triangleright)$ states that $\circ$ is the left adjoint of $\triangleright$, as expected; similarly for $|$ and $\blacktriangleright$.

Due to lack of space we cannot show many interesting corollaries; see [11].

Compartments The rules for reasoning about compartments are similar to those about compartments in Ambient Logic; the main difference is that now boundaries are structured and not only names. Clearly, these rules do not apply to membrane logic, since membranes are not structured in compartments.

─────── **Rules for Compartments** ───────

$(\!|A|\!)\neg\diamond)\quad \dfrac{\mathcal{A} \vdash \neg\diamond}{\mathcal{M}(\!|\mathcal{A}|\!) \vdash \neg\diamond}$

$(0(\!|\diamond|\!))\quad \dfrac{}{0(\!|\diamond|\!) \dashv\vdash \diamond}$

$(\mathcal{M}(\!|\vdash)\quad \dfrac{\mathcal{A} \vdash \mathcal{B} \quad \mathcal{M} \vdash \mathcal{N}}{\mathcal{M}(\!|\mathcal{A}|\!) \vdash \mathcal{N}(\!|\mathcal{B}|\!)}$

$(\mathcal{M}(\!|@)\quad \dfrac{\mathcal{M}(\!|\mathcal{A}|\!) \vdash \mathcal{B}}{\mathcal{A} \vdash \mathcal{B}@\mathcal{M}}$

$(\mathcal{M}(\!|\neg\diamond)\quad \dfrac{\mathcal{M} \vdash \neg 0}{\mathcal{M}(\!|\mathcal{A}|\!) \vdash \neg\diamond}$

$(\mathcal{M}(\!|\neg\diamond)\quad \dfrac{}{\mathcal{M}(\!|\mathcal{A}|\!) \vdash \neg(\neg\diamond \circ \neg\diamond)}$

$(\mathcal{M}(\!|\wedge)\quad \dfrac{}{\mathcal{M}(\!|\mathcal{A}|\!) \wedge \mathcal{M}(\!|\mathcal{B}|\!) \vdash \mathcal{M}(\!|\mathcal{A} \wedge \mathcal{B}|\!)}$

$(\mathcal{M}(\!|\vee)\quad \dfrac{}{\mathcal{M}(\!|\mathcal{A} \vee \mathcal{B}|\!) \vdash \mathcal{M}(\!|\mathcal{A}|\!) \vee \mathcal{M}(\!|\mathcal{B}|\!)}$

$(\neg@)\quad \dfrac{}{\mathcal{A}@\mathcal{M} \dashv\vdash \neg(\neg(\mathcal{A})@\mathcal{M})}$

The first two rules state that a compartment cannot be considered non-existent if the membrane is not empty or the contained system is not empty. The third rule states that an inactive membrane enclosing an empty system is logically equivalent to an empty system. The fourth rule states that a single compartment cannot be decomposed into two non-trivial systems. The rule $(\mathcal{M}(\!|\mathcal{D}@)$ shows that $\mathcal{A}@\mathcal{B}$ and $\mathcal{M}(\!|\mathcal{A}\mathcal{D}$ are adjoints, and the rule $(\neg @)$ that the compartment adjoint @ is self-dual.

The fragment about compartment is particularly simple to handle, because all rules (with assumptions) are bidirectional: $(\mathcal{M}(\!|\mathcal{D}\vdash)$ holds in both directions, and the inverses of $(\mathcal{M}(\!|\mathcal{D}\wedge)$ and $(\mathcal{M}(\!|\mathcal{D}\vee)$ are derivable.

See [11] for some corollaries about compartments.

Time and space modalities Let us now discuss the logical rules and properties about spatial and temporal modalities.

┌──────── **Some rules for spatial and temporal modalities in systems** ────────┐

$$(\Diamond\mathcal{M}(\!|\mathcal{D}) \frac{}{\mathcal{M}(\!|\Diamond\mathcal{A}\mathcal{D}\vdash\Diamond\mathcal{M}(\!|\mathcal{A}\mathcal{D}} \qquad (\Diamondblack\mathcal{M}(\!|\mathcal{D}) \frac{}{\mathcal{M}(\!|\Diamondblack\mathcal{A}\mathcal{D}\vdash\Diamondblack\mathcal{A}}$$

$$(\Diamond\circ) \frac{}{\Diamond\mathcal{A}\circ\Diamond\mathcal{B}\vdash\Diamond(\mathcal{A}\circ\mathcal{B})} \qquad (\Diamondblack\circ) \frac{}{\Diamondblack\mathcal{A}\circ\mathcal{B}\vdash\Diamondblack(\mathcal{A}\circ\mathbf{T})}$$

$$(\Diamondblack\Diamond) \frac{}{\Diamondblack\Diamond\mathcal{A}\vdash\Diamond\Diamondblack\mathcal{A}}$$

└──┘

The rules for these constructors are very similar to those of ambient logic [5]. The modalities $\Diamond$ and $\Diamondblack$ obey the rules of S4 modalities, but are not S5 modalities [9]. The last rule shows that the two modalities permute in one direction. The other direction does not hold; consider, e.g., the formula $\mathcal{A} = \langle\circlearrowleft_k\rangle(\!|\mathcal{D}$ and the system $P = \circlearrowleft^\perp_m(\!|\circlearrowleft_m(\!|\circlearrowleft_n(\!|\mathcal{D}\mathcal{D}\mathcal{D}\circ\circlearrowleft^\perp_n(\circlearrowleft_k)(\!|\mathcal{D}$. Then, $P\vDash\Diamondblack\Diamond\mathcal{A}$, but $P\nvDash\Diamond\Diamondblack\mathcal{A}$ because neither P nor any of its subsystems will ever exhibit the action $\circlearrowleft_k$.

On the other hand, the action modality $\langle\alpha\rangle\mathcal{M}$ of membranes does not satisfy the laws of S4 modality, because the relation $\xrightarrow{\alpha}$ is neither reflexive nor transitive. Nevertheless, it satisfies the laws of any Kripke modality [9].

┌──────────────── **Rules for action modality** ────────────────┐

$$(\langle\alpha\rangle) \frac{}{\langle\alpha\rangle\mathcal{M}\vdash\neg[\alpha]\neg\mathcal{M}}$$

$$([\alpha]\,K) \frac{}{[\alpha]\,(\mathcal{M}\Rightarrow\mathcal{N})\vdash[\alpha]\,\mathcal{M}\Rightarrow[\alpha]\,\mathcal{N}} \qquad ([\alpha]\vdash) \frac{\mathcal{M}\vdash\mathcal{N}}{[\alpha]\,\mathcal{M}\vdash[\alpha]\,\mathcal{N}}$$

└──┘

┌──────────────── **Some corollaries about action modality** ────────────────┐

$$([\alpha]) \frac{}{[\alpha]\,\mathcal{M}\vdash\neg\langle\alpha\rangle\neg\mathcal{M}} \qquad (\langle\alpha\rangle K) \frac{}{\langle\alpha\rangle\mathcal{M}\Rightarrow\langle\alpha\rangle\mathcal{N}\vdash\langle\alpha\rangle(\mathcal{M}\Rightarrow\mathcal{N})}$$

$$(\langle\alpha\rangle\vdash) \frac{\mathcal{M}\vdash\mathcal{N}}{\langle\alpha\rangle\mathcal{M}\vdash\langle\alpha\rangle\mathcal{N}} \qquad ([\alpha]\wedge) \frac{}{[\alpha]\,(\mathcal{M}\wedge\mathcal{N})\dashv\vdash[\alpha]\,\mathcal{M}\wedge[\alpha]\,\mathcal{N}}$$

$$([\alpha]\langle\alpha\rangle) \frac{}{[\alpha]\,\mathcal{M}\vdash\langle\alpha\rangle\mathcal{M}} \qquad (\langle\alpha\rangle\vee) \frac{}{\langle\alpha\rangle(\mathcal{M}\vee\mathcal{N})\vdash\langle\alpha\rangle\mathcal{M}\vee\langle\alpha\rangle\mathcal{N}}$$

└──┘

A quite expressive set of rules can be obtained by *reflecting* at the logical level the operational behaviour of systems and membranes. The next table shows some of these rules, which can be validated using the reaction of the calculus.

Logical rules for reactions

$(\langle\circlearrowright\rangle)$ $\dfrac{}{\langle\circlearrowright_n\rangle\mathcal{M}\langle\!\mid\!\mathcal{A}\!\mid\!\rangle \circ \langle\circlearrowright_n^{\perp}(\mathcal{K})\rangle\mathcal{N}\langle\!\mid\!\mathcal{B}\!\mid\!\rangle \vdash \Diamond \mathcal{N}\langle\!\mid\!\mathcal{K}\langle\!\mid\!\mathcal{M}\langle\!\mid\!\mathcal{A}\!\mid\!\rangle\!\mid\!\rangle \circ \mathcal{B}\!\mid\!\rangle}$

$(\langle\circlearrowright\rangle)$ $\dfrac{}{\langle\circlearrowright_n^{\perp}\rangle\mathcal{N}\langle\!\mid\!\langle\circlearrowright_n\rangle\mathcal{M}\langle\!\mid\!\mathcal{A}\!\mid\!\rangle \circ \mathcal{B}\!\mid\!\rangle \vdash \Diamond(\mathcal{M}|\mathcal{N}\langle\!\mid\!\mathcal{B}\!\mid\!\rangle \circ \mathcal{A})}$

$(\langle\circledcirc\rangle)$ $\dfrac{}{\langle\circledcirc(\mathcal{N})\rangle\mathcal{M}\langle\!\mid\!\mathcal{A}\!\mid\!\rangle \vdash \Diamond \mathcal{M}\langle\!\mid\!\mathcal{N}\langle\!\mid\!\diamond\mathcal{D} \circ \mathcal{A}\!\mid\!\rangle}$

Some corollaries about reactions

$(\langle\!\langle \text{mate}\rangle\!\rangle)$ $\dfrac{}{\langle\text{mate}_n\rangle\mathcal{M}\langle\!\mid\!\mathcal{A}\!\mid\!\rangle \circ \langle\text{mate}_n^{\perp}\rangle\mathcal{N}\langle\!\mid\!\mathcal{B}\!\mid\!\rangle \vdash \Diamond \mathcal{M}|\mathcal{N}\langle\!\mid\!\mathcal{A} \circ \mathcal{B}\!\mid\!\rangle}$

$(\langle\!\langle \text{bud}\rangle\!\rangle)$ $\dfrac{}{\langle\text{bud}_n^{\perp}(\mathcal{K})\rangle\mathcal{N}\langle\!\mid\!\langle\text{bud}_n\rangle\mathcal{M}\langle\!\mid\!\mathcal{A}\!\mid\!\rangle \circ \mathcal{B}\!\mid\!\rangle \vdash \Diamond(\mathcal{K}\langle\!\mid\!\mathcal{M}\langle\!\mid\!\mathcal{A}\!\mid\!\rangle\!\mid\!\rangle \circ \mathcal{N}\langle\!\mid\!\mathcal{B}\!\mid\!\rangle)}$

$(\langle\!\langle \text{drip}\rangle\!\rangle)$ $\dfrac{}{\langle\text{drip}_n(\mathcal{N})\rangle\mathcal{M}\langle\!\mid\!\mathcal{A}\!\mid\!\rangle \vdash \Diamond(\mathcal{N}\langle\!\mid\!\diamond\mathcal{D} \circ \mathcal{M}\langle\!\mid\!\mathcal{A}\!\mid\!\rangle)}$

These rules show the connections between action modalities $\langle a\rangle$ (in the logic of membranes) and temporal modalities $\Diamond$ (in the logic of systems). These rules are very useful in verifying dynamic properties of systems and membranes.

Predicates We need to extend the notion of validity to open formulas. Let $\mathsf{FV}(\mathcal{A}) = \{x_1 \ldots x_k\}$ be the set of free variables of a formula $\mathcal{A}$, and $\phi \in \mathsf{FV}(\mathcal{A}) \to \Lambda$ a substitution of names for variables; $\mathcal{A}_\phi$ denotes the formula $\mathcal{A}\{x_1 \leftarrow \phi(x_1), \ldots, x_n \leftarrow \phi(x_k)\}$ obtained by applying the substitution ϕ. Then,

$$\mathbf{vld}(\mathcal{A}) \triangleq \forall \phi \in \mathsf{FV}(\mathcal{A}) \to \Lambda. \forall P \in \Pi. P \vDash \mathcal{A}_\phi$$

Using this notion of validity of formulas, the definitions of sequents and rules do not need to be changed. Then, the rules for the quantifiers are the usual ones:

Rules for the universal quantifier

$(\forall L)\dfrac{\mathcal{A}\{x \leftarrow \eta\} \vdash \mathcal{B}}{\forall x.\mathcal{A} \vdash \mathcal{B}}$ $(\forall R)\dfrac{\mathcal{A} \vdash \mathcal{B}}{\mathcal{A} \vdash \forall x.\mathcal{B}}$ $(x \notin \mathsf{FV}(\mathcal{A}))$

With respect to Ambient Logic, name quantification has a slightly different meaning. In the Brane Calculus, different names are intended to denote different proteine complexes on membranes; an action and a coaction can trigger a reaction only if they are using matching complexes, i.e., names. Given this interpretation, using the quantifiers we can express properties which are schematic with respect to the names involved, that is, they do not depend on the specific complexes. For instance, $\forall x.(\langle\circlearrowright_x^{\perp}\rangle\langle\!\mid\!\langle\circlearrowright_x\rangle\langle\!\mid\!\diamond\mathcal{D}\!\mid\!\rangle\!\mid\!\rangle \Rightarrow \Diamond\diamond)$ means "if, for any given complexes, the system exhibits a matching exo and co-exo capabilities in the right places, then it can evolve (into the empty system)".

Name equality We can encode name equality just using logical constructors, and in particular the adjoint of compartment:

$$\eta = \mu \triangleq \langle\circlearrowright_\eta\rangle\langle\!\mid\!\mathbf{T}\!\mid\!\rangle @ \langle\circlearrowright_\mu\rangle$$

Proposition 2. $\forall \phi \in \mathsf{FV}(\eta, \mu) \to \Lambda. \forall P \in \Pi. P \vDash (\eta = \mu)_\phi \iff \phi(\eta) = \phi(\mu)$

As an example application, the formula

$$\forall x. \forall y. (\mathfrak{d}_x) \mathbf{T} (\mathbf{T}) \circ (\mathfrak{d}_y^\perp(\mathbf{T})) \mathbf{T} (\mathbf{T}) \circ \mathbf{T} \Rightarrow \neg x = y$$

means "no pair of membranes exhibit matching action and coaction for a phagocitosis", which can be seen as a safety property (think, e.g., of a virus trying to enter a cell, and looking for the right complexes on its surface).

Substitution The next result provides a substitution principle for validity of predicates; this will allow us to replace logically equivalent formulas inside formula contexts. Let $\mathcal{B}\{-\}$ be a formula with a hole, and let $\mathcal{B}\{\mathcal{A}\}$ the formula obtained by filling the hole with $\mathcal{A}$.

Lemma 1 (Substitution). $\mathrm{vld}(\mathcal{A}' \iff \mathcal{A}'') \Rightarrow \mathrm{vld}(\mathcal{B}\{\mathcal{A}'\} \iff \mathcal{B}\{\mathcal{A}''\})$.

Corollary 1 (Principle of substitution). $\mathcal{A}' \dashv\vdash \mathcal{A}'' \Rightarrow \mathcal{B}\{\mathcal{A}'\} \dashv\vdash \mathcal{B}\{\mathcal{A}''\}$.

4.3 From validity of Propositions to validity of Predicates

We can take advantage of (name) equality to lift validity of propositions to validity of quantified formulas. As a consequence, all the rules and corollaries we have given so far for propositional validity, can be lifted to predicate validity.

To this end, we need to prove the following proposition:

Proposition 3 (Lifting propositional validity). *Let $\mathcal{A}$ be a closed valid formula. For any injective function $\psi \in \mathsf{FN}(\mathcal{A}) \to \vartheta$ mapping names to variables, the formula $(dfn(\mathcal{A}) \Rightarrow \mathcal{A})_\psi$ is valid, where $dfn(\mathcal{A}) \triangleq \bigwedge\limits_{n,m \in \mathsf{FN}(\mathcal{A}), n \neq m} \neg(n = m)$.*

For instance, the valid proposition $[\mathfrak{d}_n] \mathcal{M} \Rightarrow \neg(\mathfrak{d}_m) \mathcal{M}$ is mapped into the valid predicate $\neg x = y \Rightarrow ([\mathfrak{d}_x] \mathcal{M} \Rightarrow \neg(\mathfrak{d}_y) \mathcal{M})$. Notice that without the inequalities between variables denoting different names, the result would not hold.

The proof of Proposition 3 relies on some *injective renaming* lemmata. This kind of lemmata, stating that the relevant meta-logical properties are preserved by name permutations, is quite common among calculi with names (they occur, e.g., in π-calculus, ambient calculus,...); the general technique for their proof is to proceed by induction on the syntax of formulas.

Lemma 2 (Fresh renaming preserves satisfaction)

1. *Let $\mathcal{M}$ be a closed membrane formula, σ a membrane and m, m' names such that $m' \notin \mathsf{FN}(\sigma) \cup \mathsf{FN}(\mathcal{M})$. Then, $\sigma \vDash \mathcal{M} \iff \sigma\{m \leftarrow m'\} \vDash \mathcal{M}\{m \leftarrow m'\}$.*
2. *Let $\mathcal{A}$ be a closed system formula, P a system and m, m' names such that $m' \notin \mathsf{FN}(P) \cup \mathsf{FN}(\mathcal{A})$. Then, $P \vDash \mathcal{A} \iff P\{m \leftarrow m'\} \vDash \mathcal{A}\{m \leftarrow m'\}$.*

Lemma 3 (Fresh renaming preserves validity). *Let $\mathcal{A}$ be a valid closed formula.*

1. *If m' is a name such that $m' \notin \mathsf{FN}(\mathcal{A})$, then $A\{m \leftarrow m'\}$ is closed and valid.*
2. *If $\phi \in \mathsf{FN}(\mathcal{A}) \to \Lambda$ is an injective renaming, then $\mathcal{A}_\phi$ is closed and valid.*

4.4 Example: Viral Infection

As an example of the expressivity of Brane Logic, we give the formulas describing a viral infection. We borrow the example of the Semliki Forest virus in [3].

━━━━━━━━━━━━━━━━━━━━ Viral infection system ━━━━━━━━━━━━━━━━━━

$$\textbf{virus} \triangleq \circlearrowright_n.\circlearrowright_k(\!|\text{nucap}|\!)$$

$$\textbf{cell} \triangleq \textbf{membrane}(\!|\text{cytosol}|\!)$$

$$\textbf{membrane} \triangleq !\circlearrowright_n^{\perp}(\text{mate}_m)|\!|!\circlearrowright_w^{\perp}$$

$$\textbf{cytosol} \triangleq \textbf{endosome} \circ Z$$

$$\textbf{endosome} \triangleq !\text{mate}_m^{\perp}|\!|!\circlearrowright_k^{\perp}(\!|\,|\!)$$

$$\textbf{infected cell} \triangleq \textbf{membrane}(\!|\text{nucap} \circ \text{cytosol}|\!)$$

It is simple to show that **cell**, if placed next to **virus**, evolves into **infected cell**

$$\textbf{virus} \circ \textbf{cell} \rightarrowtail^* \textbf{infected cell}$$

The system describe in detail an infection of the Semliki Forest virus; however, it is almost impossible to abstract from the structure of the system, for instance if we are interested only in its dynamic behaviour. There are entire subsystems (e.g. Z) or parts of mebranes (e.g. $!\circlearrowright_w$) in **cell** that are not involved in the infection process. These are only a burden in explaining what happens in the infection process. The logic can help us to abstract from these irrelevant details: the formulas describe only what is really needed for the viral attack to take place. This kind of abstraction is very important in more complex systems or for focusing only about certain aspects of their evolution.

$$Virus \triangleq \langle\circlearrowright_n\rangle\langle\circlearrowright_k\rangle\mathbf{T}(\!|Nucap|\!)$$

$$InfectableCell \triangleq \exists x.Membrane(x)(\!|Endosome(x)^{\exists}|\!)$$

$$Membrane(x) \triangleq \langle\circlearrowright_n^{\perp}(\langle\text{mate}_x\rangle T)\rangle\mathbf{T}$$

$$Endosome(x) \triangleq \langle\text{mate}_x^{\perp}\rangle\mathbf{T}|\langle\circlearrowright_k^{\perp}\rangle\mathbf{T}(\!|\mathbf{T}|\!)$$

$$InfectedCell \triangleq \mathbf{T}(\!|Nucap^{\exists}|\!)$$

A system satisfies *Virus* if and only if it can be phagocitated by cells revealing a co-phago action with key n on their surface, and, after that, it can release its nucleocapsid if enveloped in a membrane revealing a co-exo action with key k. An infectable cell is a cell containing an endosome, such that their respective membranes have matching mate and mate$^{\perp}$ actions and which exhibit the keys requested by $\circlearrowright$ and $\circlearrowright$ actions of the virus. Notice that the existential quantifier allow us to abstract from the specific key x in the membrane and the endosome: it is not important which is the specific key, only that it is the same.

Using the logical rules, we can derive that "an infectable cell can become infected if it gets close to a virus":

$$InfectableCell \vdash Virus \triangleright \Diamond InfectedCell$$

5   A Decidable Sublogic

In this section we describe a simple model checker for a decidable fragment of the Brane Logic. On the basis of undecidability results for model checking of Ambient Logic [6], we expect that the statement "$P \vDash \mathcal{A}$" is undecidable. There are several reasons for this. First, replication allows to define infinitary systems and membranes. Restricting to replication-free processes and membranes does not suffice either; in fact, following [6], it should be possible to reduce the finite model problem of first order logic to model checking of replication-free systems against first order formulas extended with compartements, composition and compositionadjoint. However, it is possible to consider fragments of the logic, where model checking is decidable. In this section, we describe a model checker for replication-free systems against adjoint-free formulas. Although this logic is not very expressive, it allows to point out the differences respect to the model checker presented in [5], especially in the verification of membrane satisfaction.

5.1   Deciding Satisfaction of Membrane Formulas

Let us consider first the problem of deciding "$\sigma \vDash \mathcal{M}$", where σ is a !-free membrane and $\mathcal{M}$ is an ▶-free membrane formula. This problem can be solved without checking system formulas. As a first step, every !-free membrane can be put in a normal form, given by a finite multiset of *prime membranes*.

Normalization of a replication-free membrane

$$\xi ::= \mathbf{0} \mid a.\sigma \qquad \text{(prime membranes)}$$

$$\mathsf{Norm}(\mathbf{0}) \triangleq [] \qquad \mathsf{Norm}(a.\sigma) \triangleq [a.\sigma]$$

$$\mathsf{Norm}(\sigma|\tau) \triangleq [\xi_1,\ldots,\xi_k,\xi'_1,\ldots,\xi'_l],$$
$$\text{where } \mathsf{Norm}(\sigma) = [\xi_1,\ldots,\xi_k] \text{ and } \mathsf{Norm}(\tau) = [\xi'_1,\ldots,\xi'_l]$$

Lemma 4. *If* $\mathsf{Norm}(\sigma) = [\xi_1,\ldots,\xi_k]$ *then* $\sigma \equiv \prod_{i=1\ldots k} \xi_i$.

The model checker algorithm for membranes consists of three mutually recursive functions: the model checker $\mathsf{Check} : \Sigma \times \Omega \to Bool$, an auxiliary checker $\mathsf{Check} : \Xi \times \Theta \to Bool$ for checking action formulas, and a function $\mathsf{Next} : \Sigma \times \Theta \to \mathcal{P}_f(\Xi)$. Intuitively, $\mathsf{Next}(\sigma,\alpha)$ is the (finite) set of residuals of σ after performing an action satisfying α.

Checking whether membrane σ satisfies closed formula $\mathcal{M}$

$$\mathsf{Check}(\sigma,\mathbf{T}) \triangleq \mathbf{T}$$

$$\mathsf{Check}(\sigma,\neg\mathcal{M}) \triangleq \neg\mathsf{Check}(\sigma,\mathcal{M})$$

$$\mathsf{Check}(\sigma,\mathcal{M}\vee\mathcal{N}) \triangleq \mathsf{Check}(\sigma,\mathcal{M}) \vee \mathsf{Check}(\sigma,\mathcal{N})$$

$$\mathsf{Check}(\sigma,\mathbf{0}) \triangleq \mathsf{Norm}(\sigma) = []$$

$$\mathsf{Check}(\sigma,\mathcal{M}|\mathcal{N}) \triangleq \text{let } \mathsf{Norm}(\sigma) = [\xi_1,\ldots,\xi_k] \text{ in}$$
$$\exists I, J.I \cup J = \{1,\ldots,k\} \wedge I \cap J = \emptyset \wedge$$
$$\mathsf{Check}(\textstyle\prod_{i\in I} \xi_i, \mathcal{M}) \wedge \mathsf{Check}(\textstyle\prod_{j\in J} \xi_j, \mathcal{N})$$

$$\mathsf{Check}(\sigma, \langle\alpha\rangle\mathcal{M}) \triangleq \exists \tau \in \mathsf{Next}(\sigma,\alpha).\mathsf{Check}(\tau,\mathcal{M})$$

$$\mathsf{Next}(\mathbf{0}, \alpha) \triangleq \emptyset$$

$$\mathsf{Next}(\sigma | \tau, \alpha) \triangleq \mathsf{Next}(\sigma, \alpha) \cup \mathsf{Next}(\tau, \alpha)$$

$$\mathsf{Next}(a.\sigma, \alpha) \triangleq \text{if } \mathsf{Check}(a, \alpha) \text{ then } \{\sigma\} \text{ else } \emptyset$$

$$\mathsf{Check}(\circlearrowright_n, \circlearrowright_m) \triangleq n = m \qquad \mathsf{Check}(\circlearrowright_n^{\perp}(\sigma), \circlearrowright_m^{\perp}(\mathcal{M})) \triangleq n = m \wedge \mathsf{Check}(\sigma, \mathcal{M})$$

$$\mathsf{Check}(\circlearrowleft_n, \circlearrowleft_m) \triangleq n = m \qquad \mathsf{Check}(\circledcirc_n(\sigma), \circledcirc_m(\mathcal{M})) \triangleq n = m \wedge \mathsf{Check}(\sigma, \mathcal{M})$$

$$\mathsf{Check}(\circlearrowleft_n^{\perp}, \circlearrowleft_m^{\perp}) \triangleq n = m \quad \mathsf{Check}(\mathsf{wrap}_n(\sigma), \mathsf{wrap}_m(\mathcal{M})) \triangleq n = m \wedge \mathsf{Check}(\sigma, \mathcal{M})$$

$$\mathsf{Check}(a, \alpha) \triangleq \mathbf{F} \text{ otherwise}$$

The algorithm always terminates, because each recursive call is on formulas and membranes smaller than the original ones.

Proposition 4. *For all !-free membranes σ and $\blacktriangleright$-free closed membrane formulas $\mathcal{M}$, $\sigma \vDash \mathcal{M}$ iff $\mathsf{Check}(\sigma, \mathcal{M}) = \mathbf{T}$.*

5.2 Deciding Satisfaction of System Formulas

The model checker for system formulas relies on the model checker for membranes. First we have to define a normalization function for systems into multisets of *prime systems*.

─── **Normalization of a replication-free system** ───

$$\pi ::= \diamond \mid \sigma \langle\!\!\langle P \rangle\!\!\rangle \qquad \text{(prime systems)}$$

$$\mathsf{Norm}(\diamond) \triangleq [] \qquad \mathsf{Norm}(\sigma \langle\!\!\langle P \rangle\!\!\rangle) \triangleq [\sigma \langle\!\!\langle P \rangle\!\!\rangle]$$

$$\mathsf{Norm}(P \circ Q) \triangleq [\pi_1, \ldots, \pi_k, \pi_1', \ldots, \pi_l'],$$

$$\text{where } \mathsf{Norm}(P) = [\pi_1, \ldots, \pi_k] \text{ and } \mathsf{Norm}(Q) = [\pi_1', \ldots, \pi_l']$$

Lemma 5. *If $\mathsf{Norm}(P) = [\pi_1, \ldots, \pi_k]$ then $P \equiv \prod_{i=1\ldots k} \pi_i$.*

As for many modal logics, we need two auxiliary functions $\mathsf{Reach}, \mathsf{SubLoc} : \Pi \to \mathcal{P}_f(\Pi)$ for checking the two modalities. Their specification is the following:

$$Q \in \mathsf{Reach}(P) \Rightarrow P \longmapsto^* Q \qquad \forall P'.P \longmapsto^* P' \Rightarrow \exists Q \in \mathsf{Reach}(P).P' \equiv Q$$

$$Q \in \mathsf{SubLoc}(P) \Rightarrow P \downarrow^* Q \qquad \forall P'.P \downarrow^* P' \Rightarrow \exists Q \in \mathsf{SubLoc}(P).P' \equiv Q$$

Due to lack of space, we omit their (easy) definitions.

─── **Checking whether system P satisfies closed formula $\mathcal{A}$** ───

$$\mathsf{Check}(P, \mathbf{T}) \triangleq \mathbf{T}$$

$$\mathsf{Check}(P, \neg \mathcal{A}) \triangleq \neg \mathsf{Check}(P, \mathcal{A})$$

$$\mathsf{Check}(P, \mathcal{A} \vee \mathcal{B}) \triangleq \mathsf{Check}(P, \mathcal{A}) \vee \mathsf{Check}(P, \mathcal{B})$$

$$\mathsf{Check}(P, \mathbf{0}) \triangleq \mathsf{Norm}(P) = []$$

$\mathsf{Check}(P, \mathcal{A}|\mathcal{B}) \triangleq \mathsf{let} \ \mathsf{Norm}(P) = [\pi_1, \ldots, \pi_k] \ \mathsf{in}$
$$\exists I, J.I \cup J = \{1, \ldots, k\} \wedge I \cap J = \emptyset \wedge$$
$$\mathsf{Check}(\textstyle\prod_{i \in I} \pi_i, \mathcal{A}) \wedge \mathsf{Check}(\textstyle\prod_{j \in J} \pi_j, \mathcal{B})$$

$\mathsf{Check}(P, \mathcal{M}(\!|\mathcal{A}|\!)) \triangleq \exists \sigma, Q.\mathsf{Norm}(P) = [\sigma(\!|Q|\!)] \wedge \mathsf{Check}(\sigma, \mathcal{M}) \wedge \mathsf{Check}(Q, \mathcal{A})$

$\mathsf{Check}(P, \forall x.\mathcal{A}) \triangleq \mathsf{let} \ m \notin \mathsf{FN}(P) \cup \mathsf{FN}(\mathcal{A}) \ \mathsf{in}$
$$\forall n \in \mathsf{FN}(P) \cup \mathsf{FN}(\mathcal{A}) \cup \{m\}.\mathsf{Check}(P, \mathcal{A}\{x \leftarrow m\})$$

$\mathsf{Check}(P, \Diamond \mathcal{A}) \triangleq \exists Q \in \mathsf{Reach}(P).\mathsf{Check}(Q, \mathcal{A})$

$\mathsf{Check}(P, \diamondsuit \mathcal{A}) \triangleq \exists Q \in \mathsf{SubLoc}(P).\mathsf{Check}(Q, \mathcal{A})$

Also this algorithm always terminates, because each recursive call is on formulas and processes smaller than the original ones. Notice that in the case of compartment, we execute the model checker over membranes defined above.

Proposition 5. *For all !-free systems P and ($\rhd\blacktriangleright$@)-free closed system formulas $\mathcal{A}$, $P \vDash \mathcal{A}$ iff $\mathsf{Check}(P, \mathcal{A}) = \boldsymbol{T}$.*

6   Conclusions

In this paper we have introduced a modal logic for describing spatial and temporal properties of biological systems represented as nested membranes, with particular attention to the computational activity which takes place *on* membranes. The logic is quite expressive, since it can describe in a easy but formal way a large range of biological situations at the abstraction level of membrane machines. For a decidable sublogic, we have given a model-checking algorithm, which is a useful tool for automatic verification of properties (e.g., vulnerabilities) of biological systems.

The work presented in this paper is intended to be the basis for further developments, in many directions. First, we can consider logics for more expressive brane calculi, e.g. with communication cross/on-membranes and protein complexes logic formulas. Suitable corresponding logical constructors can be added to the logic of actions. Also, the logic can be easily adapted to other variants of the Brane Calculus, such as the Projective Brane Calculus [7] (e.g., a system formula like $\langle \mathcal{M}; \mathcal{N} \rangle (\!|\mathcal{A}|\!)$ would carry a formula for each face of the membrane).

Another interesting aspect to investigate is the notion of logical equivalence induced by the logic. This should be similar to the equivalences induced by Hennessy-Milner logic extended with spatial connectives (for membranes) and of Ambient Logic (for systems). We think that the methodologies and results developed in [15] can be extended to our logic.

Moreover, it would be interesting to extend the decidability result to a larger class of formulas. We plan to extend the model checker algorithm to formulas without quantifiers but with the guarantees operators (i.e., the adjoints of compositions), along the lines of [6]. On a different direction, it is interesting to consider also *epistemic logics* [10], where the role of the guarantee operator is played by an epistemic operator, while maintaining decidability.

Acknowledgments. The authors wish to thank Luca Cardelli for useful discussions and for kindly providing the fancy font of the actions of Brane Calculus.

References

1. B. Alberts, D. Bray, J. Lewis, M. Raff, K. Roberts, and J. D. Watson. *Molecular biology of the cell*. Garland, second edition, 1989.
2. L. Caires. Behavioral and spatial observations in a logic for the pi-calculus. In I. Walukiewicz, editor, *FoSSaCS*, volume 2987 of *Lecture Notes in Computer Science*, pages 72–89. Springer, 2004.
3. L. Cardelli. Brane calculi. In V. Danos and V. Schachter, editors, *CMSB*, volume 3082 of *Lecture Notes in Computer Science*, pages 257–278. Springer, 2004.
4. L. Cardelli. Abstract machines of systems biology. *T. Comp. Sys. Biology*, 3737:145–168, 2005.
5. L. Cardelli and A. D. Gordon. Anytime, anywhere: Modal logics for mobile ambients. In *Proc. POPL*, pages 365–377, 2000.
6. W. Charatonik, S. Dal-Zilio, A. D. Gordon, S. Mukhopadhyay, and J.-M. Talbot. Model checking mobile ambients. *Theor. Comput. Sci.*, 308(1-3):277–331, 2003.
7. V. Danos and S. Pradalier. Projective brane calculus. In V. Danos and V. Schächter, editors, *CMSB*, volume 3082 of *Lecture Notes in Computer Science*, pages 134–148. Springer, 2004.
8. M. Hennessy and R. Milner. Algebraic laws for nondeterminism and concurrency. *J. ACM*, 32(1):137–161, 1985.
9. G. E. Hughes and M. J. Cresswell. *A companion to Modal Logic*. Methuen, London, 1984.
10. R. Mardare and C. Priami. A decidable extension of hennessy-milner logic with spatial operators. Technical Report DIT-06-009, Dipartimento di Informatica e Telecomunicazioni, University of Trento, 2006.
11. M. Miculan and G. Bacci. Modal logics for brane calculus. Technical Report UDMI/08/2006/RR, Dept. of Mathematics and Computer Science, Univ. of Udine, 2006. http://www.dimi.uniud.it/miculan/Papers/UDMI082006.pdf.
12. R. Milner. *Communication and Concurrency*. Prentice-Hall, 1989.
13. A. Regev, W. Silverman, and E. Y. Shapiro. Representation and simulation of biochemical processes using the pi-calculus process algebra. In *Pacific Symposium on Biocomputing*, pages 459–470, 2001.
14. J. C. Reynolds. Separation logic: A logic for shared mutable data structures. In *LICS*, pages 55–74. IEEE Computer Society, 2002.
15. D. Sangiorgi. Extensionality and intensionality of the ambient logics. In *Proc. POPL*, pages 4–13, 2001.

Deciding Behavioural Properties in Brane Calculi

Nadia Busi

Dipartimento di Scienze dell'Informazione, Università di Bologna,
Mura A. Zamboni 7, I-40127 Bologna, Italy
busi@cs.unibo.it

Abstract. Brane calculi are a family of biologically inspired process calculi proposed in [5] for modeling the interactions of dynamically nested membranes and small molecules.

Building on the decidability of divergence for the fragment with *mate*, *bud* and *drip* operations in [1], in this paper we extend the decidability results to a broader class of properties and to larger set of interaction primitives. More precisely, we provide the decidability of divergence, control state maintainabiliy, inevitability and boundedness properties for the calculus with molecules and without the *phago* operation.

1 Introduction

Brane calculi [5] are a family of process calculi proposed for modeling the behavior of biological membranes. The formal investigation of biological membranes has been initiated by G. Păun [13,12], in the field of automata and formal language theory, with the definition of P systems. In a process algebraic setting, the notions of membranes and compartments are explicitly represented in BioAmbients [14], a variant of Mobile Ambients [7] based on a set of biologically inspired primitives of interaction. Brane calculi represent an evolution of BioAmbients: the main difference w.r.t. previous approaches consists in the fact that the active entities reside on membranes, and not inside membranes. In [5] two basic instances of Brane Calculi are defined: the Phago/Exo/Pino (PEP) and the Mate/Bud/Drip (MBD) calculi.

The interaction primitives of PEP are inspired by *endocytosis* (the process of incorporating external material into a cell by engulfing it with the cell membrane) and *exocytosis* (the reverse process). A relevant feature of such primitives is *bitonality*, a property ensuring that there will never be a mixing of what is inside a membrane with what is outside, although external entities can be brought inside if safely wrapped by another membrane. As endocytosis can engulf an arbitrary number of membranes, it turns out to be a rather uncontrollable process. Hence, it is replaced by two simpler operations: *phagocytosis*, that is engulfing of just one external membrane, and *pinocytosis*, that is engulfing zero external membranes. In [1] we show that a fragment of PEP, namely, the calculus comprising only the phago and exo primitives, is Turing powerful.

C. Priami (Ed.): CMSB 2006, LNBI 4210, pp. 17–31, 2006.
© Springer-Verlag Berlin Heidelberg 2006

The primitives of MBD are inspired by membrane fusion (mate) and fission (mito). Because membrane fission is an uncontrollable process that can split a membrane at an arbitrary place, it is replaced by two simpler operations: *budding*, that is splitting off one internal membrane, and *dripping*, that consists in splitting off zero internal membranes. In [1] we show that the existence of a divergent computation is a decidable property. The proof of the decidability of divergence is based on the theory of well-structured transition systems [8].

The aim of this paper is to extend the decidabillity result of [1] to a larger class of interaction primitives and to a broader set of properties.

After the introduction of the two basic brane calculi PEP and MBD, containing only membranes and membrane interaction primitives, in [5] the calculus is extended with small molecules, freely floating either in the external environment or inside a membrane, and with a molecule–membrane interaction primitive. Biological membranes contain catalysts that can cause molecules, floating respectively inside and outside the membrane, to interact each other without crossing the membrane. Membranes can bind molecules on either sides of their surface, and can release molecules on either sides of their surface. Usually, such an operation occurs in an atomic (all-or-nothing) way. The *bind&release* operation permits to simultaneously bind and release multiple molecules. In this paper we extend the decidability results to the calculus with molecules, and with all the molecule–membrane and membrane–membrane interaction primitives, but the *phago* operation.

Regarding the set of decidability properties, besides providing a constructive method for deciding divergence, the theory of well-structured transition systems [8] also provides methods for deciding the following properties: control state maintainabiliy, inevitability and boundedness. We show that these methods can be fruitfully applied to the full brane calculus (without the *phago* operation) to obtain the decidability of behavioural properties.

The paper is organized as follows: in Section 2 we present the syntax and the semantics of the Full Brane Calculus, and in Section 3 we recall the theory of well-structured transition systems. The decidability results are contained in Section 4. Section 5 reports some conclusive remarks.

2   Full Brane Calculus: Syntax and Semantics

In this section we recall the syntax and the semantics of the Full Brane Calculus [5].

2.1   Syntax and Semantics of Systems and Processes

A system consists of nested membranes, and a process is associated to each membrane. Besides containing other membranes, a membrane can also contain some (small) molecules. As done in [5], We assume that small molecules do not change, do not have internal structure, and do not interact among themselves.

Definition 1. *Let Mol be an infinite set of names for molecules, ranged over by m, m',.... The set of systems is defined by the following grammar:*

$$P, Q ::= \ \diamond \mid P \circ Q \mid !P \mid \sigma(\!|P|\!) \mid m$$

The set of (finite) multisets of molecules is defined by the following grammar:

$$p, q \quad ::= \quad \diamond \mid p \circ q \mid m$$

The set of brane processes is defined by the following grammar:

$$\sigma, \tau \quad ::= \quad 0 \mid \sigma | \tau \mid !\sigma \mid a.\sigma$$

Variables a, b range over actions.

The term $\diamond$ represents the empty system; the parallel composition operator on systems is $\circ$. The replication operator $!$ denotes the parallel composition of an unbounded number of instances of a system. The term $\sigma(\!|P|\!)$ denotes the brane that performs process σ and contains system P. The term m represents a single molecule.

Multisets of molecules will be used used below to define the operation of interaction between membranes and molecules.

The term 0 denotes the empty process, whereas $|$ is the parallel composition of processes; with $!\sigma$ we denote the parallel composition of an unbounded number of instances of process σ. Term $a.\sigma$ is a guarded process: after performing the action a, the process behaves as σ.

We adopt the following abbreviations: with a we denote $a.0$, with $(\!|P|\!)$ we denote $0(\!|P|\!)$, and with $\sigma(\!|\ |\!)$ we denote $\sigma(\!|\diamond|\!)$.

The structural congruence relation on systems and processes is defined as follows:[1]

Definition 2. *The structural congruence $\equiv$ is the least congruence relation satisfying the following axioms:*

$$
\begin{aligned}
P \circ Q &\equiv Q \circ P & \sigma \mid \tau &\equiv \tau \mid \sigma \\
P \circ (Q \circ R) &\equiv (P \circ Q) \circ R & \sigma \mid (\tau \mid \rho) &\equiv (\sigma \mid \tau) \mid \rho \\
P \circ \diamond &\equiv P & \sigma \mid 0 &\equiv \sigma
\end{aligned}
$$

$$
\begin{aligned}
!\diamond &\equiv \diamond & !0 &\equiv 0 \\
!(P \circ Q) &\equiv !P \circ !Q & !(\sigma \mid \tau) &\equiv !\sigma \mid !\tau \\
!!P &\equiv !P & !!\sigma &\equiv !\sigma \\
P \circ !P &\equiv !P & \sigma \mid !\sigma &\equiv !\sigma
\end{aligned}
$$

$$0(\!|\diamond|\!) \equiv \diamond$$

Note that the set of multisets of a molecules is a subset of the set of systems; hence, the first three structural congruence axioms for systems (i.e., the axioms for commutative monoids) also hold for multisets of molecules.

---

[1] With abuse of notation we use $\equiv$ to denote both structural congruence on systems and structural congruence on processes.

Definition 3. *The basic reaction rules are the following:*

$$\text{(par)} \quad \frac{P \to Q}{P \circ R \to Q \circ R} \qquad\qquad \text{(brane)} \quad \frac{P \to Q}{\sigma(\!|P|\!) \to \sigma(\!|Q|\!)}$$

$$\text{(strucong)} \quad \frac{P' \equiv P \quad P \to Q \quad Q \equiv Q'}{P' \to Q'}$$

Rules (par) and (brane) are the contextual rules that respectively permit to a system to execute also if it is in parallel with another process or if it is inside a membrane, respectively. Rule (strucong) ensures that two structurally congruent systems have the same reactions.

With $\to^*$ we denote the reflexive and transitive closure of a relation $\to$. Given a reduction relation $\to$, we say that a system P has a *divergent computation* (or infinite computation) if there exists an infinite sequence of systems $P_0, P_1, \ldots, P_i, \ldots$ such that $P = P_0$ and $\forall i \geq 0 : P_i \to P_{i+1}$. We say that a system P *universally terminates* if it has no divergent computations. We say that P is *deterministic* iff for all P', P'': if $P \to P'$ and $P \to P''$ then $P' \equiv P''$. We say that P has a *terminating computation* (or a deadlock) if there exists Q such that $P \to^* Q$ and $Q \nrightarrow$. A system P satisfies the universal termination property if P has no divergent computations. A system P satisfies the existential termination property if P has a deadlock. Note that the existential termination and the universal termination properties are equivalent on deterministic systems.

The system P' is a *derivative* of the system P if $P \to^* P'$; the set of *derivatives* of a system P is denoted by $Deriv(P)$.

We use $\prod$ (resp. $\bigcirc$) to denote the parallel composition of a set of processes (resp. systems), i.e., $\prod_{i\in\{1,\ldots,n\}} \sigma_i = \sigma_1 \mid \ldots \mid \sigma_n$ and $\bigcirc_{i\in\{1,\ldots,n\}} P_i = P_1 \circ \ldots \circ P_n$. Moreover, $\prod_{i\in\emptyset} \sigma_i = 0$ and $\bigcirc_{i\in\emptyset} P_i = \diamond$.

2.2 Syntax and Semantics of Actions

The set of actions introduced in [5] comprises both operations representing membranes interactions and operations for interactions between molecules and membranes.

In [5] two basic calculi for membrane interactions are investigated. The first calculus (called PEP in [1]) is inspired by endocytosis/exocytosis. Endocytosis is the process of incorporating external material into a cell by "engulfing" it with the cell membrane, while exocytosis is the reverse process. As endocytosis can engulf an arbitrary amount of material, giving rise to an uncontrollable process, in [5] two more basic operations are used: *phagocytosis*, engulfing just one external membrane, and *pinocytosis*, engulfing zero external membranes.

The second basic calculus proposed in [5] (called MBD in [1]) is inspired by membrane fusion and splitting. To make membrane splitting more controllable, in [5] two more basic operations are used: *budding*, consisting in splitting off one internal membrane, and *dripping*, consisting in splitting off zero internal membranes. Membrane fusion, or merging, is called *mating*.

Regarding the interaction beween molecules and membranes, [5] observes that membranes contain catalysts that can cause molecules, floating respectively inside and outside the membrane, to interact each other without crossing the membrane. Membranes can bind molecules on either sides of their surface, and can release molecules on either sides of their surface. Usually, coordinated bindings and releases happen completely or not at all. Hence, the ability of a membrane to bind and release multiple molecules simultaneously is represented by a single *bind&release* operation.

Definition 4. *Let Name be a denumerable set of ambient names, ranged over by $n, m, \ldots$. The set of actions of the Full Brane Calculus is defined by the following grammar:*

$$a \quad ::= \quad \circlearrowright_n \mid \circlearrowright_n^\perp(\sigma) \mid \circlearrowleft_n \mid \circlearrowleft_n^\perp \mid \odot(\sigma)$$
$$mate_n \mid mate_n^\perp \mid bud_n \mid bud_n^\perp(\sigma) \mid drip(\sigma)$$
$$p(q) \rightrightarrows p'(q')$$

Action $\circlearrowright_n$ denotes phagocytosis; the co-action $\circlearrowright_n^\perp$ is meant to synchronize with $\circlearrowright_n$; names n are used to pair-up related actions and co-actions. The co-phago action is equipped with a process σ, this process will be associated to the new membrane that engulfs the external membrane. Action $\circlearrowleft_n$ denotes exocytosis, and synchronizes with the co-action $\circlearrowleft_n^\perp$. Exocytosis causes an irreversible mixing of membranes. Action $\odot$ denotes pinocytosis. The pino action is equipped with a process σ: this process will be associated to the new membrane, that is created inside the brane performing the pino action.

Actions $mate_n$ and $mate_n^\perp$ will synchronize to obtain membrane fusion. Action bud_n permits to split one internal membrane, and synchronizes with the co-action $bud_n^\perp$. Action $drip$ permits to split off zero internal membranes. Actions $bud^\perp$ and $drip$ are equipped with a process σ, that will be associated to the new membrane created by the brane performing the action.

The action $p(q) \rightrightarrows p'(q')$ binds, in general, the multiset p of molecules outside the membrane and the multiset q of molecules inside the membrane if that is possible, it instantly releases the multiset p' of molecules outside the membrane and the multiset q' of molecules inside the membrane.

Definition 5. *The reaction relation for the Full Brane Calculus is the least relation containing the axioms in Table 1, and satisfying the rules in Definition 3.*

In [5] it is shown that the operations of mating, budding and dripping can be encoded in PEP.

3   Well-Structured Transition System

The decidability results presented in this paper are based on the theory of well-structured transition systems [8]. Such a theory permits to show the decidability of some behavioural properties, such as, e.g., the universal termination, boundedness, coverability for finitely branching transition systems, provided that the

Table 1. The set of axioms of the reduction rule for the Full Brane Calculus

(phago) $\circlearrowright_n.\sigma|\sigma_0(\!|P|\!) \circ \circlearrowright_n^{\perp}(\rho).\tau|\tau_0(\!|Q|\!) \to \tau|\tau_0(\!|\rho(\!|\sigma|\sigma_0(\!|P|\!)|\!) \circ Q|\!)$

(exo) $\circlearrowright_n^{\perp}.\tau|\tau_0(\!|\circlearrowright_n.\sigma|\sigma_0(\!|P|\!) \circ Q|\!) \to P \circ \sigma|\sigma_0|\tau|\tau_0(\!|Q|\!)$

(pino) $\circledcirc(\rho).\sigma|\sigma_0(\!|P|\!) \to \sigma|\sigma_0(\!|\rho(\!|\,|\!) \circ P|\!)$

(mate) $mate_n.\sigma|\sigma_0(\!|P|\!) \circ mate_n^{\perp}.\tau|\tau_0(\!|Q|\!) \to \sigma|\sigma_0|\tau|\tau_0(\!|P \circ Q|\!)$

(bud) $bud_n^{\perp}(\rho).\tau|\tau_0(\!|bud_n.\sigma|\sigma_0(\!|P|\!) \circ Q|\!) \to \rho(\!|\sigma|\sigma_0(\!|P|\!)|\!) \circ \tau|\tau_0(\!|Q|\!)$

(drip) $drip(\rho).\sigma|\sigma_0(\!|P|\!) \to \rho(\!|\,|\!) \circ \sigma|\sigma_0(\!|P|\!)$

(B&R) $p \circ p(q) \rightrightarrows p'(q').\sigma|\sigma_0(\!|q \circ P|\!) \to p' \circ \sigma|\sigma_0(\!|q' \circ P|\!)$

set of states can be equipped with a well-quasi-ordering, i.e., a quasi-ordering relation which is compatible with the transition relation and such that each infinite sequence of states admits an increasing subsequence.

We start by recalling some basic definitions and results from [8] concerning well-structured transition systems, as well as on well-quasi-orderings on sequences of elements belonging to a well-quasi-ordered set, that will be used in the following sections.

A *quasi-ordering* (qo) is a reflexive and transitive relation.

A *partial-ordering* $\leq$ is a quasi-ordering satisfying the following property: if $x \leq y$ and $y \leq x$ then $x = y$.

Definition 6. *A* well-quasi-ordering *(wqo) is a quasi-ordering* $\leq$ *over a set X such that, for any infinite sequence $x_0, x_1, x_2, \ldots$ in X, there exist indexes $i < j$ such that $x_i \leq x_j$.*

Note that, if $\leq$ is a wqo, then any infinite sequence $x_0, x_1, x_2, \ldots$ contains an infinite increasing subsequence $x_{i_0}, x_{i_1}, x_{i_2}, \ldots$ (with $i_0 < i_1 < i_2 < \ldots$).

Definition 7. *Let $\leq$ be a wqo over a set X, and let $I \subseteq X$.*
The set I is upward closed *if the following holds: $\forall x, y : x \leq y \land x \in I$ imply $y \in I$.*

Transition systems can be formally defined as follows.

Definition 8. *A* transition system *is a structure $TS = (S, \to)$, where S is a set of states and $\to \subseteq S \times S$ is a set of transitions.*
We write $Succ(s)$ to denote the set $\{s' \in S \mid s \to s'\}$ of immediate successors of $s \in S$.
TS is finitely branching *if $\forall s \in S : Succ(s)$ is finite. We restrict to finitely branching transition systems.*

Well-structured transition systems, defined as follows, provide the key tool to decide properties of computations.

Definition 9. *A well-structured transition system (with strong compatibility) is a transition system $TS = (S, \rightarrow)$, equipped with a quasi-ordering $\leq$ on S, also written $TS = (S, \rightarrow, \leq)$, such that the two following conditions hold:*

1. **well-quasi-ordering***: $\leq$ is a well-quasi-ordering, and*
2. **strong compatibility***: $\leq$ is (upward) compatible with $\rightarrow$, i.e., for all $s_1 \leq t_1$ and all transitions $s_1 \rightarrow s_2$, there exists a state t_2 such that $t_1 \rightarrow t_2$ and $s_2 \leq t_2$.*

The following theorems (most of them are special cases of results in [8]) will be used to obtain our decidability results.

Theorem 1. *Let $TS = (S, \rightarrow, \leq)$ be a finitely branching, well-structured transition system with decidable $\leq$ and computable Succ. The existence of an infinite computation starting from a state $s \in S$ is decidable.*

Theorem 2. *Let $TS = (S, \rightarrow, \leq)$ be a finitely branching, well-structured transition system with decidable $\leq$ and computable Succ. Let $I \subseteq S$ be an upward closed set of states. It is decidable if there exists a computation, starting from a state $s \in S$, such that all states reached during the computation belong to I.*

The theorem above provides the decidability of the *control state maintainability problem* and the *inevitability problem*.

Given an initial state s and a finite set $X = \{s_1, \ldots, s_n\}$ of states, the control state maintainability problem consists in checking if there exists a computation, starting from s, where all states cover one of the s_i (i.e., for all states s' reachable during the computation, there exists $i \in \{1, \ldots, n\}$ such that $s_i \leq s'$).

The inevitability problem is the dual problem of the control state maintainability problem, and consists in checking if all computations starting from an initial state s eventually visit a state not covering one of the s_i.

The *boundedness problem* consists in checking if the set of states reachable from an initial state s is finite.

Theorem 3. *Let $TS = (S, \rightarrow, \leq)$ be a finitely branching, well-structured transition system with decidable $\leq$ and computable Succ. If $\leq$ is also a partial order, then the boundedness problem is decidable.*

To show that the quasi-ordering relation we will define on MBD systems is a well-quasi-ordering we need the following result, due to Higman [9] and stating that the set of the finite sequences over a set equipped with a wqo is well-quasi-ordered.

Given a set S, with S^* we denote the set of finite sequences of elements in S.

Definition 10. *Let S be a set and $\leq$ a wqo over S. The relation $\leq_*$ over S^* is defined as follows. Let $t, u \in S^*$, with $t = t_1 t_2 \ldots t_m$ and $u = u_1 u_2 \ldots u_n$. We have that $t \leq_* u$ iff there exists an injection f from $\{1, 2, \ldots, m\}$ to $\{1, 2, \ldots, n\}$ such that $t_i \leq u_{f(i)}$ and $i \leq f(i)$ for $i = 1, \ldots, m$.*

Note that relation $\leq_*$ is a quasi-ordering over S^*.

Lemma 1. **[Higman]** *Let S be a set and $\leq$ a wqo over S. Then, the relation $\leq_*$ is a wqo over S^*.*

Also the following propositions will be used to prove that the relation on systems is a well-quasi-ordering:

Proposition 1. *Let S be a finite set. Then the equality is a wqo over S.*

Proposition 2. *Let S, T be sets and $\leq_S$, $\leq_T$ be wqo over S and T, respectively. The relation $\leq$ over $S \times T$ is defined as follows: $(s_1, t_1) \leq (s_2, t_2)$ iff ($s_1 \leq_S s_2$ and $t_1 \leq_T t_2$). The relation $\leq$ is a wqo over $S \times T$.*

4   Decidability of Properties in Brane Calculi

In this section we exploit the theory of well-structured transition systems to investigate the decidability of properties in Brane Calculi.

A first step in this direction has been carried out in [1], where we showed that universal termination is decidable for the MBD basic Brane Calculus. In this work we extend such a technique to deal with a larger fragment of the Full Brane Calculus, as well as with other properties of systems.

In [1] we proved that the PEP basic brane calculus (more precisely, the PEP fragment with only phago and exo actions) is Turing powerful. More precisely, we provide a deterministic encoding of a Random Access Machine (RAM) [16,11] satisfying the following property: all the computations of the encoding of a RAM terminate if and only if the RAM terminates. This means that there is no hope to decide universal termination on a calculus that extends the PEP calculus.

To understand to which fragment of the Full Brane Calculus we can extend the decidability results, we recall some crucial points on decidability of universal termination in MBD. The proof that the quasi-ordering defined in [1] for MBD systems turns out to be a well-quasi-ordering is based on the existence of an upper bound to the maximum nesting level of the set of derivatives of a system. A key property of MBD systems, observed in [5], is the following: the reduction reactions in MBD do not increase the maximum nesting levels of membranes in a system. Hence, the nesting level of membranes in a system P provides an upper bound to the nesting level of membranes in the set of the derivatives of P.

Clearly, the key property of MBD systems no longer holds when moving to PEP systems, as both the pino and the phago actions can increase the nesting level of the system. Whereas there is no hope to provide an upper bound to the maximum nesting level of the derivatives of systems containing the phago operation (as witnessed by the system $!(\circlearrowleft_n^\perp(0).\circlearrowright_n(\!|\;|\!)) \circ \circlearrowright_n(\!|\;|\!))$, we will show that it is possible to provide an upper limit to the nesting level even in presence of the pino operation.

To this aim, we define the calculus BC^{-phago} as the fragment of the Full Brane Calculus obtained by dropping the *phago* operation from the set of actions. The results presented in this section hold for the calculus BC^{-phago}.

We recall that our decidability results are based on the theory of well-structured transition systems [8]. Such a theory provides decidability techniques for properties of systems, provided that the transition system is finitely branching and that the set of states of a system can be equipped with a well-quasi-ordering, i.e., a quasi-ordering relation which is compatible with the transition relation and such that each infinite sequence of states admits an increasing subsequence.

Hence, to provide decidability of properties for BC^{-phago}, we start by providing an alternative semantics that is equivalent w.r.t. termination to the one presented in Section 2, but which is based on a finitely branching transition system and permits to define a well-quasi-ordering on the derivatives of a given system (i.e., the set of systems reachable from a given initial system). Then, by exploiting the theory developed in [8], we show that divergence, control state maintainability, inevitability, boundedness are decidable properties for BC^{-phago} systems.

4.1   A Finitely Branching Semantics for BC^{-phago} Systems

The finitely branching semantics provided in this section is essentially an extension to BC^{-phago} of the finitely branching semantics of MBD provided in [1]. Here we recall the main issues.

Because of the structural congruence rules, the reaction transition system for BC^{-phago} is not finitely branching. To obtain a finitely branching transition system (with the same behavior w.r.t. termination), we take the transition system whose states are the equivalence classes of structural congruence.

Technically, it is possible to define a normal form for systems, up to the commutative and associative laws for the $\circ$ and $|$ operators.

In a system in normal form, the presence of a replicated version of a sequential process $!a.\sigma$ (resp. system $!(\sigma(\!|P|\!))$ or molecule $!m$) forbids the presence of any occurrence of the nonreplicated version of the same process (resp. system or molecule), as well as of other occurrences of the replicated version of the process (resp. system or molecule). Moreover, replication is distributed over the components of parallel composition operators, and redundant replications and empty systems and terms are removed.

Definition 11. *Let $\overset{ca}{=}$ be the least congruence on systems satisfying the commutative and associative rules for $\circ$ and $|$.*

A brane process σ is in normal form if $\sigma \overset{ca}{=} \prod_{i \in I} a_i.\sigma_i \mid \prod_{j \in J} !a'_j.\sigma'_j$, where

- *σ_i and σ'_j are in normal form for $i \in I$ and $j \in J$;*
- *if $a_i = bud_n^{\perp}(\rho)$ or $a_i = drip(\rho)$ or $a_i = \textcircled{\circ}(\rho)$ then ρ is in normal form;*
  if $a'_j = bud_n^{\perp}(\rho)$ or $a'_j = drip(\rho)$ or $a'_j = \textcircled{\circ}(\rho)$ then ρ is in normal form;
- *if $\sigma_i \overset{ca}{=} \sigma'_j$ then $a_i \overset{ca}{\neq} a'_j$;*
- *if $\sigma'_i \overset{ca}{=} \sigma'_j$ and $a'_i \overset{ca}{=} a'_j$ then $i = j$.*

A system P is in normal form if $P \overset{ca}{=} \bigcirc_{i \in I} \sigma_i(\!|P_i|\!) \circ \bigcirc_{j \in J} !(\sigma'_j(\!|P'_j|\!)) \circ \bigcirc_{h \in H} m_h \circ \bigcirc_{k \in K} !(m'_k)$, where

- σ_i, P_i, σ'_j and P'_j are in normal form for $i \in I$ and $j \in J$;
- if $P_i \stackrel{ca}{=} P'_j$ then $\sigma_i \stackrel{ca}{\neq} \sigma'_j$;
- if $P'_i \stackrel{ca}{=} P'_j$ and $\sigma'_i \stackrel{ca}{=} \sigma'_j$ then $i = j$;
- $m_h \neq m'_k$ for all $h \in H$ and $k \in K$.

The function nf produces the normal form of a process or a system:

Definition 12. *The normal form of a process is defined inductively as follows:*

$$
\begin{aligned}
nf(0) &= 0 \\
nf(a.\sigma) &= a.nf(\sigma) && a \in \{mate_n, mate_n^\perp, bud_n, \eth_n, \eth_n^\perp\} \\
nf(a(\rho).\sigma) &= a(nf(\rho)).nf(\sigma) && a \in \{bud_n^\perp, drip, \circledcirc\} \\
nf(p(q) \rightrightarrows p'(q')) &= nf(p)(nf(q)) \rightrightarrows nf(p')(nf(q'))
\end{aligned}
$$

Let $nf(\sigma) = \prod_{i \in I} a_i.\sigma_i \mid \prod_{j \in J}!a'_j.\sigma'_j$ and $nf(\tau) = \prod_{h \in H} b_h.\tau_h \mid \prod_{k \in K}!b'_k.\tau'_k$. Then

$$
\begin{aligned}
nf(\sigma \mid \tau) = &\prod\{a_i.\sigma_i \mid i \in I \wedge \forall k \in K : a_i.\sigma_i \stackrel{ca}{\neq} b'_k.\tau'_k\} \mid \\
&\prod\{b_h.\tau_h \mid h \in H \wedge \forall j \in J : b_h.\tau_h \stackrel{ca}{\neq} a'_j.\sigma'_j\} \mid \\
&\prod\{!a'_j.\sigma'_j \mid j \in J\} \mid \\
&\prod\{!b'_k.\tau'_k \mid k \in K \wedge \forall j \in J : b'_k.\tau'_k \stackrel{ca}{\neq} a'_j.\sigma'_j\}
\end{aligned}
$$

and

$$
nf(!\sigma) = \prod\{!a_i.\sigma_i \mid i \in I\} \mid \prod\{!a'_j.\sigma'_j \mid j \in J\}
$$

Definition 13. *The normal form of a system is defined inductively as follows:*

$$
\begin{aligned}
nf(\diamond) &= \quad \diamond \\
nf(m) &= \quad m \\
nf(\sigma(\!|P|\!)) &= nf(\sigma)(\!|nf(P)|\!)
\end{aligned}
$$

Let $nf(P) = \bigcirc_{i \in I} \sigma_i(\!|P_i|\!) \circ \bigcirc_{j \in J}!(\sigma'_j(\!|P'_j|\!)) \circ \bigcirc_{u \in U} m_u \circ \bigcirc_{v \in V} m'_v$ and $nf(Q) = \bigcirc_{h \in H} \tau_h(\!|Q_h,|\!) \circ \bigcirc_{k \in K}!(\tau'_k(\!|Q'_k|\!)) \circ \bigcirc_{w \in W} n_w \circ \bigcirc_{z \in Z} n'_z$. Then

$$
\begin{aligned}
nf(P \circ Q) = &\bigcirc\{\sigma_i(\!|P_i|\!) \mid i \in I \wedge \forall k \in K : \sigma_i(\!|P_i|\!) \stackrel{ca}{\neq} \tau'_k(\!|Q'_k|\!)\} \circ \\
&\bigcirc\{\tau_h(\!|Q_h|\!) \mid h \in H \wedge \forall j \in J : \tau_h(\!|Q_h|\!) \stackrel{ca}{\neq} \sigma'_j(\!|P'_j|\!)\} \circ \\
&\bigcirc\{!\sigma'_j(\!|P'_j|\!) \mid j \in J\} \circ \\
&\bigcirc\{!\tau'_k(\!|Q'_k|\!) \mid k \in K \wedge \forall j \in J : \tau'_k(\!|Q'_k|\!) \stackrel{ca}{\neq} \sigma'_j(\!|P'_j|\!)\} \circ \\
&\bigcirc\{m_u \mid u \in U \wedge \forall z \in Z : m_u \neq n'_z\} \circ \\
&\bigcirc\{n_w \mid w \in W \wedge \forall v \in V : n_w \neq m'_v\} \circ \\
&\bigcirc\{!m'_v \mid v \in V\} \circ \\
&\bigcirc\{!n'_z \mid z \in Z \wedge \forall v \in V : n'_z \neq m'_v\}
\end{aligned}
$$

$$
\begin{aligned}
nf(!P) = &\bigcirc\{!\sigma_i(\!|P_i|\!) \mid i \in I\} \circ \\
&\bigcirc\{!\sigma'_j(\!|P'_j|\!) \mid j \in J\} \circ \\
&\bigcirc\{!m_u \mid u \in U\} \circ \\
&\bigcirc\{!m'_v \mid v \in V\}
\end{aligned}
$$

We need an alternative, finitely branching semantics for systems in normal form, denoted by $\mapsto$, that is equivalent to the semantics of Section 2. We do not report the rules here for the lack of space; the definition of such a semantics for the MBD calculus can be found in [1].

The following result, relating the reduction relations $\to$ and $\mapsto$, holds:

Lemma 2. *Let P, Q be BC^{-phago} systems. If $P \to P'$ then $nf(P) \mapsto nf(P')$. If $nf(P) \mapsto Q$ then $P \to Q$.*

4.2 Decidability of Termination for MBD Systems

Let us consider a system P in normal form. In this section we provide a quasi-order on the derivatives of P (and a quasi-order on brane processes) that turns out to be a wqo compatible with $\mapsto$. Hence, exploiting the results in section 3, we obtain decidability of termination.

We note that each system (resp. process) in normal form is essentially a finite sequence of objects of kind $\sigma(\!|Q|\!)$ or $!(\sigma(\!|Q|\!))$ (resp. of objects of kind $a.\sigma$ or $!a.\sigma$). If we consider the nesting level of membranes, we note that each subsystem Q contained in a subterm $\sigma(\!|Q|\!)$ or $!(\sigma(\!|Q|\!))$ of a system R is simpler than R. More precisely, the maximum nesting level of membranes in Q is strictly smaller than the maximum nesting level of membranes in R. As already observed in [6], the reactions in MBD preserve the nesting level of membranes. The only operation that can increase the nesting level of membranes is *pino*. However, we note that the number of pino operations nested one inside the other in the processes of a system is bounded.

Hence, the sum of the nesting level of membranes in a system P with the nesting depth of the pino operation in the processes of P turns out to be an upper bound to the nesting level of membranes in the set of the (normal forms of the) derivatives of P.

Definition 14. *The nesting level of a system is defined inductively as follows:*

$$
\begin{aligned}
nl(\diamond) &= 0 \\
nl(m) &= 0 \\
nl(\sigma(\!|P|\!)) &= nl(P) + 1 \\
nl(P \circ Q) &= max\{nl(P), nl(Q)\} \\
nl(!P) &= nl(P)
\end{aligned}
$$

Definition 15. *The nesting depth of the pino operation in a process is defined inductively as follows:*

$$
\begin{aligned}
ndpino(0) &= 0 \\
ndpino(a.\sigma) &= ndpino(\sigma) & a &\in \{mate_n, mate_n^\perp, bud_n, \circlearrowright_n, \\
& & & \quad \circlearrowright_n^\perp, p(q) \Rightarrow p'(q')\} \\
ndpino(a(\rho).\sigma) &= max\{ndpino(\rho), ndpino(\sigma)\} & a &\in \{bud_n^\perp, drip\} \\
ndpino(\odot(\rho).\sigma) &= max\{1 + ndpino(\rho), ndpino(\sigma)\} \\
ndpino(\sigma \mid \tau) &= max\{ndpino(\sigma), ndpino(\tau)\} \\
ndpino(!\sigma) &= ndpino(\sigma)
\end{aligned}
$$

The nesting depth of the pino operation in a system is defined inductively as follows:

$$ndpino(\diamond) = 0$$
$$ndpino(P \circ Q) = max\{ndpino(P), ndpino(Q)\}$$
$$ndpino(!P) = ndpino(P)$$
$$ndpino(\sigma(\!| P |\!)) = max\{ndpino(\sigma), ndpino(P)\}$$
$$ndpino(m) = 0$$

Thanks to normal forms, we have that the set of processes of kind $a.\sigma$ or $!a.\sigma$ that occur as subterms in the derivatives (w.r.t. $\mapsto$) of a process in normal form is finite. This fact will be used to show that the quasi-orders on processes and on systems are wqo.

Definition 16. *Let P be a system in normal form. The set of derivatives of P w.r.t. $\mapsto$ is defined as follows: $nfDeriv(P) = \{P' \mid P \mapsto^* P'\}$.*

The following lemma provides an upper bound to the nesting level of the derivatives of a system P:

Lemma 3. *Let P be a systems in normal form and let $P' \in nfDeriv(P)$. Then $nl(P') \leq nl(P) + ndpino(P)$.*

We introduce a quasi-order $\preceq_{proc}$ on processes in normal form such that $\sigma \preceq_{proc} \tau$ if

- for each occurrence of a replicated guarded process at top-level in σ there is a corresponding occurrence of the same process at top-level in τ;
- for each occurrence of a guarded process at top-level in σ there is either a corresponding occurrence of the same process or an occurrence of the replicated version of the process at top-level in τ.

Definition 17. *Let σ and τ be two processes in normal form. Let $\sigma = \prod_{i \in I} a_i.\sigma_i \mid \prod_{j \in J} !a'_j.\sigma'_j$ and $\tau = \prod_{h \in H} b_h.\tau_h \mid \prod_{k \in K} !b'_k.\tau'_k$, and $H \cap K = \emptyset$. We say that $\sigma \preceq_{proc} \tau$ if there exists a pair of functions (f, g) such that:*

- $f : I \to H \cup K$ and $g : J \to K$
- $\forall i, i' \in I :$ if $f(i) = f(i')$ and $f(i) \in H$ then $i = i'$
- $\forall i \in I :$ if $f(i) \in H$ then $b_{f(i)}.\tau_{f(i)} \overset{ca}{=} a_i.\sigma_i$
- $\forall i \in I :$ if $f(i) \in K$ then $b'_{f(i)}.\tau'_{f(i)} \overset{ca}{=} a_i.\sigma_i$
- $\forall j \in J : b'_{g(j)}.\tau'_{g(j)} \overset{ca}{=} a'_j.\sigma'_j$

We define a quasi-order on systems such that $R \preceq_{sys} S$ if

- for each replicated molecule $!m$ at top level in R there is a corresponding replicated molecule $!m$ at top level in S;
- for each replicated membrane $!(\rho(\!| R_1 |\!))$ at top-level in R there is a corresponding replicated membrane $!(\sigma(\!| S_1 |\!))$ at top-level in S such that ρ is smaller than σ and R_1 is smaller than S_1;
- for each occurrence of a molecule m at top-level in R there is

- either a corresponding occurrence of molecule m at top-level in S
- or an occurrence of a replicated molecule $!m$ at top-level in S;
- for each occurrence of a membrane $\rho(\!|R_1|\!)$ at top-level in R there is
  - either a corresponding occurrence of a membrane $\sigma(\!|S_1|\!)$ at top-level in S such that ρ is smaller than σ and R_1 is smaller than S_1
  - or an occurrence of a replicated membrane $!(\rho(\!|R_1|\!))$ at top-level in S.

Definition 18. *Let P, Q be systems. Let $P = \bigcirc_{i \in I} \sigma_i(\!|P_i|\!) \circ \bigcirc_{j \in J} !(\sigma'_j(\!|P'_j|\!)) \circ \bigcirc_{u \in U} m_u \circ \bigcirc_{v \in V} m'_v$ and $Q = \bigcirc_{h \in H} \tau_h(\!|Q_h,|\!) \circ \bigcirc_{k \in K} !(\tau'_k(\!|Q'_k|\!)) \circ \bigcirc_{w \in W} n_w \circ \bigcirc_{z \in Z} n'_z$. Suppose that the sets H, K, W and Z are pairwise disjoint.*

We say that $P \preceq_{sys} Q$ if there exists a tuple of functions (f_1, g_1, f_2, g_2) such that:

- $f_1 : I \to H \cup K$ *and* $g_1 : J \to K$
- $\forall i, i' \in I :$ *if* $f_1(i) = f_1(i')$ *and* $f_1(i) \in H$ *then* $i = i'$
- $\forall i \in I :$ *if* $f_1(i) \in H$ *then* $\sigma_i \preceq_{proc} \tau_{f_1(i)}$ *and* $P_i \preceq_{sys} Q_{f_1(i)}$
- $\forall i \in I :$ *if* $f_1(i) \in K$ *then* $\sigma_i \preceq_{proc} \tau'_{f_1(i)}$ *and* $P_i \preceq_{sys} Q'_{f_1(i)}$
- $\forall j \in J : \sigma'_j \preceq_{proc} \tau'_{g_1(j)}$ *and* $P'_j \preceq_{sys} Q'_{g_1(j)}$
- $f_2 : U \to W \cup Z$ *and* $g_2 : V \to Z$
- $\forall u, u' \in U :$ *if* $f_2(u) = f_2(u')$ *and* $f_2(u) \in W$ *then* $u = u'$
- $\forall u \in U :$ *if* $f_2(u) \in W$ *then* $m_u = n_{f_2(u)}$
- $\forall u \in U :$ *if* $f_2(u) \in Z$ *then* $m_u = n'_{f_2(u)}$
- $\forall v \in V : m'_v = n'_{g_2(v)}$

It is easy to see that $\preceq_{proc}$ and $\preceq_{sys}$ are partial orders.

The relation $\preceq_{sys}$ is strongly compatible with $\mapsto$:

Theorem 4. *Let P, P', Q be systems in normal form. If $P \mapsto P'$ and $P \preceq_{sys} Q$ then there exists Q' in normal form such that $Q \mapsto Q'$ and $Q \preceq_{sys} Q'$.*

By Higman lemma and Proposition 1 it easy to prove that

Lemma 4. *Let P be a system in normal form. The relation $\preceq_{proc}$ is a wqo over the set of processes that can appear as subterms in the derivatives of P.*

The relation $\preceq_{sys}$ is a wqo over a subset of derivatives whose elements have a nesting level smaller than a given natural number. The proof proceeds by induction on the nesting level of membranes, and makes use of Higman's Lemma, of Lemma 4 and of Proposition 2.

Theorem 5. *Let P be a system in normal form and $n \geq 0$. The relation $\preceq_{sys}$ is a wqo over the set of systems appearing as subsystems in the derivatives of P, and whose nesting level is not greater than n.*

The following result can be deduced from Lemma 3 and Theorem 5:

Theorem 6. *Let P be a system in normal form. The relation $\preceq_{sys}$ is a wqo over the set $nfDeriv(P)$.*

The following theorem ensures that the hypothesis of Theorem 1 are satisfied.

Theorem 7. *Let P be a system in normal form. Then the transition system $(nfDeriv(P), \mapsto, \preceq_{sys})$ is a well-structured transition system with decidable $\preceq_{sys}$ and computable Succ. Moreover, $\preceq_{sys}$ is a partial-ordering relation.*

By the above theorem and Theorems 1, 2 and 3 we get the following

Corollary 1. *Let P be a BC^{-phago} system. The following properties are decidable for P: divergence, control state maintainability, inevitability, boundedness.*

Control state maintainability can be used to check safety properties, such as, e.g., the fact that all the derivatives of a system contain at least one occurrence of a given molecule (or at least two occurrences of molecules belonging to some specified set). Inevitability can be used to check, e.g., if in all the computation a state is eventually reached that does contain no occurrences of a given molecule. Boundedness can be used to check if the number of membranes or of molecules can arbitrarily grow during the computation.

5 Conclusion

In this paper we showed the decidability of a set of properties for the Brane Calculus with molecules but without the *phago* operation. We conjecture that the results presented in this paper also hold for systems that can perform a bounded number of phago operations. A synctactical characterization of a subset of systems satisfing this requirement consists in forbidding the presence of a phago operation inside the subsystems (and subprocesses) of kind $!P$ (resp. $!\sigma$).

We plan to extend the results presented in this paper to the analysis of other properties. We claim that the technique adopted to decide the existence of a divergent computation in well-structured transition systems can be adapted to check the presence of some cyclic behaviour in the system.

In the present paper we exploit the so-called *tree saturation methods* for well-structured transition systems: such a class of methods essentially consists in representing (an approximation of) all the computations in a finite tree-like structure. Another class of methods, called *set saturation methods*, is based on the following property of well-quasi-orderings: any infinite, increasing sequence of upward-closed sets $I_1 \subseteq I_2 \subseteq \ldots$ eventually stabilizes (i.e., there exists k s.t. $I_k = I_{k+1}$). We plan to exploit set saturation methods to investigate the decidability of other properties.

The decidability results for well-structured transition systems are all constructive, i.e., they provide a computable procedure for deciding the systems properties. We plan to develop a tool for the animation and the analysis of Brane Calculus systems, also based on the results presented in this work.

In [1] we provided a deterministic encoding of Random Access Machines in the PEP fragment with only *phago* and *exo* operations. A byproduct of the results presented in this paper is the fact that the PEP fragment with only *exo* and *pino* operations is not expressive enough to provide a deterministic encoding of a RAM.

In [2] we provide an encoding of a Random Access Machine in the MBD calculus which preserves the existence of a terminating computation. This means that deadlock is not decidable for MBD. A direct consequence is the undecidability of deadlock also for BC^{-phago}. It could be worthwhile to investigate the (un)decidability of the reachability and liveness properties – which turn out to be equivalent to deadlock in, e.g., Place/Transition Petri nets [15] – for (fragments) of Brane Calculi.

In [4] we modeled the LDL cholesterol degradation pathway [10] in Full Brane Calculus (with mate, pino, exo, drip and bind&release actions), and we showed how to apply the techniques illustrated in the present paper for the analysis of properties of such a biological pathway.

References

1. N. Busi and R. Gorrieri. On the Computational Power of Brane Calculi. In Proc. Computational Methods in System Biology 2005 (CMSB 2005), *Transactions on Computational Systems Biology*, LNCS, Springer, to appear.
2. N. Busi. On the computational power of the Mate/Bud/Drip Brane Calculus: interleaving vs. maximal parallelism. Proc 6th International Workshop on Membrane Computing (WMC6), LNCS 3850, Springer, 2006.
3. N. Busi and G. Zavattaro. On the expressive power of movement and restriction in pure mobile ambients. *Theoretical Computer Science*, 322:477–515, 2004.
4. N. Busi and C. Zandron. Modeling and Analysis of Biological Processes by Mem(Brane) Calculi and Systems. In *Proc. Winter Simulation Conference 2006*, to appear.
5. L. Cardelli. Brane Calculi - Interactions of biological membranes. Proc. Computational Methods in System Biology 2004 (CMSB 2004), LNCS 3082, Springer, 2005.
6. L. Cardelli. Abstract Machines for System Biology. Draft, 2005.
7. L. Cardelli and A.D. Gordon. Mobile Ambients. *Theoretical Computer Science*, 240(1):177–213, 2000.
8. A. Finkel and Ph. Schnoebelen. Well-Structured Transition Systems Everywhere! *Theoretical Computer Science*, 256:63–92, Elsevier, 2001.
9. G. Higman. Ordering by divisibility in abstract algebras. In *Proc. London Math. Soc.*, vol. 2, pages 236–366, 1952.
10. H. Lodish, A Berk, P. Matsudaira, C. A. Kaiser, M. Krieger, M. P. Scott, S. L. Zipursky, and J. Darnell. Molecular Cell Biology. W.H. Freeman and Company, 4th edition, 1999.
11. M.L. Minsky. *Computation: finite and infinite machines*. Prentice-Hall, 1967.
12. G. Păun. *Membrane Computing. An Introduction*. Springer, 2002.
13. G. Păun. Computing with membranes. *Journal of Computer and System Sciences*, 61(1):108–143, 2000.
14. A. Regev, E. M. Panina, W. Silverman, L. Cardelli, E. Shapiro. BioAmbients: An Abstraction for Biological Compartments. *Theoretical Computer Science*, 325(1):141–167, Elsevier, 2004.
15. W. Reisig. *Petri nets: An Introduction*. EATCS Monographs in Computer Science, Springer, 1985.
16. J.C. Shepherdson and J.E. Sturgis. Computability of recursive functions. *Journal of the ACM*, 10:217–255, 1963.

Probabilistic Model Checking of Complex Biological Pathways[*]

J. Heath[1], M. Kwiatkowska[2], G. Norman[2], D. Parker[2], and O. Tymchyshyn[2]

[1]School of Biosciences
[2]School of Computer Science
University of Birmingham, Birmingham, B15 2TT, UK

Abstract. Probabilistic model checking is a formal verification technique that has been successfully applied to the analysis of systems from a broad range of domains, including security and communication protocols, distributed algorithms and power management. In this paper we illustrate its applicability to a complex biological system: the FGF (Fibroblast Growth Factor) signalling pathway. We give a detailed description of how this case study can be modelled in the probabilistic model checker PRISM, discussing some of the issues that arise in doing so, and show how we can thus examine a rich selection of quantitative properties of this model. We present experimental results for the case study under several different scenarios and provide a detailed analysis, illustrating how this approach can be used to yield a better understanding of the dynamics of the pathway.

1 Introduction

There has been considerable success recently in adapting approaches from computer science to the analysis of biological systems and, in particular, biochemical pathways. The majority of this work has relied on simulation-based techniques developed for discrete stochastic models [7]. These allow modelling of the evolution of individual molecules, whose rates of interaction are controlled by exponential distributions. The principal alternative modelling paradigm, using ordinary differential equations, differs in that it reasons about how the average concentrations of the molecules evolve over time. In this paper, as in [4,3], we adopt the stochastic modelling approach, but employ methods which allow calculation of *exact* quantitative measures of the model under study.

We use probabilistic model checking [19] and the probabilistic model checker PRISM [9,14] as a framework for the modelling and analysis of biological pathways. This approach is motivated by the success of previous work which has demonstrated the applicability of these techniques to the analysis of a wide variety of complex systems [11]. One benefit of this is the ability to employ the existing efficient implementations and tool support developed in this area. Additionally, we enjoy the advantages of model checking, for example, the use of

[*] Supported in part by EPSRC grants GR/S72023/01, GR/S11107 and GR/S46727 and Microsoft Research Cambridge contract MRL 2005-44.

both a formal model and specification of the system under study and the fact that the approach is exhaustive, that is, all possible behaviours of the system are analysed. Our intention is that the methods in this paper should be used in conjunction with the classical simulation and differential equation based approaches to provide greater insight into the complex interactions of biological pathways. This paper provides a detailed illustration of the applicability of probabilistic model checking to this domain through the analysis of a complex biological pathway called FGF (Fibroblast Growth Factor).

Related Work. The closest approach to that presented here is [4], where the probabilistic model checker PRISM is used to model the RKIP inhibited ERK pathway. The main difference is that in [4] the authors consider a "population" based approach to modelling using approximate techniques where concentrations are modelled by discrete abstract quantities. In addition, here we demonstrate how a larger class of temporal properties including reward-based measures are applicable to the study biological systems. Also related to the RKIP inhibited ERK pathway is [3], where it is demonstrated how the stochastic process algebra PEPA [8] can be used to model biological systems. The stochastic π-calculus [15] has been proposed as a model language for biological systems [18,16]; this approach has so far been used in conjunction with stochastic simulation, for example through the tools BioSpi [16] and SPiM [12].

In parallel with the development of the PRISM model of the FGF pathway presented in this paper, we have constructed a separate π-calculus model [22,13] and applied stochastic simulation through BioSpi. Although currently these works focus on different aspects of the pathway, in the future we aim to use this complex case study as a basis for investigating the advantages of stochastic simulation and probabilistic model checking.

2   Probabilistic Model Checking and PRISM

Probabilistic model checking is a formal verification technique for the modelling and analysis of systems which exhibit stochastic behaviour. This technique is a variant of *model checking*, a well-established and widely used formal method for ascertaining the correctness of real-life systems. Model checking requires two inputs: a description of the system in some high-level modelling formalism (such as a Petri net or process algebra), and specification of one or more desired properties of that system in temporal logic (e.g. CTL or LTL). From these, one can construct a model of the system, typically a labelled state-transition system in which each state represents a possible configuration and the transitions represent the evolution of the system from one configuration to another over time. It is then possible to automatically verify whether or not each property is satisfied, based on a systematic and exhaustive exploration of the model.

In probabilistic model checking, the models are augmented with quantitative information regarding the likelihood that transitions occur and the times at which they do so. In practice, these models are typically Markov chains or Markov decision processes. In this paper, it suffices to consider *continuous-time*

```
Reactions:
1. A+B ⟷ A:B    (complexation)
2. A    ⟶       (degradation)

Reaction rates:
- complexation      : r₁
- decomplexation    : r₂
- degradation       : r₃
```

(a) System of reactions

```
module M
  ab : [0..2] init 1;

  // 0: a degraded, b free 1: a,b free 2: a,b bound
  [] ab=1 → r₁ : (ab'=2); // bind
  [] ab=2 → r₂ : (ab'=1); // unbind
  [] ab=1 → r₃ : (ab'=0); // degrade
endmodule
```

(b) PRISM encoding 1

```
module A
  a : [0..1] init 1;

  [bind] a=1 → r₁ : (a'=0);
  [rel]  a=0 → r₂ : (a'=1);
  []     a=1 → r₃ : (a'=0);
endmodule
```

```
module B
  b : [0..1] init 1;

  [bind] b=1 → (b'=0);
  [rel]  b=0 → (b'=1);
endmodule
```

```
module AB
  ab : [0..1] init 0;

  [bind] ab=0 → (ab'=1);
  [rel]  ab=1 → (ab'=0);
endmodule
```

(c) PRISM encoding 2

```
rewards a=1 : 1; endrewards
```

(d) Reward structure 1

```
rewards [bind] true : 1; endrewards
```

(e) Reward structure 2

Fig. 1. Simple example and possible PRISM representations

Markov chains (CTMCs), in which transitions between states are assigned (positive, real-valued) rates, which are interpreted as the rates of negative exponential distributions. The model is augmented with rewards associated with states and transitions. Rewards associated with states (*cumulated rewards*) are incremented in proportion to the time spent in the state, while rewards associated with transitions (*impulse rewards*) are incremented each time the transition is taken.

Properties of these models, while still expressed in temporal logic, are now quantitative in nature. For example, rather than verifying that "the protein always eventually degrades", we may ask "what is the probability that the protein eventually degrades?" or "what is the probability that the protein degrades within T hours?". Reward-based properties include "what is the expected energy dissipation within the first T time units?" and "what is the expected number of complexation reactions before relocation occurs?".

PRISM [9,14] is a probabilistic checking tool developed at the University of Birmingham. Models are specified in a simple state-based language based on Reactive Modules. An extension of the temporal logic CSL [1,2] is used to specify properties of CTMC models augmented with rewards. The tool employs state-of-the-art symbolic approaches using data structures based on binary decision diagrams [10]. Also of interest, the tool includes support for PEPA [8] and has recently been extended to allow for simulation-based analysis using Monte-Carlo methods and discrete event simulation. For further details, see [14].

3   Modelling a Simple Biological System in PRISM

We now illustrate PRISM's modelling and specification languages through an example: the simple set of biological reactions given in Figure 1(a). We consider two proteins A and B which can undergo complexation with rate r_1 and decomplexation with rate r_2. In addition, A can degrade with rate r_3.

We give two alternative approaches for modelling these reactions in PRISM, shown in Figures 1(b) and 1(c), respectively. A model described in the PRISM language comprises a set of *modules*, the state of each being represented by a set of finite-ranging *variables*. In approach 1 (Figure 1(b)) we use a single module with one variable, representing the (three) possible states of the whole system (which are listed in the italicised comments in the figure). The behaviour of this module, i.e. the changes in states which it can undergo, is specified by a number of *guarded commands* of the form $[] \; g \rightarrow r : u$, with the interpretation that if the predicate (guard) g is true, then the system is updated according to u (where $x' = \ldots$ denotes how the value of variable x is changed). The rate at which this occurs is r, i.e. this is the value that will be attached to the corresponding transition in the underlying CTMC.

In approach 2 (Figure 1(c)) we represent the different possible forms that the proteins can take (A, B and $A{:}B$) as separate modules, each with a single variable taking value 0 or 1, representing its absence or presence, respectively. To model interactions where the state of several modules changes simultaneously, we use *synchronisation*, denoted by attaching action labels to guarded commands (placed inside the square brackets). For example, when the *bind* action occurs, variables a and b in modules A and B change from 1 to 0 and variable ab in module AB changes from 0 to 1. In this example, the rate of each combined transition is fully specified in module A and we have omitted the rates from the other modules. More precisely, PRISM assigns a rate of 1 to any command for which none is specified and computes the rate of a combined transition as the product of the rates for each command. Note that independent transitions, involving only a single module, can also be included, as shown by the modelling of degradation (which only involves A), by omitting the action label.

In general, a combination of the above two modelling approaches is used. In simple cases it is possible to use a single variable, but as the system becomes more complex the use of separate variables and synchronisation becomes more desirable. We will see this later in the paper.

Properties of CTMCs are specified in PRISM using an extension of the temporal logic CSL. We now give a number of examples for the model in Figure 1(c).

- *What is the probability that the protein A is bound to the protein B at time instant T?* ($\mathcal{P}_{=?}[\text{true } \mathcal{U}^{[T,T]} \; ab{=}1]$);
- *What is the probability that the protein A degrades before binding to the protein B?* ($\mathcal{P}_{=?}[ab{=}0 \; \mathcal{U} \; (a{=}0 \wedge ab{=}0)]$);
- *During the first T time units, what is the expected time that the protein A spends free?* ($\mathcal{R}_{\leq 0.5 \cdot T}[\mathcal{C}^{\leq T}]$, assuming a reward structure which associates reward 1 with states where the variable a equals 1 - see Figure 1(d));
- *What is the expected number of times that the proteins A and B bind before A degrades?* ($\mathcal{R}_{=?}[\mathcal{F} \; (a{=}0 \wedge ab{=}0)]$, assuming a reward of 1 is associated with any transition labelled by *bind* - see Figure 1(e)).

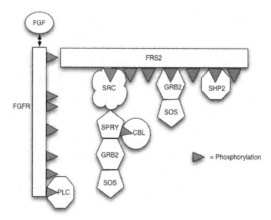

Fig. 2. Diagram showing the different possible bindings in the pathway

4   Case Study: FGF

Fibroblast Growth Factors (FGF) are a family of proteins which play a key role in the process of cell signalling in a variety of contexts, for example wound healing. The mechanisms of the FGF signalling pathway are complex and not yet fully understood. In this section, we present a model of the pathway which is based on literature-derived information regarding the early stages of FGF signal propagation and which incorporates several features that have been reported to negatively regulate this propagation [6,21,5,20].

Our model incorporates protein-protein interactions (including competition for partners), phosphorylation and dephosphorylation, protein complex relocation and protein complex degradation (via ubiquitin-mediated proteolysis). Figure 2 illustrates the different components in the pathway and their possible bindings. Below is a list of the reactions included in the model. Further details are provided in Figure 3.

1. An FGF ligand binds to an FGF receptor (FGFR) creating a complex of FGF and FGFR.
2. The existence of this FGF:FGFR dimer leads to phosphorylation of FGFR on two residues Y653 and Y654 in the activation loop of the receptor.
3. The dual Y653/654 form of the receptor leads to phosphorylation of other FGFR receptor residues: Y663, Y583, Y585, Y766 (in this model we only consider Y766 further).
4. **and 5.** The dual Y653/654 form of the receptor also leads to phosphorylation of the FGFR substrate FRS2, which binds to both the phosphorylated and dephosphorylated forms of the FGFR.
6. FRS2 can also be dephosphorylated by a phosphotase, denoted Shp2.
7. A number of effector proteins interact with the phosphorylated form of FRS2. In this model we include Src, Grb2:Sos and Shp2.

1. FGF binds to FGFR
   FGF+FGFR $\leftrightarrow$ FGFR:FGF ($k_{on} = 5e+8\text{M}^{-1}\text{s}^{-1}$, $k_{off}=1e-1\text{s}^{-1}$)
2. Whilst FGFR:FGF exists
   FGFR Y653 $\rightarrow$ FGFR Y653P ($k_{cat}=0.1\text{s}^{-1}$)
   FGFR Y654 $\rightarrow$ FGFR Y654P ($k_{cat}=0.1\text{s}^{-1}$)
3. When FGFR Y653P and FGFR Y654P
   FGFR Y463 $\rightarrow$ FGFR Y463P ($k_{cat}=70\text{s}^{-1}$)
   FGFR Y583 $\rightarrow$ FGFR Y583P ($k_{cat}=70\text{s}^{-1}$)
   FGFR Y585 $\rightarrow$ FGFR Y585P ($k_{cat}=70\text{s}^{-1}$)
   FGFR Y766 $\rightarrow$ FGFR Y766P ($k_{cat}=70\text{s}^{-1}$)
4. FGFR binds FRS2
   FGFR+ FRS2 $\leftrightarrow$ FGFR:FRS2 ($k_{on} = 1e+6\text{M}^{-1}\text{s}^{-1}$, $k_{off}=2e-2\text{s}^{-1}$)
5. When FGFR Y653P, FGFR Y654P and FGFR:FRS2
   FRS2 Y196 $\rightarrow$ FRS2 Y196P ($k_{cat}=0.2\text{s}^{-1}$)
   FRS2 Y290 $\rightarrow$ FRS2 Y290P ($k_{cat}=0.2\text{s}^{-1}$)
   FRS2 Y306 $\rightarrow$ FRS2 Y306P ($k_{cat}=0.2\text{s}^{-1}$)
   FRS2 Y382 $\rightarrow$ FRS2 Y382P ($k_{cat}=0.2\text{s}^{-1}$)
   FRS2 Y392 $\rightarrow$ FRS2 Y392P ($k_{cat}=0.2\text{s}^{-1}$)
   FRS2 Y436 $\rightarrow$ FRS2 Y436P ($k_{cat}=0.2\text{s}^{-1}$)
   FRS2 Y471 $\rightarrow$ FRS2 Y471P ($k_{cat}=0.2\text{s}^{-1}$)
6. Reverse when Shp2 bound to FRS2:
   FRS2 Y196P $\rightarrow$ FRS2 Y196 ($k_{cat}=12\text{s}^{-1}$)
   FRS2 Y290P $\rightarrow$ FRS2 Y290 ($k_{cat}=12\text{s}^{-1}$)
   FRS2 Y306P $\rightarrow$ FRS2 Y306 ($k_{cat}=12\text{s}^{-1}$)
   FRS2 Y382P $\rightarrow$ FRS2 Y382 ($k_{cat}=12\text{s}^{-1}$)
   FRS2 Y436P $\rightarrow$ FRS2 Y436 ($k_{cat}=12\text{s}^{-1}$)
   FRS2 Y471P $\rightarrow$ FRS2 Y471 ($k_{cat}=12\text{s}^{-1}$)
   FRS2 Y392P $\rightarrow$ FRS2 Y392 ($k_{cat}=12\text{s}^{-1}$)
7. FRS2 effectors bind phosphoFRS2:
   Src+FRS2 Y196P $\leftrightarrow$ Src:FRS2 Y2196P ($k_{on} = 1e+6\text{M}^{-1}\text{s}^{-1}$, $k_{off}=2e-2\text{s}^{-1}$)
   Grb2+FRS2 Y306P $\leftrightarrow$ Grb2:FRS2 Y306P($k_{on} = 1e+6\text{M}^{-1}\text{s}^{-1}$, $k_{off}=2e-2\text{s}^{-1}$)
   Shp2+FRS2 Y471P $\leftrightarrow$ Shp2:FRS2 Y471P($k_{on} = 1e+6\text{M}^{-1}\text{s}^{-1}$, $k_{off}=2e-2\text{s}^{-1}$)
8. When Src:FRS2 we relocate/remove
   Src:FRS2 $\rightarrow$ relocate out ($t_{1/2}=15\text{min}$)
9. When Plc:FGFR it degrades FGFR
   PLC+FGFRY 766 $\leftrightarrow$ PLC:FGFR 766 ($k_{on} = 1e+6\text{M}^{-1}\text{s}^{-1}$, $k_{off}=2e-2\text{s}^{-1}$)
   PLC:FGFR $\rightarrow$ degFGFR   ($t_{1/2}=60\text{min}$)
10. Spry appears in time-dependent manner:
    $\rightarrow$ Spry ($t_{1/2}=15\text{min}$)
11. Spry binds Src and is phosphorylated:
    Spry+Src $\leftrightarrow$ Spry Y55:Src   ($k_{on} = 1e+5\text{M}^{-1}\text{s}^{-1}$, $k_{off}=1e-4\text{s}^{-1}$)
    Spry Y55:Src $\rightarrow$ Spry Y55P:Src ($k_{cat}=10\text{s}^{-1}$)
    Spry Y55P+Src $\leftrightarrow$ Spry Y55P:Src ($k_{on} = 1e+5\text{M}^{-1}\text{s}^{-1}$, $k_{off}=1e-4\text{s}^{-1}$)
    Spry Y55P+Cbl $\leftrightarrow$ Spry Y55P:Cbl ($k_{on} = 1e+5\text{M}^{-1}\text{s}^{-1}$, $k_{off}=1e-4\text{s}^{-1}$)
    Spry Y55P+Grb2 $\leftrightarrow$ Spry Y55P:Grb2($k_{on} = 1e+5\text{M}^{-1}\text{s}^{-1}$, $k_{off}=1e-4\text{s}^{-1}$)
12. phosphoSpry binds Cbl which degrades/removes FRS2
    Spry Y55P:Cbl+FRS2 $\leftrightarrow$ FRS-Ubi ($k_{cat}=8.5e-4\text{s}^{-1}$)
    FRS2-Ubi $\rightarrow$ degFrs2 ($t_{1/2}=5\text{min}$)
13. Spry is dephosphorylated by Shp2: (when Shp2 bound to FRS2)
    Spry Y55P $\rightarrow$ Spry Y55 ($k_{cat}=12\text{s}^{-1}$)
14. Grb2 binds Sos
    Grb2+Sos $\leftrightarrow$ Grb2:Sos ($k_{on} = 1e+5\text{M}^{-1}\text{s}^{-1}$, $k_{off}=1e-4\text{s}^{-1}$)

Fig. 3. Reaction rules for the pathway

8. and 9. These are two methods of attenuating signal propagation by removal (i.e. relocation) of components. In step **8.** if Src is associated with the phosphorylated FRS2 Y219, this leads to relocation (i.e. endocytosis and/or degradation of FGFR:FRS2). In step **9.** if Plc is bound to Y766 of FGFR, this leads to relocation/degradation of FGFR.

10. The signal attenuator Spry is a known inhibitor of FGFR signalling and is synthesised in response to FGFR signalling. Here we include a variable to regulate the concentration of Spry protein in a time dependent manner.
11. We incorporate the association of Spry with Src and concomitant phosphorylation of Spry residue Y55.
12. The Y55 phosphorylated form of Spry binds with Cbl, which leads to ubiquitin modification of FRS2 and a degradation of FRS2 through ubiquitin-mediated proteolysis.
13. The Y55P form of Spry is dephosphorylated by Shp2 bound to FRS2 Y247P.
14. Grb2 binds to the Y55P form of Spry. In our model Spry competes with FRS2 for Grb2 as has been suggested from some studies in the literature.

Note that this model is not intended to, and cannot be, a fully accurate representation of a real-world FGF signalling pathway. Its primary purpose at this stage of development is as a tool to evaluate biological hypotheses that are not easily obtained by intuition or manual methods. To this end, the model is an abstraction as argued in [17], created to facilitate predictive "in silico" experiments for a range of scenarios. Results of such "in silico genetics" experiments based on simulations of a stochastic π-calculus model of the above set of reactions are described in [22] (see also [13]).

We explicitly draw attention to the following issues. The reactions selected are based upon their current biological interest rather than complete understanding of the components of FGF signalling. Indeed, at this stage we have ignored many reactions that could prove significant in regulation of FGFR signalling in real cells. However, the design permits the incorporation of further modifications to the core model as biological understanding advances. The model is idealised in that it does not take into account variations in composition, affinities or rate constants that might occur in different cell types or physiological conditions. However, a useful computational modelling approach should accommodate future quantitative or qualitative modifications to the core model.

5   Modelling in PRISM

We now describe the specification in PRISM of the FGF model from the previous section. We employ a combination of the two approaches discussed in Section 3. Each of the basic elements of the pathway, including all possible compounds and receptors residues (FGF, FGFR, FRS2, Plc, Src, Spry, Sos, Grb2, Cbl and Shp2) is represented by a separate PRISM module. Synchronisation between modules is used to model reactions involving interactions of multiple elements. However, the different forms which each can take (for example, which other compounds it is bound to) are represented by one or more variables within the module.

Our model represents a single instance of the pathway, i.e. there can be at most one of each compound. This has the advantage that the resulting state space is relatively small (80,616 states); however, the model is highly complex due to the large number of different interactions that can occur in the pathway (there are over 560,000 transitions between states). Furthermore, as will

```
formula Frs = relocFrs2=0 ∧ degFrs2=0; // FRS2 not relocated or degraded
module FRS2
    FrsUbi      : [0..1] init 0; // ubiquitin modification of FRS2
    relocFrs2   : [0..1] init 0; // FRS2 relocated
    degFrs2     : [0..1] init 0; // FRS2 degraded
    Y196P       : [0..1] init 0;... Y471P : [0..1] init 0; // phosporilation of receptors
    // compounds bound to FRS2
    FrsFgfr     : [0..1] init 0; // 0: FGFR not bound, 1: FGFR bound
    FrsGrb      : [0..2] init 0; // 0: Grb2 not bound, 1: Grb2 bound, 2: Grb2:Sos bound
    FrsShp      : [0..1] init 0; // 0: Shp2 not bound, 1: Shp2 bound
    FrsSrc      : [0..8] init 0;
    // 0: Src not bound      1: Src bound,         2: Src:Spry
    // 3: Src:SpryP,         4: Src:SpryP:Cbl,     5: Src:SpryP:Grb
    // 6: Src:SpryP:Grb:Cbl, 7: Src:SpryP:Grb:Sos, 8: Src:SpryP:Grb:Sos:Cbl
                                    . . .
    // phosporilation of receptors (5)
    [] Frs∧ Y653P=1∧ Y654P=1∧FrsFgfr=1∧ Y196P=0 → 0.2 : (Y196P'=1); // Y196
                                    . . .
    [] Frs∧ Y653P=1∧ Y654P=1∧FrsFgfr=1∧ Y471P=0 → 0.2 : (Y471P'=1); // Y471
    // dephosporilation of Y196 (6) - remove Src if bound
    []          Frs∧ FrsShp=1∧ Y196P=1∧FrsSrc=0 → 12 : (Y196P'=0);
    [src_rel] Frs∧ FrsShp=1∧ Y196P=1∧ FrsSrc>0 → 12 : (Y196P'=0)∧(FrsSrc'=0);
                                    . . .
    // dephosporilation of Y471 (6) - remove Shp2 since bound
    [shp_rel] Frs∧ FrsShp=1∧ Y471P=1 → 12 : (Y471P'=0)∧(FrsShp'=0);
                                    . . .
    // Src:FRS2→ degFRS2 [8]
    [] Frs∧ FrsSrc>0 → 1/(15*60) : (relocFrs2'=1);
                                    . . .
    // Spry55p:Cbl+FRS2→ Frs-Ubi [12]
    [] Frs∧ FrsSrc=4,6,8 ∧ FrsUbi=0 → 0.00085 : (FrsUbi'=1);
    // FRS2-Ubi→ degFRS2 [12]
    [] Frs∧ FrsUbi=1 → 1/(5*60) : (degFrs2'=1);
                                    . . .
    // Grb2+Sos↔ Grb2:Sos [14]
    [sos_bind_frs] Frs∧ FrsGrb=1 → 1        : (FrsGrb'=2);      // Grb:FRS2
    [sos_bind_frs] Frs∧ FrsSrc=5,6→ 1        : (FrsSrc'=FrsSrc+2); // Grb:SpryP:Src:FRS2
    [sos_rel_frs]  Frs∧ FrsGrb=2 → 0.0001 : (FrsGrb'=1);        // Grb:FRS2
    [sos_rel_frs]  Frs∧ FrsSrc=7,8→ 0.0001 : (FrsSrc'=FrsSrc-2); // Grb:SpryP:Src:FRS2
                                    . . .

endmodule
```

Fig. 4. Fragment of the PRISM module for FRS2 and related compounds

be demonstrated later in the paper, the model is sufficiently rich to explain the roles of the components in the pathway and how they interact. The study of a single instance of the pathway is also motivated by the fact that the same signal dynamics (Figure 7(a)) were obtained in [22,13] for a model where the number of molecules of each type were initially set to 100. Fragments of the PRISM code for the modules representing FRS2, Src and Sos are given in Figures 4, 5 and 6, respectively. The full version is available from the PRISM web page [14].

Figure 4 shows the module for FRS2. It contains variables representing whether FRS2 is currently: undergoing ubiquitin modification ($FrsUbi$); relocated ($relocFrs2$); degraded ($degFrs2$); and bound to other compounds ($FrsFgfr$, $FrsGrb$, $FrsShp$ and $FrsSrc$). It also has variables representing the phosphorylation status of each of FRS's receptors ($Y196P, \ldots, Y471P$).

The first set of commands given in Figure 4 correspond to the phosphorylation of receptors in FRS (reaction **5** in Figure 3). Since the only variables that are

```
module SRC
   Src : [0..8] init 1;
   // 0: Src bound to FRS2,  1: Src not bound,        2: Src:Spry
   // 3: Src:SpryP,           4: Src:SpryP:Cbl,        5: Src:SpryP:Grb
   // 6: Src:SpryP:Grb:Cbl,   7: Src:SpryP:Grb:Sos,    8: Src:SpryP:Grb:Sos:Cbl

   // Src+FRS2196P↔Src:FRS2 (7)
   [src_bind] Src>0 → (Src'=0);
   [src_rel]  Src=0 → (Src'=FrsSrc);
   // Spry+Src→Spry55:Src or Spry55P+Src→Spry55P:Src (11)
   [spry_bind] Src=1 → 1 : (Src'=Spry+1);
   // Spry+Src←Spry55:Src (11)
   [spry_rel] Src=2 → 0.01 : (Src'=1);
   // Spry55P+Src←Spry55P:Sr (11)c
   [spry_rel] Src>2 → 0.0001 : (Src'=1);
   // Spry55:Src→Spry55P:Src (11)
   [] Src=2 → 10 : (Src'=3);
   // SpryP+Cbl↔SpryP:Cbl (11)
   [cbl_bind_src] Src=3,5,7→ 1        : (Src'=Src+1);
   [cbl_rel_src]  Src=4,6,8→ 0.0001 : (Src'=Src-1);
   // SpryP+Grb↔SpryP:Grb (11)
   [grb_bind_src] Src=3,4  → 1        : (Src'=Src+2*Grb);
   [grb_rel_src]  Src=5,6  → 0.0001 : (Src'=Src-2); // SOS not bound
   [grb_rel_src]  Src=7,8  → 0.0001 : (Src'=Src-4); // SOS bound
                           . . .
endmodule
```

Fig. 5. PRISM module for Src and related compounds

```
module SOS
   Sos : [0..1] init 1;

   // Grb2+Sos↔Grb:Sos
   [sos_bind]     Sos=1 → (Sos'=0); // Grb2 free
   [sos_bind_frs]Sos=1 → (Sos'=0); // Grb2:FRS2 or to Grb2:SpryP:SRC:FRS2
   [sos_rel]      Sos=0 → (Sos'=1); // Grb2 free
   [sos_rel_frs]  Sos=0 → (Sos'=1); // Grb2:FRS2 or to Grb2:SpryP:SRC:FRS2
                       . . .
endmodule
```

Fig. 6. PRISM module for Sos

updated are local to this module, the commands have no action label, i.e. we do not require any other module to synchronise on these commands. The guards of these commands incorporate dependencies on the current state both of FRS2 itself and of other compounds. More precisely, FGFR must be bound to FRS2 and certain receptors of FGFR must have already been phosphorylated.

Elsewhere, in Figure 4, we see commands that use synchronisation to model interactions with other compounds, e.g. the release of Src (the commands labelled *src_rel*) and the binding and release of Sos (the commands labelled *sos_bind_frs* and *sos_rel_frs*). Note the corresponding commands in modules *SRC* (Figure 5) and *SOS* (Figure 6). In each of these cases, as discussed in Section 3, the rate of the combined interaction is specified in the *FRS2* module and is hence omitted from the corresponding commands in *SRC* and *SOS*. Also, in the module for Sos (Figure 6), there are different action labels for the binding and release of

Sos with Grb2; this is because Grb2 can be either free or bound to a number of different compounds when it interacts with Sos. For example, Grb2 can be bound to Frs2 (through reaction **7**) or Spry (through reaction **11**), and Spry can in turn be bound to Src, which can also be bound to FRS2.

Notice how, in the commands for binding and unbinding of Src with FRS2 in Figure 4 (labelled *sos_bind_frs* and *sos_rel_frs*), we can use the value of *FrsSrc* to update the value of *Src*, rather than separating each case into individual commands. Also worthy of note are the updates to *Src* in Figure 5 when either Grb2 or Grb2:Sos bind to Src. To simplify the code, we have used a single command for each of these possible reactions, and therefore updates which either increment or decrement the variable *Src* by 2 or 4 (the variable *Grb* takes value 1 if Grb2 is not bound to Sos and value 2 if Sos is bound).

6   Property Specification

Our primary goal in this case study is to analyse the various mechanisms previously reported to negatively regulate signalling. Since the binding of Grb2 to FRS2 serves as the primary link between FGFR activation and ERK signalling, we examine the amount of Grb2 bound to FRS2 as the system evolves. In addition, we investigate the different causes of degradation which, based on the system description, can be caused by one of the following reactions occurring:

- when Src:FRS2 is present, FRS2 is relocated (reaction **8**);
- when Plc:FGFR is present, it degrades FGFR (reaction **9**);
- when phosphoSpry binds to Cbl, it degrades FRS2 (reaction **12**).

Below, we present a list of the various properties of the model that we have analysed, and the form in which they are supplied to the PRISM tool. For the latter, we define a number of *atomic propositions*, essentially predicates over the variables in the PRISM model, which can be used to identify states of the model that have certain properties of interest. These include a_{grb2}, which indicates that Grb2 is bound to FRS2 (i.e. those states where the variable *FrsGrb* of Figure 4 is greater than zero), and a_{src}, a_{plc} and a_{spry}, corresponding to the different causes of degradation/relocation given above. For properties using expected rewards (with the $\mathcal{R}_{=?}[\cdot]$ operator), we also explain the reward structure used.

A. *What is the probability that Grb2 is bound to FRS2 at the time instant T?* $(\mathcal{P}_{=?}[\mathbf{true}\ \mathcal{U}^{[T,T]}\ a_{grb2}])$;

B. *What is the expected number of times that Grb2 binds to FRS2 by time T?* $(\mathcal{R}_{=?}[\mathcal{C}^{\leq T}]$, where a reward of 1 is assigned to all transitions involving Grb2 binding to FRS2);

C. *What is the expected time that Grb2 spends bound to FRS2 within the first T time units?* $(\mathcal{R}_{=?}[\mathcal{C}^{\leq T}]$, where a reward of 1 is assigned to states where Grb2 is bound to FRS2, i.e. those satisfying atomic proposition $a_{grb2})$;

D. *What is the long-run probability that Grb2 is bound to FRS2?* $(\mathcal{S}_{=?}[a_{grb2}])$;

E. *What is the expected number of times Grb2 binds to FRS2 before degradation or relocation occurs?* $(\mathcal{R}_{=?}[F\ (a_{src}\vee a_{plc}\vee a_{spry})])$, with rewards as for **B**;

F. *What is the expected time Grb2 spends bound to FRS2 before degradation or relocation occurs?* ($\mathcal{R}_{=?}[F\ (a_{src} \vee a_{plc} \vee a_{spry})]$, with rewards as for **C**);

G. *What is the probability that each possible cause of degradation/relocation has occurred by time T?* (e.g. $\mathcal{P}_{=?}[\neg(a_{src} \vee a_{plc} \vee a_{spry})\ \mathcal{U}^{[0,T]}\ a_{src}]$ in the case Src causes relocation);

H. *What is the probability that each possible cause of degradation/relocation occurs first?* (e.g. $\mathcal{P}_{=?}[\neg(a_{src} \vee a_{plc} \vee a_{spry})\ \mathcal{U}\ a_{plc}]$ in the case when Plc causes degradation);

I. *What is the expected time until degradation or relocation occurs in the pathway?* ($\mathcal{R}_{=?}[\mathcal{F}\ (a_{src} \vee a_{plc} \vee a_{spry})]$ where all states have reward 1).

7   Results and Analysis

We used PRISM to construct the FGF model described in Section 5 and analyse the set of properties listed in Section 6. This was done for a range of different scenarios. First, we developed a base model, representing the full system, in which we suppose that initially FGF, unbound and unphosphorylated FGFR, unphosphorylated FRS2, unbound Src, Grb2, Cbl, Plc and Sos are all present in the system (Spry arrives into the system with the half-time of 10 minutes).

Subsequently, we performed a series of "in silico genetics" experiments on the model designed to investigate the roles of the various components of the activated receptor complex in controlling signalling dynamics. This involves deriving a series of modified models of the pathway where certain components are omitted (Shp2, Src, Spry or Plc), and is easily achieved in a PRISM model by just changing the initial value of the component under study. For example, to remove Src from the system we just need to change the initial value of the variable *Src* from 1 to 0 (see Figure 5).

For each property we include the statistics for 5 cases: for the full pathway and for the pathway when either Shp2, Src, Spry or Plc is removed. Figures 7(a)–(c) show the transient behaviour (i.e. at each time instant T) of the signal (binding of Grb2 to FRS2) for the first 60 minutes, namely properties **A**, **B** and **C** from the previous section. Table 1 gives the the long-run behaviour of the signal, i.e. properties **D**, **E** and **F**. The latter three results can be regarded as the values of the first three in "the limit", i.e. as either T tends to infinity or degradation occurs. Figures 7(d)–(f) show the transient probability of each of the possible causes of relocation or degradation occurring (property **G**). Table 2 shows the results relating to degradation in the long-run (properties **H** and **I**).

We begin with an analysis of the signal (binding of Grb2 to FRS2) in the full model, i.e. see the first plot ("full model") in Figure 7 and the first lines of Tables 1 and 2. The results presented demonstrate that the probability of the signal being present (Figure 7(a)) shows a rapid increase, reaching its maximum level at about 1 to 2 minutes. The peak is followed by a gradual decrease in the signal, which then levels off at a small non-zero value. In this time interval Grb2 repeatedly binds to FRS2 (Figure 7(b)) and, as time passes, Grb2 spends a smaller proportion of time bound to FRS2 (Figure 7(c)).

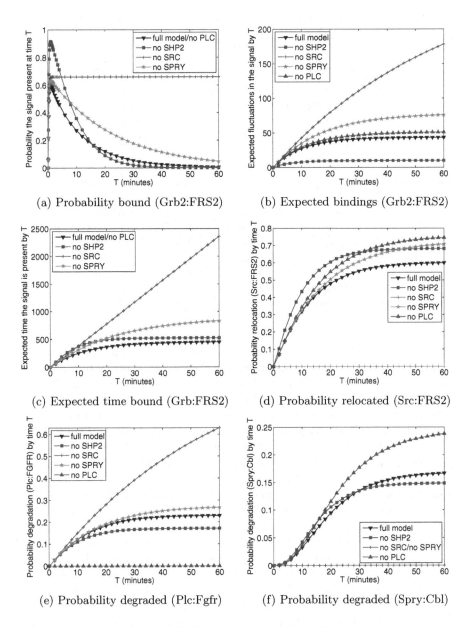

(a) Probability bound (Grb2:FRS2)

(b) Expected bindings (Grb2:FRS2)

(c) Expected time bound (Grb:FRS2)

(d) Probability relocated (Src:FRS2)

(e) Probability degraded (Plc:Fgfr)

(f) Probability degraded (Spry:Cbl)

Fig. 7. Transient numerical results

The rapid increase in the signal is due the relevant reactions (the binding of Grb2 to FRS2 triggered by phosphorylation of FRS2, which requires activated FGFR to first bind to FRS2) all occurring at very fast rates. On the other hand, the decline in the signal is caused either by dephosphorylation of FRS2 (due to Shp2 being bound to FRS2) or by relocation/degradation of FRS2.

Table 1. Long run and expected reachability properties for the signal

	probability bound	expected no. of bindings	expected time bound (min)
full model	7.54e-7	43.1027	6.27042
no Shp2	3.29e-9	10.0510	7.78927
no Src	0.659460	283.233	39.6102
no Spry	4.6e-6	78.3314	10.8791
no Plc	0.0	51.5475	7.56241

Table 2. Probability and expected time until degradation/relocation in the long run

	probability of degradation/relocation			expected time (min)
	Src:FRS2	Plc:FGFR	Spry:Cbl	
full model	0.602356	0.229107	0.168536	14.0258
no Shp2	0.679102	0.176693	0.149742	10.5418
no Src	-	1.0	0.0	60.3719
no Spry	0.724590	0.275410	-	16.8096
no Plc	0.756113	-	0.243887	17.5277

Dephosphorylation of FRS2 is both fast and allows Grb2 to rebind (as FRS2 can become phosphorylated again). The overall decline in signal is due to relocation of FRS2 caused by bound Src which takes a relatively long time to occur (Table 2 and Figure 7(d)). Degradation caused by Spry has little impact since it is not present from the start and, by the time it appears, it is more likely that Grb2 is no longer bound or Src has caused relocation (Table 2, Figure 7(d) and Figure 7(f)).

The fact that the signal levels out at a non-zero value (Table 1) is caused by Plc degrading the FGF receptor bound to FRS2 and Grb2. More precisely, after FGFR is degraded by Plc, no phosphorylation of partner FRS2 residues is possible. The signal stays non-zero since neither Src-mediated relocation and degradation, nor Shp-mediated dephosphorylation, are possible when respective FRS2 residues are not active. The non-zero value is very small because it is more likely that Src has caused relocation (Table 2). The repeated binding of Grb2 to FRS2 (Figure 7(b)) is caused by the dephosphorylation of FRS2, which is soon phosphorylated again and allows Grb2 to rebind. The decrease in the proportion of time that Grb2 is bound to FRS2 is due to the probability of FRS2 becoming relocated/degraded increasing as time passes (Figure 7(d)–(f)).

Next, we further illustrate the role of the components by analysing models in which different elements of the pathway are not present.

Shp2. Figure 7(a) shows that the peak in the signal is significantly larger than that seen under normal conditions. By removing Shp2 we have removed, as explained above, the fast reaction for the release of Grb2 from FRS2, and this justifies the larger peak. The faster decline in the signal is due to there being a greater chance of Src being bound (as Shp2 causes the dephosphorylation of FRS2, it also causes the release of Src from FRS2), and hence the increased chance relocation (Figure 7(d) and Table 2). These observations are also the cause for the decrease in the time until degradation/relocation when Shp2 is removed (Table 2) and the fact that the other causes of degradation/relocation are less likely (Figures 7(e)–(f) and Table 2). Dephosphorylation due to bound Shp2 was responsible for the large number of times that Grb2 and FRS2 bind (and unbind) in the original model; we do not see such a large number of bindings once Shp2 is removed (Figure 7(b) and Table 1).

Src. As Figure 7(a) demonstrates, the suppression of Src is predicted to have a major impact on signalling dynamics: after a fast increase, the signal fails to decrease substantially. This is supported by the results presented in both Figures 7(d)–(f) and Table 2 which show that Src is the main cause of signal degradation, and by removing Src the time until degradation or relocation greatly increases. The failure of Spry to degrade the signal (Figure 7(f) and Table 2) is attributed to its activation being downstream of Src. Note that, this also means that Plc is the only remaining cause of degradation.

Spry. The model fails to reproduce the role of Spry in inhibiting the activation of the ERK pathway by competition for Grb2:Sos. More precisely, our results show that the suppression of Spry does not result in signal reduction. This can be explained by the differences in system designs: under laboratory conditions the action of Spry is measured after Spry is over-expressed, whereas, under normal physiological conditions, Spry is known to arrive slowly into the system. Removing Spry removes one of the causes of degradation, and therefore increases the other causes of degradation/relocation (Figures 7(d)–(e) and Table 2). Moreover, the increase in the probability of Plc causing degradation/relocation leads to an increase in the chance of Grb2 and FRS2 remaining bound (Table 2).

Plc. While having a modest effect on transient signal expression, the main action of Plc removal is to cause the signal to stabilise at zero (Table 1). This is due to Plc being the only causes of degradation/relocation not relating to FRS2. The increase in time until degradation (Table 2) is also attributed to the fact that, by removing Plc, we have eliminated one of the possible causes of degradation. This also has the effect that the other causes of relocation/degradation are more likely (Figure 7(d), Figure 7(f) and Table 2).

8   Conclusions

In this paper we have shown that probabilistic model checking can be a useful tool in the analysis of biological pathways. The technique's key strength is that it allows the calculation of exact quantitative properties for system events occurring over time, and can therefore support a detailed, quantitative analysis of the interactions between the pathway components. By developing a model of a complex, realistic signalling pathway that is not yet well understood, we were able to demonstrate, firstly, that the model is robust and that its predictions agree with biological data [22,13] and, secondly, that probabilistic model checking can be used to obtain a wide range of quantitative measures of system dynamics, thus resulting in deeper understanding of the pathway.

We intend to perform further analysis of the FGF pathway, including an investigation into the effect that changes to reaction rates and initial concentrations will have on the pathway's dynamics. Future work will involve both comparing this probabilistic model checking approach with simulation and ODEs, and also investigation of how to scale the methodology yet further.

References

1. A. Aziz, K. Sanwal, V. Singhal, and R. Brayton. Verifying continuous time Markov chains. In *Proc. CAV'96*, volume 1102 of *LNCS*, pages 269–276. Springer, 1996.
2. C. Baier, B. Haverkort, H. Hermanns, and J.-P. Katoen. Model checking continuous-time Markov chains by transient analysis. In *Proc. CAV'00*, volume 1855 of *LNCS*, pages 358–372. Springer, 2000.
3. M. Calder, S. Gilmore, and J. Hillston. Modelling the influence of RKIP on the ERK signalling pathway using the stochastic process algebra PEPA. *Transactions on Computational Systems Biology*, 2006. To appear.
4. M. Calder, V. Vyshemirsky, D. Gilbert, and R. Orton. Analysis of signalling pathways using continuous time Markov chains. *Transactions on Computational Systems Biology*, 2006. To appear.
5. I. Dikic and S. Giordano. Negative receptor signalling. *Curr Opin Cell Biol.*, 15:128–135, 2003.
6. V. Eswarakumar, I. Lax, and J. Schlessinger. Cellular signaling by fibroblast growth factor receptors. *Cytokine Growth Factor Rev.*, 16(2):139–149, 2005.
7. D. Gillespie. Exact stochastic simulation of coupled chemical reactions. *Journal of Physical Chemistry*, 81(25):2340–2361, 1977.
8. J. Hillston. *A Compositional Approach to Performance Modelling*. Cambridge University Press, 1996.
9. A. Hinton, M. Kwiatkowska, G. Norman, and D. Parker. PRISM: A tool for automatic verification of probabilistic systems. In *Proc. TACAS'06*, volume 3920 of *LNCS*, pages 441–444. Springer, 2006.
10. M. Kwiatkowska, G. Norman, and D. Parker. Probabilistic symbolic model checking with PRISM: A hybrid approach. *International Journal on Software Tools for Technology Transfer (STTT)*, 6(2):128–142, 2004.
11. M. Kwiatkowska, G. Norman, and D. Parker. Probabilistic model checking in practice: Case studies with PRISM. *ACM SIGMETRICS Performance Evaluation Review*, 32(4):16–21, 2005.
12. A. Phillips and L. Cardelli. A correct abstract machine for the stochastic pi-calculus. In *Proc.BioCONCUR'04*, ENTCS. Elsevier, 2004.
13. www.cs.bham.ac.uk/~oxt/fgfmap.html.
14. PRISM web site. www.cs.bham.ac.uk/~dxp/prism.
15. C. Priami. Stochastic π-calculus. *The Computer Journal*, 38(7):578–589, 1995.
16. C. Priami, A. Regev, W. Silverman, and E. Shapiro. Application of a stochastic name passing calculus to representation and simulation of molecular processes. *Information Processing Letters*, 80:25–31, 2001.
17. A. Regev and E. Shapiro. Cellular abstractions: Cells as computation. *Nature*, 419(6905):343, 2002.
18. A. Regev, W. Silverman, and E. Shapiro. Representation and simulation of biochemical processes using the pi- calculus process algebra. In *Pacific Symposium on Biocomputing*, volume 6, pages 459–470. World Scientific Press, 2001.

19. J. Rutten, M. Kwiatkowska, G. Norman, and D. Parker. *Mathematical Techniques for Analyzing Concurrent and Probabilistic Systems*, volume 23 of *CRM Monograph Series*. AMS, 2004.
20. J. Schlessinger. Epidermal growth factor receptor pathway. Sci. STKE (Connections Map), http://stke.sciencemag.org/cgi/cm/stkecm;CMP_14987.
21. M. Tsang and I. Dawid. Promotion and attenuation of FGF signaling through the Ras-MAPK pathway. *Science STKE*, pe17, 2004.
22. O. Tymchyshyn, G. Norman, J. Heath, and M. Kwiatkowska. Computer assisted biological reasoning: The simulation and analysis of FGF signalling pathway dynamics. Submitted for publication.

Type Inference in Systems Biology

François Fages and Sylvain Soliman

Projet Contraintes, INRIA Rocquencourt,
BP105, 78153 Le Chesnay Cedex, France
Firstname.Lastname@inria.fr
http://contraintes.inria.fr

Abstract. Type checking and type inference are important concepts and methods of programming languages and software engineering. Type checking is a way to ensure some level of consistency, depending on the type system, in large programs and in complex assemblies of software components. Type inference provides powerful static analyses of pre-existing programs without types, and facilitates the use of type systems by freeing the user from entering type information. In this paper, we investigate the application of these concepts to systems biology. More specifically, we consider the Systems Biology Markup Language SBML and the Biochemical Abstract Machine BIOCHAM with their repositories of models of biochemical systems. We study three type systems: one for checking or inferring the functions of proteins in a reaction model, one for checking or inferring the activation and inhibition effects of proteins in a reaction model, and another one for checking or inferring the topology of compartments or locations. We show that the framework of abstract interpretation elegantly applies to the formalization of these abstractions and to the implementation of linear time type checking as well as type inference algorithms. Through some examples, we show that the analysis of biochemical models by type inference provides accurate and useful information. Interestingly, such a mathematical formalization of the abstractions used in systems biology already provides some guidelines for the extensions of biochemical reaction rule languages.

1 Introduction

Type checking and type inference are important concepts and methods of programming languages and software engineering [1]. Type checking is a way to ensure some level of consistency, depending on the type system, in large programs and in complex assemblies of software components. Type inference provides powerful static analyzes of pre-existing programs without types, and facilitates the use of type systems by freeing the user from entering type information.

In this paper, we investigate the application of these concepts to systems biology. More specifically, we consider the Systems Biology Markup Language SBML [2] and the Biochemical Abstract Machine BIOCHAM [3]. In both of these languages, the biochemical models are described through a set of reaction rules. We study three type systems:

C. Priami (Ed.): CMSB 2006, LNBI 4210, pp. 48–62, 2006.
© Springer-Verlag Berlin Heidelberg 2006

1. one for checking or inferring the protein functions in a reaction model,
2. one for checking or inferring the activation and inhibition effects in a reaction model,
3. and another one for checking or inferring the topology of compartments or locations in reaction models with space considerations.

To this end, the formal framework of abstract interpretation will be used to provide type systems with a precise mathematical definition. Abstract interpretation is a theory of abstraction introduced by Cousot and Cousot in [4] as a framework for reasoning about programs, their semantics, and for designing static analysers, among which type inference systems [5]. Although not strictly necessary to the presentation of the type inference methods considered in this paper, we believe that that formal framework is very relevant to systems biology, as a formalism for providing a mathematical sense to modeling issues concerning multiple abstraction levels and their formal relationship.

We show that the framework of abstract interpretation elegantly applies to the formalization of the three abstractions considered in this paper and to the implementation of linear time type checking as well as type inference algorithms. Through examples of biochemical systems coming from the biomodels.net and BIOCHAM repositories of models, we show that the static analysis of reaction models by type inference provides both accurate and useful information. Interestingly, we show that these considerations also provide some guidelines concerning the extensions of biochemical reaction rule-based languages.

2   Preliminaries on Abstract Interpretation, Type Checking and Type Inference

2.1   Concrete Domain of Reaction Models

Following SBML and BIOCHAM conventions, a model of a biochemical system is a set of reaction rules of the form e for S => S' where S is a set of molecules given with their stoichiometric coefficient, called a *solution*, S' is the transformed solution, and e is a kinetic expression involving the concentrations of molecules (which are not strictly required to appear in S). The set of molecules is noted $\mathcal{M}$. We will use the BIOCHAM operators + and * to denote solutions as 2*A + B, as well as the syntax of catalyzed reactions e for S =[C]=> S' as an abbreviation for e for S+C => S'+C.

A set of reaction rules like $\{e_i$ for S_i => $S_i'\}_{i=1,\ldots,n}$ over molecular concentration variables $\{x_1, \ldots, x_m\}$, can be interpreted under different semantics. The traditional *differential semantics* interpret the rules by the following system of Ordinary Differential Equations (ODE):

$$dx_k/dt = \sum_{i=1}^{n} r_i(x_k) * e_i - \sum_{j=1}^{n} l_j(x_k) * e_j$$

where $r_i(x_k)$ (resp. l_i) is the stoichiometric coefficient of x_k in the right (resp. left) member of rule i. Thanks to its wide range of mathematical tools, this semantics is the most commonly used, when the data is available and the system of a reasonable size. The *stochastic semantics* interpret the kinetic expressions as transition probabilities (see for instance [6]), while the *boolean semantics* forget the kinetic expressions and interpret the rules as a non-deterministic (asynchronous) transition system over boolean states representing the absence or presence of molecules. In BIOCHAM these three semantics are implemented [7], while in the SBML exchange format, no particular semantics are defined.

For the simple analyzes considered in this paper, the concrete domain of reaction models will be the syntactic domain of formal reaction rules, with no other semantics than a data structure. A reaction model is thus a set of reaction rules, and the domain of reaction models is ordered by set inclusion, i.e. by the information ordering.

Definition 1. *The universe of reactions is the set of possible rules*
$$\mathcal{R} = \{e \text{ for } S \Rightarrow S' \mid e \text{ is a kinetic expression,}$$
$$\text{and } S \text{ and } S' \text{ are solutions } \}.$$
The concrete domain $\mathcal{D}_\mathcal{R}$ of reaction models is the power-set of reaction rules ordered by inclusion $\mathcal{D}_\mathcal{R} = (\mathcal{P}(\mathcal{R}), \subseteq)$.

2.2   Abstract Domains, Abstractions and Galois Connections

In the general setting of abstract interpretation, an abstract domain is a lattice $L(\sqsubseteq, \bot, \top, \sqcup, \sqcap)$ defined by the set L and the partial order $\sqsubseteq$, and where $\bot$, $\top$, $\sqcup$, $\sqcap$ denote the least element, the greatest element, the least upper bound and the greatest lower bound respectively.

As often the case in program analysis, the concrete domain and the abstract domains considered for analyzing biochemical models, are power-sets, that is set lattices $\mathcal{P}(\mathcal{S})(\subseteq, \emptyset, \mathcal{S}, \cup, \cap)$ ordered by inclusion, with the empty set as $\bot$ element, and the base set $\mathcal{S}$ (such as the universe of reaction rules here) as $\top$ element. An abstraction is formalized by a Galois connection as follows [4]:

Definition 2. *A Galois connection $C \to_\alpha A$ between two lattices C and A is defined by abstraction and concretization functions $\alpha : C \to A$ and $\gamma : A \to C$ that satisfy $\forall c \in C, \forall y \in A : x \sqsubseteq_C \gamma(y) \Leftrightarrow \alpha(x) \sqsubseteq_A y$.*

For any Galois connection, we have the following properties:

1. $\gamma \circ \alpha$ is extensive (i.e. $x \sqsubseteq_C \gamma \circ \alpha(x)$) and represents the information lost by the abstractions;
2. α preserves $\sqcup$, and γ preserves $\sqcap$;
3. α is one-to-one *iff* γ is onto *iff* $\gamma \circ \alpha$ is the identity.

If $\gamma \circ \alpha$ is the identity, the abstraction α loses no information, and C and A are isomorphic from the information standpoint (although γ may not be one-to-one).

We will consider three abstract domains: one for protein functions, where molecules are abstracted into categories such as kinases and phosphatases, one for the influence graph, where the biochemical reaction rules are abstracted in activation and inhibition binary relations between molecules, and one for location topologies, where the reaction (and transport) rules are abstracted retaining only the neighborhood information between locations.

2.3 Type Checking and Type Inference by Abstract Interpretation

In this setting, a type system A for a concrete domain C is simply a Galois connection $C \rightarrow_\alpha A$. The *type inference* problem is, given a concrete element $x \in C$ (e.g. a reaction model) to compute $\alpha(x)$ (e.g. the protein functions that can be inferred from the reactions). The *type checking* problem is, given a concrete element $x \in C$ and a typing $y \in A$ (e.g. a set of protein functions), to determine whether $x \sqsubseteq_C \gamma(y)$ (i.e. whether the reactions provide less and compatible information on the protein functions) which is equivalent to $\alpha(x) \sqsubseteq_A y$ (i.e. whether the typing contains the inferred types).

The simple type systems considered in this paper will be implemented with type checking and type inference algorithms that basically browse the reactions, and check or collect the type information for each rule independently and in linear time.

3 A Type System for Protein Functions

To investigate the use of type inference in the domain of protein functions we first restrict ourselves to two simple functions: kinase and phosphatase. These correspond to the action of adding (resp. removing) a phosphate group to (resp. from) a compound.

For the sake of simplicity, we do not consider other categories such as protease (in degradation rules), acetylase and deacetylase (in modification rules), etc. This choice is in accordance with the BIOCHAM syntax which allows to mark the modified sites of a protein with the operator ~, as in P~{p,q} without distinguishing however between a phosphorylation and an acetylation for instance. We thus consider BIOCHAM models containing compounds with different levels of phosphorylation or acetylation, without distinguishing the different forms of modification, and call them phosphorylation, by abuse of terminology.

The analysis of protein functions in a reaction model is interesting for several reasons. First, the kind of information (kinase activity) collected on proteins can be checked using online databases like GO, the Gene Ontology [8]. Second, in the context of the machine learning techniques implemented in BIOCHAM for completing or revising a model w.r.t. a temporal logic specification [7], the information that an enzyme acts as a kinase or as a phosphatase drastically reduce the search space for reaction additions, and help find more biologically plausible model revisions.

3.1 Abstract Domain of Protein Functions

Definition 3. *The abstract domain of protein functions $\mathcal{D_F}$ is the domain of functions from molecules $\mathcal{M}$ to pairs of booleans, representing "has kinase function" (true/false) and "has phosphatase function" (true/false).*

Definition 4. $\alpha : \mathcal{D_R} \rightarrow \mathcal{D_F}$ *is defined for each molecule as the disjunction of α on each single rule and each pair of rules:*

α(A =[B]=> C) = *where* C *is more phosphorylated than* A *(i.e. its set of active phosphorylation sites strictly includes that of* A*) is abstracted as* B *has kinase function.*

α(A =[B]=> C) = *where, on the contrary,* A *is more phosphorylated than* C, *we abstract that* B *has phosphatase function.*

α(A + B => A-B, A-B => C + B) = *where* C *is more phosphorylated than* A *is abstracted as* B *has kinase function.*

α(A + B => A-B, A-B => C + B) = *where, on the contrary,* A *is more phosphorylated than* C, *we abstract that* B *has phosphatase function.*

3.2   Evaluation Results

MAPK model. On a simple example of the MAPK cascade extracted from the SBML repository and originally based on [9], the type inference algorithm determines that RAFK, RAF~{p1} and MEK~{p1,p2} have a kinase function; RAFPH, MEKPH and MAPKPH have a phosphatase function; and the other compounds have no function inferred.

If we wanted to type-check such a model, we would correctly check all phosphatases but would miss an example of the kinase function of MAPK~{p1,p2}, since its action is not visible in the above model.

Kohn's Map. Kohn's map of the mammalian cell cycle control [10] has been transcribed in BIOCHAM to serve as a large benchmarking example of 500 species and 800 rules [11]. To check if this abstraction scales up we tried it on this model, and indeed obtain the answer in less than one second CPU time (on a PC 1,7GHz). Here is an excerpt of the output of the type inference:

```
cdk7-cycH is a kinase
Wee1 is a kinase
Myt1 is a kinase
cdc25C~{p1} is a phosphatase
cdc25C~{p1,p2} is a phosphatase
Chk1 is a kinase
C-TAK1 is a kinase
Raf1 is a kinase
cdc25A~{p1} is a phosphatase
cycA-cdk1~{p3} is a kinase
cycA-cdk2~{p2} is a kinase
cycE-cdk2~{p2} is a kinase
cdk2~{p2}-cycE~{p1} is a kinase
cycD-cdk46~{p3} is a kinase
cdk46~{p3}-cycD~{p1} is a kinase
cycA-cdk1~{p3} is a kinase
cycB-cdk1~{p3} is a kinase
cycA-cdk2~{p2} is a kinase
cycD-cdk46~{p3} is a kinase
cdk46~{p3}-cycD~{p1} is a kinase
Plk1 is a kinase
```

```
pCAF is a kinase
p300 is a kinase
HDAC1 is a phosphatase
```

It is worth noticing that in these results no compound is both a kinase and a phosphatase. The `cdc25 A` and `C` are the only phosphatases found in the whole map with `HDAC1`). The type inference also tells us that the cyclin-dependant kinases have a kinase function when in complex with a cyclin. Finally the acetylases `pCAF`, `p300` and the deacetylase `HDAC1` are detected but identified to kinases and phosphatases respectively, since the BIOCHAM syntax does not distinguish between phosphorylation and acetylation.

4   A Type System for Activation and Inhibitory Influences

4.1   Abstract Domain of Influences

Influence networks for activation and inhibition have been introduced for the analysis of gene expression in the setting of gene regulatory networks [12]. Such influence networks are in fact an abstraction of complex reaction networks, and can be applied as such to protein interaction networks. However the distinction between the influence network and the reaction network is crucial to the application of Thomas's conditions of multistationarity and oscillations [12,13] to protein interaction network, and there has been some confusion between the two kinds of networks [14]. Here we precisely define influence networks as an abstraction of (or a type system for) reaction networks.

Definition 5. *The abstract domain of influences is the powerset of the binary relations of activation and inhibition between compounds* $\mathcal{D}_\mathcal{I} = \mathcal{P}(\{A$ *activates* $B \mid A, B \in \mathcal{M}\} \cup \{A$ *inhibits* $B \mid A, B \in \mathcal{M}\})$.
   The influence abstraction $\alpha : \mathcal{D}_\mathcal{R} \to \mathcal{D}_\mathcal{I}$ *is the function*
$$\alpha(x) = \{A \text{ inhibits } B \quad \mid \exists(e_i \text{ for } S_i \Rightarrow S'_i) \in x,$$
$$l_i(A) > 0 \text{ and } r_i(B) - l_i(B) < 0\}$$
$$\cup \{A \text{ activates } B \mid \exists(e_i \text{ for } S_i \Rightarrow S'_i) \in x,$$
$$l_i(A) > 0 \text{ and } r_i(B) - l_i(B) > 0\}$$

In particular, we have the following influences for elementary reactions of complexation, modification, synthesis and degradation:

$\alpha(\{A + B => C\}) = \{$ A inhibits B, A inhibits A, B inhibits A,
                                       B inhibits B, A activates C, B activates C$\}$
$\alpha(\{A = [C] => B\}) = \{$ C inhibits A, A inhibits A, A activates B, C activates B$\}$
$\alpha(\{A = [B] => \_\}) = \{$ B inhibits A, A inhibits A$\}$
$\alpha(\{\_ = [B] => A\}) = \{$ B activates A$\}$

The inhibition loops on the reactants are justified by the negative sign in the Jacobian matrix of the differential semantics of such reactions. It is worth noting however that they are often omitted in the influence graphs considered in the literature, as well as with some other influences, according to functionality, kinetic and non-linearity considerations.

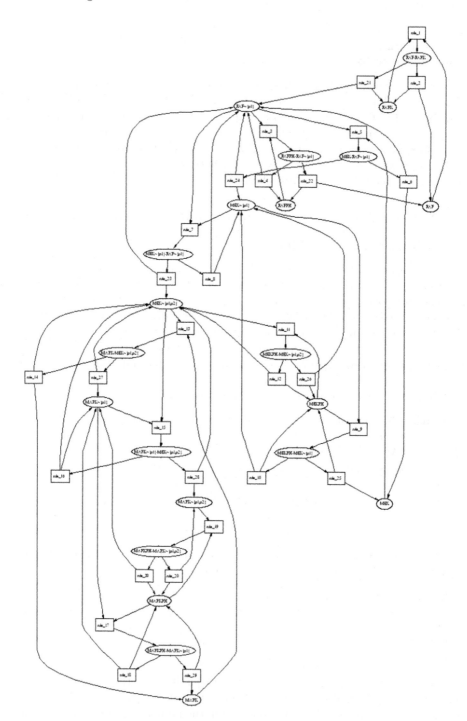

Fig. 1. Reaction graph of the MAPK model

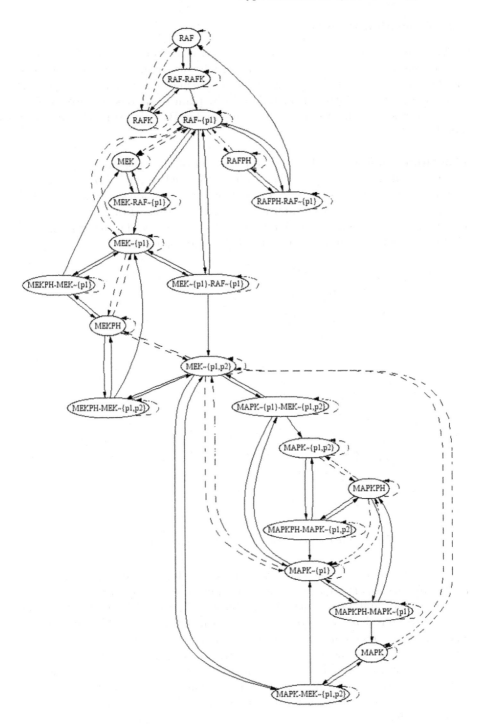

Fig. 2. Inferred influence graph of the MAPK model

4.2   Evaluation Results

MAPK model. Let us first consider the MAPK signalling model of [9]. Fig. 1 depicts the reaction graph as a bipartite graph with round boxes for molecules and rectangular boxes for rules. Fig. 2 depicts the inferred influence graph, where activation (resp. inhibition) is materialized by plain (resp. dashed) arrows. The graph layouts of the figures have been computed in BIOCHAM by the Graphviz suite[1].

p53-Mdm2 model. In the p53-Mdm2 model of [15], the protein $Mdm2$ is localized explicitly in two possible locations: the nucleus and in the cytoplasm, and transport rules are considered. Fig. 4 depicts the reaction graph of the model.

Fig. 3 depicts the inferred influence graph. Note that $Mdm2$ in the nucleus has both an activation and an inhibitory effect on $p53$ u. This corresponds to different influences in different regions of the phase space.

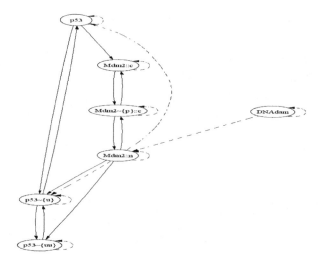

Fig. 3. Inferred influence graph of the p53-Mdm2 model

Fig. 5 depicts the core influence graph considered for the logical analysis of this model [16]. In the core influence graph, some influence are neglected, as expected, however some inhibitions, such the inhibitory effect of $p53$ on $Mdm2$ in the nucleus, are considered while they do not appear in the inferred influence graph. The reason for these omissions is the way the reaction model is written. Some inhibitory effects are indeed expressed in the kinetic expression by subtraction of, or division by, the molecular concentration of some compounds that do not appear in the rule itself. Those inhibitions are thus missed by the type inference algorithm. An example of such a rule is the following one for the inhibition of $Mdm2$ by $p53$:

---

[1] http://www.graphviz.org/

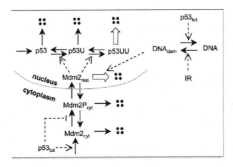

Fig. 4. Original reaction graph considered in [15] for the p53-Mdm2 model

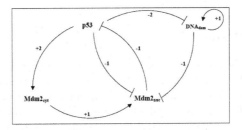

Fig. 5. Core influence graph

```
macro(p53tot,[p53]+[p53~{u}]+[p53~{uu}]).
(kph*[Mdm2::c]/(Jph+p53tot),MA(kdeph))for Mdm2::c <=> Mdm2~{p}::c.
```

Obviously, we cannot expect to infer such inhibitory effects from the kinetic expressions with all generality, however the model being written that way without fully decomposing all influences by reaction rules, a refinement of the abstraction function taking into account the kinetic expression is worth investigating. As an alternative, one could extend the syntax of reaction rules in order to indicate the inhibitors of the reaction, in a somewhat symmetric fashion to catalysts.

Kohn's Map. On Kohn's map, the type inference of activation and inhibition influences takes less than one second CPU time (on a PC 1,7GHz) for the complete model, showing again the efficiency of the type inference algorithm.

5   A Type System for Location Topologies

To date, models of biochemical systems generally abstract from space considerations. Models taking into account cell compartments and transport phenomena are thus much less common. Nevertheless, with the advent of systems biology computational tools, more and more models are refined with space considerations and transport delays, e.g. [15]. In SBML [2] level 1 version 1, locations have been introduced as purely symbolic compartments without topology. We show in this section how the topology can be inferred from the reaction rules, and checked in different models.

5.1 Abstract Domain of Location Topologies

Definition 6. *Abstract domain of neighborhood relation $\mathcal{D}_\mathcal{N}$ is a relation on pairs of molecules $\mathcal{M} \times \mathcal{M}$.*

Definition 7. $\alpha : \mathcal{D}_\mathcal{R} \to \mathcal{D}_\mathcal{N}$ *is defined by the union of its definition on single rules:*

$\alpha(\texttt{E for } \texttt{A}_1 + \cdots + \texttt{A}_n \texttt{ => } \texttt{B}_1 + \cdots + \texttt{B}_m) = $ *All A_i and all B_j are pairwise neighbors, and for all C_k such that [C_k] appears in E, C_k is a neighbor of all A_i and all B_j.*

5.2 Evaluation Results

Models from `biomodels.net`. We have taken models from the literature through the `biomodels.net` database. Of the 50 models in the current version (dated January 2006) only 13 have more than one compartment, and only 7 of those use the *outside* attribute of SBML to provide more topological insight.

The neighboring relation is inferred in these models imported in BIOCHAM, and then checked consistent with the provided *outside* relation.

For instance for calcium oscillations, we tried both the Marhl et al. model of [17] and the Borghans et al. model of [18].

In the first case (model `BIOMD0000000039.xml`), three locations are defined: the cytosol, the endoplasmic reticulum and a mitochondria, from the reactions the inferred topology is that the cytosol is neighbor of the two other locations. This correspond exactly to the information obtained from the *outside* annotations (the cytosol being marked as the outside of the two other locations).

In the second case (models BIOMD0000000043.xml to `BIOMD0000000045.xml`) we focused on the last model (*two-pool*) since it is the only one with 4 different locations: the extracellular space, the cytosol and two internal vesiculae. The location inference produces a topology where the cytosol is neighbor of all other locations. Once again this is correct w.r.t. the outside information provided in the SBML file: both vesiculae have the cytosol as outside location and the cytosol itself has the extracellular space as outside location.

These considerations show that there is some mismatch between the SBML reaction models and the choice of expressing outside vs neighborhood properties of locations. In the perspective of type checking and type inference, neighborhood relations should be preferred as they can be checked, or inferred from the reaction model, whereas the outside relation contain more information that, while helpful for the modeler as meta-data, cannot be handled automatically without abstracting it first in neighbors properties.

P53/Mdm2. The first example comes from [15]: a model of the p53/Mdm2 interaction with two locations where the transport between cytoplasm and nucleus is necessary to explain some time delays observed in the mutual repression of these proteins.

```
biocham: load_biocham('EXAMPLES/locations/p53Mdm2.bc').
...
biocham: show_neighborhood.
c and n are neighbors
```

In this precise case, the model as published does not systematically use the volume ratio in the kinetics. The transcription and type-checking of the model showed that if one wanted to keep the background degradation rate of $Mdm2$ (without DNA damage) independent of the location, one obtains different kinetics than those of the published model. In this case a formal transcription in BIOCHAM (or SBML) provided a supplementary model-validation step.

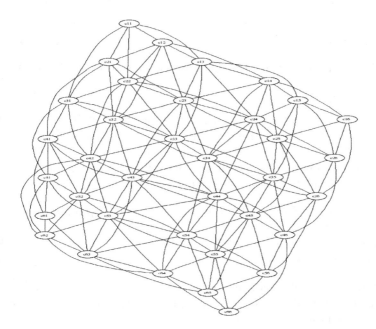

Fig. 6. Delta-Notch square cell grid inferred in a 6x6 model, with modifiers, reactants and products as pairwise neighbors

Delta and Notch Model. The next example is adapted from [19]. The Delta and Notch proteins are crucial to the cell fate in several different organisms. A population of neighboring cells (here we chose a square grid) is represented through locations and the model allows to observe the salt-and-pepper coloring (corresponding to high Delta-low Notch/low Delta-high Notch) typical of the Delta-Notch lateral inhibition based differentiation. The signaling pathways are

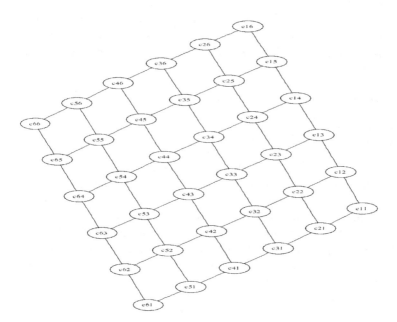

Fig. 7. Delta-Notch square cell grid inferred in a 6x6 model, without modifier-modifier neighborhood

simplified to the extreme to take into account only the direct effect of Delta and Notch expression on the local and neighboring cells. This example would thus not provide a good basis for the abstraction of section 4.

Depending on the abstraction chosen we obtain figure 6 and 7. In the first case the abstraction used is not the one given in section 5.1 but

Definition 8. $\alpha : \mathcal{D}_\mathcal{R} \to \mathcal{D}_\mathcal{N}$ *is defined by the union of its definition on single rules:*

$\alpha(\texttt{E for } A_1 + \cdots + A_n \texttt{ => } B_1 + \cdots + B_m) = $ *All A_i, all B_j, and all C_k such that $[C_k]$ appears in E, are pairwise neighbors.*

This was indeed a reasonable candidate for an abstraction, but proved too coarse on some examples since co-modifiers are often put in the kinetic expression of a single rule for simplification purposes.

6   Conclusion

We have shown that the framework of abstract interpretation applies to the formalization of some abstractions commonly used in systems biology, and to the implementation of linear-time type checking as well as type inference algorithms.

In the three type systems studied in this paper, for protein functions, activation and inhibitory influences, and location topologies respectively, the analyses are based on static information gained directly from the syntax of reaction rules, without considering their formal semantics, nor their precise dynamics. It is worth noting that this situation also occurs in program analysis where the syntax of programs may capture a sufficient part of the semantics for many analyses. Here, it is remarkable that such simple analyses already provide useful information on biological models, independently from their dynamics for which different definitions are considered (discrete, continuous, stochastic, etc.) [7].

The formal definition of the influence graph as an abstraction of the reaction model eliminates some confusion that exists in the use of Thomas's conditions [12,13] for the analysis of reaction models [14]. Such a formalization shows also that the influence graphs usually considered in the literature are further abstractions obtained by forgetting some influences, based on non-linearity considerations [20]. Some inhibitions may also be missing in the inferred influences when they are hidden in the kinetic expressions of the reactions and do not appear explicitly in the reactants. This suggests either to refine the abstraction function to take into account the kinetic expression when possible, or to extend the syntax of reactions in order to make explicit such inhibitory effects, in a symmetric fashion to catalysts for activations. In SBML there is actually an unique symmetrical notion of *Modifiers* which is not sufficient to infer the influence graph.

Similarly, the inference of protein functions and of location neighborhood have shown that the static analysis of reaction models by type inference provides both accurate and useful information. They also provide some guidelines for the extensions of biochemical reaction languages, like for instance in SBML considering neighborhood rather than outside properties, and introducing a syntax for the modification of compounds, and in BIOCHAM differentiating phosphorylation from other forms of modifications like acetylation.

Acknowledgement. This work benefited from partial support of the Network of Excellence REWERSE of the European Union.

References

1. Cardelli, L.: Typeful programming. In Neuhold, E.J., Paul, M., eds.: Formal Description of Programming Concepts. Springer-Verlag, Berlin (1991) 431–507
2. Hucka, M., et al.: The systems biology markup language (SBML): A medium for representation and exchange of biochemical network models. Bioinformatics **19** (2003) 524–531
3. Fages, F., Soliman, S., Chabrier-Rivier, N.: Modelling and querying interaction networks in the biochemical abstract machine BIOCHAM. Journal of Biological Physics and Chemistry **4** (2004) 64–73
4. Cousot, P., Cousot, R.: Abstract interpretation: A unified lattice model for static analysis of programs by construction or approximation of fixpoints. In: POPL'77: Proceedings of the 6th ACM Symposium on Principles of Programming Languages, New York, ACM Press (1977) 238–252 Los Angeles.
5. Cousot, P.: Types as abstract interpretation (invited paper). In: POPL'97: Proceedings of the 24th ACM Symposium on Principles of Programming Languages, New York, ACM Press (1997) 316–331 Paris.

6. Gillespie, D.T.: Exact stochastic simulation of coupled chemical reactions. Journal of Physical Chemistry **81** (1977) 2340–2361

7. Calzone, L., Chabrier-Rivier, N., Fages, F., Soliman, S.: Machine learning biochemical networks from temporal logic properties. Transactions on Computational Systems Biology (2006) CMSB'05 Special Issue (to appear).

8. Ashburner, M., Ball, C.A., Blake, J.A., Botstein, D., Butler, H., Cherry, J.M., Davis, A.P., Dolinski, K., Dwight, S.S., Eppig, J.T., Harris, M.A., Hill, D.P., Issel-Tarver, L., Kasarskis, A., Lewis, S., Matese, J.C., Richardson, J.E., Ringwald, M., Rubin, G.M., Sherlock, G.: Gene ontology: tool for the unification of biology. Nature Genetics **25** (2000) 25–29

9. Levchenko, A., Bruck, J., Sternberg, P.W.: Scaffold proteins may biphasically affect the levels of mitogen-activated protein kinase signaling and reduce its threshold properties. PNAS **97** (2000) 5818–5823

10. Kohn, K.W.: Molecular interaction map of the mammalian cell cycle control and DNA repair systems. Molecular Biology of the Cell **10** (1999) 2703–2734

11. Chabrier-Rivier, N., Chiaverini, M., Danos, V., Fages, F., Schächter, V.: Modeling and querying biochemical interaction networks. Theoretical Computer Science **325** (2004) 25–44

12. Thomas, R., Gathoye, A.M., Lambert, L.: A complex control circuit : regulation of immunity in temperate bacteriophages. European Journal of Biochemistry **71** (1976) 211–227

13. Soulé, C.: Graphic requirements for multistationarity. ComplexUs **1** (2003) 123–133

14. Markevich, N.I., Hoek, J.B., Kholodenko, B.N.: Signaling switches and bistability arising from multisite phosphorylation in protein kinase cascades. Journal of Cell Biology **164** (2005) 353–359

15. Ciliberto, A., Novák, B., Tyson, J.J.: Steady states and oscillations in the p53/mdm2 network. Cell Cycle **4** (2005) 488–493

16. Kaufman, M.: Private communication. (2006)

17. Marhl, M., Haberichter, T., Brumen, M., Heinrich, R.: Complex calcium oscillations and the role of mitochondria and cytosolic proteins. BioSystems **57** (2000) 75–86

18. Borghans, J., Dupont, G., Goldbeter, A.: Complex intracellular calcium oscillations: a theoretical exploration of possible mechanisms. Biophysical Chemistry **66** (1997) 25–41

19. Ghosh, R., Tomlin, C.: Lateral inhibition through delta-notch signaling: A piecewise affine hybrid model. In Springer-Verlag, ed.: Proceedings of the 4th International Workshop on Hybrid Systems: Computation and Control, HSCC'01. Volume 2034 of Lecture Notes in Computer Science., Rome, Italy (2001) 232–246

20. Thomas, R., Kaufman, M.: Multistationarity, the basis of cell differentiation and memory. Chaos **11** (2001) 170–195

Stronger Computational Modelling of Signalling Pathways Using Both Continuous and Discrete-State Methods

Muffy Calder[1], Adam Duguid[2], Stephen Gilmore[2], and Jane Hillston[2]

[1] Department of Computing Science, University of Glasgow, Glasgow G12 8QQ,
Scotland
[2] Laboratory for Foundations of Computer Science, The University of Edinburgh,
Edinburgh EH9 3JZ, Scotland

Abstract. Starting from a biochemical signalling pathway model expressed in a process algebra enriched with quantitative information we automatically derive both continuous-space and discrete-state representations suitable for numerical evaluation. We compare results obtained using implicit numerical differentiation formulae to those obtained using approximate stochastic simulation thereby exposing a flaw in the use of the differentiation procedure producing misleading results.

1   Introduction

The malfunction of cellular signalling processes has significant detrimental effects, leading to uncontrolled cell proliferation, as in cancer; or leading to other cells in the body being attacked, as in auto-immune diseases. The dynamics of cell signalling mechanisms are profoundly complex and at present are not fully understood. Computational modelling of cell signal transduction is an important intellectual tool in the scientific study of the biological processes which control and regulate cellular function.

An example of an influential computational study of intracellular signal networks is [1]. The authors develop an ordinary differential equation (ODE) model of epidermal growth factor (EGF) receptor signal pathways in order to give insight into the activation of the MAP kinase cascade through the kinases Raf, MEK and ERK-1/2. The ODE model is substantial, consisting of 94 state variables and 95 parameters. It is analysed using the numerical integration procedures of the Matlab numerical computing platform and tested using sensitivity analysis. The results increase our understanding of EGF receptor signal transduction and suggest avenues for experimental work to test hypotheses generated from the computational model. Published in 2002 the article is highly regarded and has subsequently been cited by as many as 150 other research papers.

We have previously proposed a method of investigating cell signalling pathways using a process algebra enhanced with quantitative information, PEPA [2], applied in [3] and [4]. Process algebras are well-known in theoretical computer science but are still unfamiliar to most computational biologists so we wished to

C. Priami (Ed.): CMSB 2006, LNBI 4210, pp. 63–77, 2006.
© Springer-Verlag Berlin Heidelberg 2006

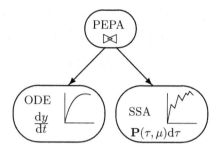

Fig. 1. A high-level model in the PEPA process algebra can be used to generate either a system of ODEs or a stochastic simulation

help to establish their relevance by reproducing the results of [1], starting from the published paper together with its supplementary material and the Matlab ODE model made available by the authors.

We were able to reproduce the results from [1] starting from our model in the PEPA process algebra but because we were starting from the vantage point of modelling in process algebra we could apply other analysis procedures, unavailable to the authors of [1] (Figure 1). To our surprise when modelling in process algebra we discovered that the computational simulation conducted by ODEs in [1] contains a systematic flaw in the analysis process which affects many of the results, some significantly. To the best of our knowledge these errors are presently unknown: at the very least they were unknown to us. Using the insights obtained from our analysis procedures we were able to return to the differential equation model, diagnose and correct the flaws in the analysis, and show agreement between the results obtained using continuous-space analysis and the results obtained using a discrete-state stochastic analysis.

Computational methods are well-understood to be complex and delicate so the relevance of this finding is not that there is an error in one particularly rich and valuable numerical study, or that modelling with ODEs is an unsatisfactory procedure, but rather that modelling in high-level languages (such as process algebras or Petri nets) may give a methodological advantage which allows an entire class of hard-to-detect errors and corner cases to be discovered and diagnosed before the results are published and promulgated to the wider scientific community.

As original contributions the present paper contains the analysis of the process used to detect the error in the earlier modelling study [1], a description of the new software tool used for integrated continuous-space and discrete-state stochastic analysis of PEPA process algebra models, and an overview of an extensive process algebra modelling study comprising 188 process definitions describing the dynamics of 95 of the reaction channels in the signalling cascade of the EGF receptor-induced MAP kinase pathway.

Structure of this paper: In Section 2 we present background material on our previous work. We follow this in Section 3 with a discussion of related work. In Section 4 we present an introduction to quantitative process algebras, considering the expressive capabilities of these languages. In Section 5 we explain how these languages are used in modelling. Section 6 presents a comparison of our analysis results and the results of other authors. In Section 7 we discuss the software tool used to perform the analysis. Finally, we present conclusions in Section 8.

2   Background

In an earlier study we made two distinct computational models of the Ras/Raf-1/MEK/ERK signalling pathway, both expressed in the PEPA process algebra. Our models were based on the deterministic model presented directly as a system of coupled ordinary differential equations in [5].

Our process algebra models adhere to two distinct modelling styles—the *reagent-centric* and *pathway* models from [3]. We interpreted these under the continuous-time Markov chain semantics for the PEPA language, and thus these gave rise to stochastic models of the pathway. We used well-known procedures of numerical linear algebra to conduct a quantitative stochastic evaluation of the pathway. We used the process algebraic reasoning apparatus of the PEPA language to establish that these two models were *strongly equivalent*, meaning that a timing-aware observer could not distinguish between them. In the extension of this work in [6] we presented automatic procedures for converting in both directions between the reagent-centric and pathway views.

We revisited the reagent-centric model in [4], mapping it to a system of ODEs. The model considered in [4] adds additional species to the model presented in [5] in order to concentrate on a detail of the pathway not considered in [5]. We applied the mapping procedure from [4] to a reduced version of the model without these additional species and were able to show that the model gave rise to exactly the same system of ODEs as studied previously in [5] establishing a precise formal equivalence between the process algebra model and the ODE model.

The deterministic and stochastic approaches to computational modelling in systems biology are often presented as alternatives; one should choose one approach or the other. Some authors have suggested that stochastic approaches are technically superior because they can expose small-scale effects which are caused by some molecular species being present in the reaction volume in very low copy numbers. We are instead in agreement with the authors of [7], who argue that the principal challenge is *choosing the appropriate framework* for the modelling study at hand. For some problems the influence of effects such as intra-cellular noise or circumstances such as low copy numbers is sufficiently great that a thorough stochastic treatment is essential. In other modelling problems no such influences are manifest and a deterministic treatment based on reaction rate equations is the correct approach.

The divergence between the stochastic behaviour exposed at low copy numbers of reactants and the deterministic approach based on reaction rate equations is due to the reliance of the ODE-based analysis on the assumption of continuity and the use of the law of mass action, essentially an empirical law derived from *in vitro* experimentation. Gillespie's Stochastic Simulation Algorithm (SSA) [8] makes no use of such an empirical law, and is instead grounded in the theory of statistical thermodynamics. In consequence it is an *exact* procedure for numerically simulating the dynamic evolution of a chemically reacting system, even at low copy numbers. However, the SSA method converges, as the number of reactants increases, to the solution computed by the ODEs so that the methods are in agreement in the limit [9].

Gillespie's exact algorithm models systems in which there are M possible reactions represented by the indexed family R_μ $(1 \leq \mu \leq M)$. It builds on a *reaction probability density function* $P(\tau, \mu \mid \mathbf{X})$ such that $P(\tau, \mu \mid \mathbf{X})d\tau$ is the probability that given the state $\mathbf{X}$ at time t, the *next* reaction in the volume will occur in the infinitesimal time interval $(t + \tau, t + \tau + d\tau)$ *and* be an R_μ reaction. Starting from an initial state, SSA randomly picks the time and type of the next reaction to occur, updates the global state to record the fact that this reaction has happened, and then repeats.

In practice, Gillespie's SSA is effective only for non-stiff systems on short time scales. An approximate acceleration procedure called "τ-leaping" was later developed by Gillespie and Petzold [10]. The "implicit τ-leaping" method [11] was developed to attack the orthogonal problem of *stiffness*, common in multi-scale modelling, where different time-scales are appropriate for reactions. Recent advances in the field include the development of *slow-scale SSA* which produces a dramatic speed-up relative to SSA by prioritising rare events [12].

A recent survey paper on stochastic simulation is [13]. A comparison paper on stochastic simulation methods and their relation to differential-equation based analysis of reaction kinetics is [9].

3   Related Work

We are not the first authors to investigate the model from [1] using stochastic simulation methods. An earlier comparison using the binomial τ-leap method appeared in [14]. However, the authors of [14] compare the solutions computed by their binomial τ-leap method with the solutions computed by Gillespie's stochastic simulation algorithm and did not compare with the results from [1]. For this reason the authors of [14] did not find the error which we uncovered by comparing the results computed by stochastic simulation with the results computed by the authors of [1] using ordinary differential equations.

In [15] the authors use the PRISM probabilistic model checker [16] to check logical formulae of Continuous Stochastic Logic (CSL) [17] against models of signalling pathways expressed as state-machines in the PRISM modelling language, comparing the result against an ODE model coded in the Matlab numerical platform.

A recent technical note [18] uses modelling in a stochastic process calculus and stochastic simulation to investigate the MAPK cascade previously studied in [19] using ordinary differential equations. [18] uses synthetic values for rate constants (all are set to 1.0) so comparison with the results of [19] is not meaningful.

4   Process Algebras

Process algebras are concise formally-defined modelling languages for the precise description of concurrent, communicating systems. Our belief is that they are well-suited to modelling cell signalling pathways and our interest here is exclusively in process algebras which are decorated with quantitative information [20]. The PEPA process algebra [2] which we use benefits from formal semantic descriptions of different characters which are appropriate for different uses. The structured operational semantics presented in [2] maps the PEPA language to a Continuous-Time Markov Chain (CTMC) representation. A continuous-space semantics maps PEPA models to a system of ordinary differential equations (ODEs) [21], admitting different solution procedures.

4.1   Expressiveness

Because we are modelling in a high-level language it is possible to apply these very different numerical evaluation procedures to compute different kinds of quantitative information from the same model. This is a freedom which we would not have if we had coded a Markov chain or a differential equation-based representation of the model directly in a numerical computing platform such as Matlab. One freedom which the use of a high-level language gives the modeller is the possibility to use either discrete-state or continuous-space analysis procedures. Another is the option of applying *both* types of analysis to the same model, and that is the approach which we have used here.

One strength of the PEPA process algebra as an expressive and practical modelling language is its support for *multi-way co-operation*; we have made use of this expressive power in all of our modelling studies in systems biology. Genuinely tri-molecular collisions occur only exceptionally rarely in dilute fluids so these do not normally arise in our modelling for this reason. Rather a collision between, say, an enzyme and a substrate to produce a compound, is expressed in PEPA as a three-way co-operation between the input enzyme and substrate (whose molecular concentrations are reduced) and the output compound (whose molecular concentration is increased). Similarly a reaction channel with two input species and two output species is represented as a four-way co-operation in PEPA. Some reaction channels may have more inputs or more outputs and so having this expressive power available in our chosen process algebra seems well-suited to the type of modelling which is undertaken in the area.

4.2   Combinators of the Language

We give only a brief introduction to the PEPA language here. The reader is referred to [2] for the definitive description.

PEPA provides a set of combinators which allow expressions to be built which define the behaviour of components via the activities that they engage in. These combinators are presented below.

Prefix $(\alpha, r).P$: Prefix is the basic mechanism by which the behaviours of components are constructed. This combinator implies that after the component has carried out activity (α, r), it behaves as component P.

Choice $P_1 + P_2$: This combinator represents a competition between components. The system may behave either as component P_1 or as P_2. All current activities of the two components are enabled. The first activity to complete distinguishes one of these components and the other is then discarded.

Cooperation: $P_1 \bowtie_L P_2$: This describes the synchronization of components P_1 and P_2 over the activities in the cooperation set L. The components may proceed independently with activities whose types do not belong to this set. A particular case of the cooperation is when $L = \emptyset$. In this case, components proceed with all activities independently. The notation $P_1 \parallel P_2$ is used as a shorthand for $P_1 \bowtie_0 P_2$. In a cooperation, the rate of a shared activity is defined as the rate of the slowest component.

Hiding: P/L This component behaves like P except that any activities of types within the set L are *hidden*, i.e. such an activity exhibits the unknown type τ and the activity can be regarded as an internal delay by the component. Such an activity cannot be carried out in cooperation with any other component: the original action type of a hidden activity is no longer externally accessible, to an observer or to another component; the duration is unaffected.

Constant: $A \stackrel{def}{=} P$ Constants are components whose meaning is given by a defining equation: $A \stackrel{def}{=} P$ gives the constant A the behaviour of the component P. This is how we assign names to components (behaviours). An explicit recursion operator is not provided but components of infinite behaviour may be readily described using sets of mutually recursive defining equations.

5   Modelling

For this system we developed a *reagent-centric* model. In this style of modelling we associate a distinct PEPA component with each reagent in the system. This is a more abstract mapping than is used in most of the work using stochastic π-calculus [22], where a distinct component is associated with each *molecule* in the system.

In the reagent-centric style, we represent the state of the system as the conjunction of the states of the components, each local state corresponding to a concentration level of an individual reagent. Concentration levels are discretized and the local states of the PEPA component records the impact of each possible reaction on the concentration level. The impact will depend on the role that the reagent plays within this particular reaction. This is summarised in Table 1.

Enzymatic reactions are possible when the enzyme is present in high concentration, and have no impact on the amount of enzyme although the current concentration of the enzyme will affect the rate of reaction. Conversely for inhibitory reactions: the inhibitor must be in low concentration and will remain low and its concentration has a regulatory effect on the rate of the reaction.

Table 1. The impact and role of reagents

Reagent role	Impact on reagent	Impact on reaction rate
Producer	decreases concentration	has a positive impact, i.e. proportional to the current concentration level
Product	increases concentration	has no impact on the rate, except at saturation
Enzyme	concentration unchanged	has a positive impact, i.e. proportional to current concentration
Inhibitor	concentration unchanged	has a negative impact, i.e. inversely proportional to current concentration

A PEPA model in this style can be thought to define a schematic for the possible reactions in the system. In the ODE mapping the local states represent the concentrations of the reagents. In the mapping to stochastic simulation, the local states indicate the types of molecules involved in the reactions and this is automatically mapped to a chemical master equation representation suitable for simulation using Gillespie's algorithm.

Figure 2 shows a small network, and the PEPA reagent-centric model that describes the graphical representation. In this example the PEPA components are A, B and C, and are tagged with H and L to designate the high and low concentrations, the coarsest possible discretization. The PEPA equations record the impact of each reaction on the concentration of that reagent.

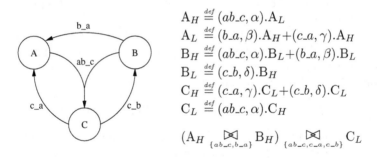

$$A_H \stackrel{def}{=} (ab\_c, \alpha).A_L$$
$$A_L \stackrel{def}{=} (b\_a, \beta).A_H + (c\_a, \gamma).A_H$$
$$B_H \stackrel{def}{=} (ab\_c, \alpha).B_L + (b\_a, \beta).B_L$$
$$B_L \stackrel{def}{=} (c\_b, \delta).B_H$$
$$C_H \stackrel{def}{=} (c\_a, \gamma).C_L + (c\_b, \delta).C_L$$
$$C_L \stackrel{def}{=} (ab\_c, \alpha).C_H$$

$$(A_H \underset{\{ab\_c,b\_a\}}{\bowtie} B_H) \underset{\{ab\_c,c\_a,c\_b\}}{\bowtie} C_L$$

Fig. 2. PEPA reagent-centric example

$$ab\_c, A + B \rightarrow C, \alpha \qquad c\_b, C \rightarrow B, \delta$$
$$b\_a, B \rightarrow A \qquad , \beta \qquad c\_a, C \rightarrow A, \gamma$$

Fig. 3. An equivalent model in chemical reaction language

The PEPA definitions in Figure 2 give rise to four reactions shown in Figure 3 in the chemical reaction language format $W, X \rightarrow Y, Z$. W is the name for the reaction, $X = \{X_1 + ... + X_n\}$ lists all the components that are consumed in this named reaction. Y is a list in the same format as X representing those components that are increased by this reaction. The last part of the reaction, Z, defines a rate constant from which the reaction rate is derived.

The reaction ab_c consists of two reactants and one product. From the PEPA definition in Fig. 2, components A and B transition from a high to low state via the activity/reaction ab_c: they are the two reactants of reaction ab_c. Similarly, component C transitions from a low to high state by reaction ab_c: it is the product of this reaction. This form of reasoning is used to transform all the PEPA equations into chemical reaction language format.

The rate of each reaction is not simply the defined constant. Where previously the reaction ab_c was defined as A + B → C, α, we take the constant α and multiply it by the number of molecules in both the A and B components (to allow for all permutations) to give a reaction rate of αAB, the mass action rate.

As outlined above, both stochastic simulation and ODE analysis are available. ODEs derived from PEPA in this manner will always respect the rules of conservation, as PEPA works on a static number of components. The inclusion of stoichiometric information outside of the PEPA model does however allow for a more powerful representation. In this case the numbers of each components required in each reaction are any valid integer i.e. ab_c requires 3 units of component A instead of 1.

5.1   Schoeberl Model in PEPA

In attempting to reproduce the model created by Schoeberl *et al.*, the main source of information came from the supplementary material to [1]. The complexity of the model highlights the issues surrounding graphical representations as can be seen in Fig. 4.

The reaction v7, highlighted in blue is a uni-directional reaction and shows one instance of internalisation. Other reactions such as v2, v3 are bi-directional yet with no obvious difference within the graphical scheme. Additional information in the form of tabled reactions and rates, for example

$$v7, \qquad [(EGF\text{-}EGFR^*)2] \rightarrow [(EGF\text{-}EGFRi^*)2]$$

can resolve some of the ambiguities, and by making joint use of these two representations the PEPA model can be constructed. Each component is taken in turn, with each reaction it participates in recorded against it. If we use (EGF-EGFR)2 (which can be seen in Fig. 4) as an example; (EGF-EGFR)2 can become phosphorylated (v3) and form (EGF-EGFR*)2, and this autophosphorylation can be reversed. This information would allow us to construct a definition such as that presented in equation (1).

$$EGF\text{-}EGFR2_H \stackrel{def}{=} (v3, k3).EGF\text{-}EGFR2_L$$
$$EGF\text{-}EGFR2_L \stackrel{def}{=} (v\text{-}3, k\text{-}3).EGF\text{-}EGFR2_H \qquad (1)$$

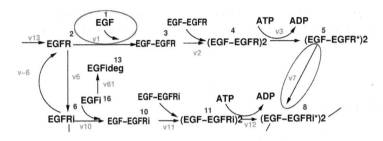

Fig. 4. An extract of the signalling pathway (reproduced from [1])

Going further, we realise that (EGF-EGFR)2 is formed from the dimerization of EGF-EGFR (v2) and that this step can also be reversed. Adding this information to the previous definitions produces the definitions shown in (2).

$$\text{EGF-EGFR2}_H \stackrel{\text{def}}{=} (v3, k3).\text{EGF-EGFR2}_L + (v\text{-}2, k\text{-}2).\text{EGF-EGFR2}_L$$

$$\text{EGF-EGFR2}_L \stackrel{\text{def}}{=} (v\text{-}3, k\text{-}3).\text{EGF-EGFR2}_H + (v2, k2).\text{EGF-EGFR2}_H \quad (2)$$

In this manner, each component can be built up to form the complete model. Some of the more complex compounds, such as (EGF-EGFR*)2-GAP-Shc*-Grb2-Sos, participate in nine reactions creating large definitions. The definitions are structurally similar, consisting of multiple choice operators for the prefixes.

This brief description can account for the majority of the model but not all. The dimerization process seen in reactions v9 and v11 currently require the addition of stoichiometric information. Through the interface to our software tool (described in Section 7) you can stipulate that two EGF-EGFR complexes form one (EGF-EGFR)2. When converting to Matlab this is translated to

$$\frac{dy(3)}{dy} = -2k_2 y(3)^2$$

and

$$\frac{dy(4)}{dy} = k_2 y(3)^2$$

where $y(3)$ is EGF-EGFR and $y(4)$ is (EGF-EGFR)2. Certain complexes can degrade such as EGFRi and EGFi, forming components that only increase in volume.

The final behaviour that requires consideration is that of EGF. EGF binds to the EGF receptors, circled in red on the left in Fig. 4. The reactions present within [1] all suggest that EGF is consumed in this binding. This is not the case and in the Matlab model the rate of change for EGF is set to zero for all reactions it is involved in. This can be likened to a reservoir: the EGF is present at a given concentration but there exists so much at this level that the reduction is negligible. In the PEPA model this must be made explicit from the start. The influence of EGF can be defined either as a secondary rate parameter, effectively increasing the rate at which the reaction will take place, or EGF can be defined as a catalyst in the relevant reactions. In the PEPA model the catalytic route was taken and so defined as $\text{EGF}_H \stackrel{def}{=} (v1, k1).\text{EGF}_H$.

6   Comparison

Figure 5 shows the time series plots for the six components highlighted in the original Schoeberl *et al.* paper. Each graph has three time series plots:

1. the solution of the original model[1] from [1] which is a Matlab program which specifies a fixed time step and solution using the `ode15s` procedure from the Matlab ODE suite [23];
2. the result of a τ-leap simulation of our PEPA model; and
3. the solution of an amended version of the original model using smaller time steps with the `ode15s` procedure.

Each form of analysis was run for the same duration (60 minutes) in order to replicate the results of the original model as closely as possible. Of the six components MEK-PP, Raf* and Ras-GTP spike in a short space of time, and so to more readily show the differences the time series were cut short once the rate of change had dropped off towards zero.

The use of the particular step within the solver is most apparent in Ras-GTP. The original model's results indicate a peak at two minutes with a value of 8000 molecules/cell. The true peak occurs earlier, reaching double the original value at 16,000 molecules/cell. As can be seen, the value at two minutes is correct, but that the speed at which this component changes means the bulk of the reaction has already taken place, and the analysis incorrectly steps over the true peak onto the negative gradient of the curve. Differences can be seen also within the Raf* and to a lesser extent MEK-PP. In all of the graphs, it is nearly impossible to distinguish the τ-leap and variable-step `ode15s` solver at this resolution.

This discrepancy only became apparent when comparing the results from the stochastic simulation and that of the ODE analysis, and we were only in a position to compare these alternative models because we generated both from a high-level process algebra description. Prior to running the τ-leap simulation, the arguments for the ODE analysis of the PEPA model had been extracted from the original model. Hence the same results were obtained, with the peaks in identical places.

The time taken to solve the ODE model using the stiff solver with smaller time steps was almost identical to the time taken to solve model with fixed larger time steps. The time taken to solve the model using the τ-leap method is longer than the time taken to solve the model using Matlab's stiff ODE solver (`ode15s`) but shorter than the time required by a standard solver such as `ode45`.

---

[1] Available on-line at `http://web.mit.edu/dllaz/egf_pap/`

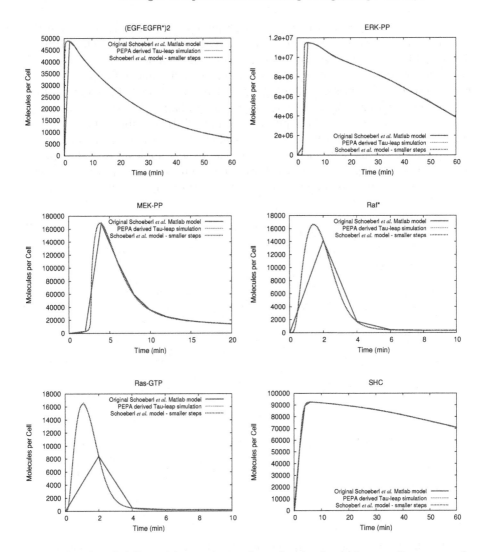

Fig. 5. Graphs of differential equation and stochastic simulation results compared. The solid red line is the solution of the original model from [1], which shows marked differences in some graphs from the solution of the PEPA-derived τ-leap simulation and (a dashed green line) and the solution of the ODE model using smaller time steps (a dotted blue line). The solution of the PEPA-derived τ-leap simulation and the solution of the ODE model using smaller time steps are virtually indistinguishable in the graphs.

7   Implementation

The reason to have a formally-defined high-level language for performance modelling is that it is possible to implement software tools which evaluate models according to the formal semantics of the language. For the present study we produced a tool platform to support the compilation of PEPA models in the reagent-centric style by extending the Choreographer platform [24] which we developed for general quantitative analysis of PEPA models.

Choreographer is an integrated development environment for process algebraic modelling, comprising a language-sensitive editor for PEPA and a toolbox of solution procedures for continuous-time Markov chains. We extended Choreographer to communicate with the publicly-available ISBJava library for stochastic simulation as used by the Dizzy [25] chemical kinetics stochastic simulation software package. We also extended Choreographer to communicate with the Matlab numerical computing platform, which we use for numerical integration of ODEs. A screenshot of our extended Choreographer platform appears in Figure 6.

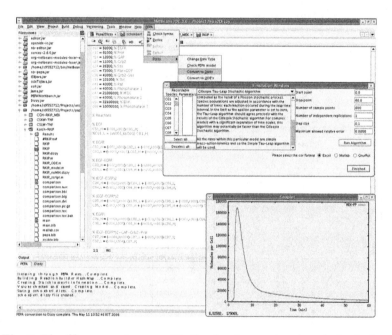

Fig. 6. The Choreographer quantitative development and analysis platform

8   Conclusions

Errors in the use of typical computing applications frequently manifest themselves as a null pointer dereference or a segmentation fault: the application tells the user that an error has occurred. Errors in the use of numerical computing routines are more insidious than errors in typical computing. No memory faults are signalled and the application often completes normally within the anticipated duration of run, delivering a plausible graph of analysis results. Without any such alarm bells being sounded the modeller must always be on guard to look for potential traps such as an over-generous step-size and it is entirely forgivable if they cannot always do this for every graph in every modelling study.

Rather than place this intellectual burden on the modeller we would prefer to use stronger computational modelling procedures which would routinely apply both continuous-state analysis methods (such as ODE solution) and discrete-state analysis (such as stochastic simulation). High-level modelling languages such as the PEPA process algebra are helpful here. Instead of coding the differential equations and the stochastic simulation directly we generate these from a single process algebra model, gaining the value of the application of both types of analysis without the expense of any re-implementation.

Using this approach we uncovered a flaw in the results presented in [1]. We had no *a priori* reason to suspect that there was a flaw; comparing the stochastic simulation results to the ODE solution identified a clear problem, at a modest computational cost. All computations were done on a single desktop PC. We believe that the insights obtained from this study stand as a good advertisement for the usefulness of high-level modelling languages for analysing complex biological processes whether process algebras, Petri nets or SBML [26].

We compared in Figure 5 the analysis results obtained by solution of the differential equations with the solutions computed by stochastic simulation. As is typical for stiff systems, some effects are best considered over different time scales. Some species (such as ERK-PP and SHC) exhibit high concentration for a period of hours. Others (such as Raf* and Ras-GTP) peak within minutes. The large time step used in the computation in [1] is not a problem for the analysis of the long-lived species but gives misleading results for those species which are short-lived.

We discovered very good agreement between the results calculated by the τ-leap method and the results calculated from the differential equations when a variable timestep is used. The solution of the variable timestep ODEs agrees almost exactly everywhere with the solution obtained from Gillespie's approximate τ-leap method: these two lines are overlapping on the plots in Figure 5.

Acknowledgements. Muffy Calder and Adam Duguid are supported by the DTI Beacon Bioscience Projects programme. Stephen Gilmore and Jane Hillston are supported by the EU IST-3-016004-IP-09 project SENSORIA. Jane Hillston is supported by the Engineering and Physical Sciences Research Council Advanced Research Fellowship EP/C543696/1 "Process Algebra Approaches to Collective Dynamics". The authors acknowledge helpful discussions with Richard Orton of the Bioinformatics Research Centre, University of Glasgow on aspects of the computational modelling of the MAPK pathway.

References

1. B. Schoeberl, C. Eichler-Jonsson, E.D. Gilles, and G. Muller. Computational modeling of the dynamics of the MAP kinase cascade activated by surface and internalized EGF receptors. *Nature Biotechnology*, 20:370–375, 2002.
2. J. Hillston. *A Compositional Approach to Performance Modelling.* Cambridge University Press, 1996.

3. Muffy Calder, Stephen Gilmore, and Jane Hillston. Modelling the influence of RKIP on the ERK signalling pathway using the stochastic process algebra PEPA. In Anna Ingolfsdottir and Hanne Riis Nielson, editors, *Proceedings of the Bio-Concur Workshop on Concurrent Models in Molecular Biology*, London, England, August 2004.

4. Muffy Calder, Stephen Gilmore, and Jane Hillston. Automatically deriving ODEs from process algebra models of signalling pathways. In Gordon Plotkin, editor, *Proceedings of Computational Methods in Systems Biology (CMSB 2005)*, pages 204–215, Edinburgh, Scotland, April 2005.

5. K.-H. Cho, S.-Y. Shin, H.-W. Kim, O. Wolkenhauer, B. McFerran, and W. Kolch. Mathematical modeling of the influence of RKIP on the ERK signaling pathway. In C. Priami, editor, *Computational Methods in Systems Biology (CSMB'03)*, volume 2602 of *LNCS*, pages 127–141. Springer-Verlag, 2003.

6. Muffy Calder, Stephen Gilmore, and Jane Hillston. Modelling the influence of RKIP on the ERK signalling pathway using the stochastic process algebra PEPA. *Transactions on Computational Systems Biology*, 2006. Extended version of [3]. To appear.

7. O. Wolkenhauer, M. Ullah, W. Kolch, and K.-H. Cho. Modelling and simulation of intracellular dynamics: Choosing an appropriate framework. *IEEE Transactions on Nanobioscience*, 3(3):200–207, September 2004.

8. D.T. Gillespie. Exact stochastic simulation of coupled chemical reactions. *Journal of Physical Chemistry*, 81(25):2340–2361, 1977.

9. T.E. Turner, S. Schnell, and K. Burrage. Stochastic approaches for modelling in vivo reactions. *Computational Biology and Chemistry*, 28:165–178, 2004.

10. D.T. Gillespie and L.R. Petzold. Improved leap-size selection for accelerated stochastic simulation. *J. Comp. Phys.*, 119:8229–8234, 2003.

11. M. Rathinam, L.R. Petzold, Y. Cao, and D.T. Gillespie. Stiffness in stochastic chemically reacting systems: The implicit tau-leaping method. *Journal of Chemical Physics*, 119(24):12784–12794, December 2003.

12. Y. Cao, D.T. Gillespie, and L. Petzold. Accelerated stochastic simulation of the stiff enzyme-substrate reaction. *Journal of Chemical Physics*, 123:144917–1 – 144917–12, 2005.

13. D. Gillespie and L. Petzold. *System Modelling in Cellular Biology*, chapter Numerical Simulation for Biochemical Kinetics. MIT Press, 2006. Ed. Z. Szallasi, J. Stelling and V. Periwal.

14. Abhijit Chatterjee, Kapil Mayawala, Jeremy S. Edwards, and Dionisios G. Vlachos. Time accelerated Monte Carlo simulations of biological networks using the binomial τ-leap method. *Bioinformatics*, 21(9):2136–2137, 2005.

15. M. Calder, V. Vyshemirsky, D. Gilbert, and R. Orton. Analysis of signalling pathways using the PRISM model checker. In Gordon Plotkin, editor, *Proceedings of Computational Methods in Systems Biology (CMSB 2005)*, Edinburgh, Scotland, April 2005.

16. M. Kwiatkowska, G. Norman, and D. Parker. PRISM: Probabilistic symbolic model checker. In A.J. Field and P.G. Harrison, editors, *Proceedings of the 12th International Conference on Modelling Tools and Techniques for Computer and Communication System Performance Evaluation*, number 2324 in Lecture Notes in Computer Science, pages 200–204, London, UK, April 2002. Springer-Verlag.

17. A. Aziz, K. Sanwal, V. Singhal, and R. Brayton. Model checking continuous time Markov chains. *ACM Transactions on Computational Logic*, 1:162–170, 2000.

18. Luca Cardelli. Mapk cascade. Microsoft Research Cambridge UK technical note. Available on-line at http://research.microsoft.com/Users/luca/Notes/Mapk Cascade.pdf, July 2005.

19. Chi-Ying F. Huang and James E. Ferrell Jr. Ultrasensitivity in the mitogen-activated protein kinase cascade. *Biochemistry*, 93(19):10078–10083, September 1996.

20. J. Hillston. Process algebras for quantitative analysis. In *Proceedings of the 20th Annual IEEE Symposium on Logic in Computer Science (LICS' 05)*, pages 239–248, Chicago, June 2005. IEEE Computer Society Press.

21. J. Hillston. Fluid flow approximation of PEPA models. In *Proceedings of the Second International Conference on the Quantitative Evaluation of Systems*, pages 33–43, Torino, Italy, September 2005. IEEE Computer Society Press.

22. C. Priami, A. Regev, W. Silverman, and E. Shapiro. Application of a stochastic name passing calculus to representation and simulation of molecular processes. *Information Processing Letters*, 80:25–31, 2001.

23. Lawrence F. Shampine and Mark W. Reichelt. The Matlab ODE suite. *SIAM J. Sci. Comput.*, 18(1):1–22, 1997.

24. Mikael Buchholtz, Stephen Gilmore, Valentin Haenel, and Carlo Montangero. End-to-end integrated security and performance analysis on the DEGAS Choreographer platform. In I.J. Hayes J.S. Fitzgerald and A. Tarlecki, editors, *Proceedings of the International Symposium of Formal Methods Europe (FM 2005)*, number 3582 in LNCS, pages 286–301. Springer-Verlag, June 2005.

25. S. Ramsey, D. Orrell, and H. Bolouri. Dizzy: stochastic simulation of large-scale genetic regulatory networks. *J. Bioinf. Comp. Biol.*, 3(2):415–436, 2005.

26. M. Hucka, A. Finney, H. M. Sauro, H. Bolouri, J. C. Doyle, and H. Kitano *et al.* The systems biology markup language (SBML): a medium for representation and exchange of biochemical network models. *Bioinformatics*, 19(4), 2003.

A Formal Approach to Molecular Docking

Davide Prandi

Dipartimento di Informatica e Telecomunicazioni, Università di Trento,
Via Sommarive 14, I-38050 Povo (TN) - Italy
prandi@dit.unitn.it

Abstract. Drugs are small molecules designed to regulate the activity of specific biological receptors. Design new drugs is long and expensive, because modifying the behavior of a receptor may have unpredicted side effects. Two paradigms aim to speed up the drug discovery process: molecular docking estimates if two molecules can bind, to predict unwanted interactions; systems biology studies the effects of pharmacological intervention from a system perspective, to identify pathways related to the disease. In this paper we start from process calculi theory to integrate information from molecular docking into systems biology paradigm. In particular, we introduce Beta-binders[D], a process calculus for representing molecular complexation driven by the shape of the ligands involved and the subsequent molecular changes.

Keywords: Formal Methods, Process Calculi, Drug Discovery, Molecular Docking, Systems Biology.

1 Introduction

A *drug* is a chemical substance designed to regulate the activity of specific biological receptors called *targets*. The process of *drug discovery* requires extensive study to determine the biological and biochemical problems that could underlie the disease. This is because, biological processes in the human body are tightly interconnected and modifying the behavior of a receptor may have dangerous side effects. Therefore drug discovery requires years of study and a large amount of money in order to find a drug for a potential target.

Molecular docking aims to predict whether one molecule will bind to another. If the geometry of a pair of molecules is complementary and involves favorable biochemical interactions, the two molecules will potentially bind *in vitro* or *in vivo*. The latest programs and algorithms (e.g. [1,2]) help researchers to find more efficient drugs, and to predict the behavior of new chemical compounds. Molecular docking impacts on costs and time consumed predicting non-specific interactions of drug molecules, and thus potential side effects.

In [3], the authors observe that *"knowing a target is not the same as knowing what the target does"*. The actual drug design process is founded on a *reductionist* approach: scientists search for a "magic bullet" that targets a specific molecule (e.g. an enzyme). If the target plays a role in different pathways, possible *on-target side effects* may emerge late in the drug discovery process. It is the

C. Priami (Ed.): CMSB 2006, LNBI 4210, pp. 78–92, 2006.
© Springer-Verlag Berlin Heidelberg 2006

case of Rofecoxib, used in the treatment of osteoarthritis and acute pain conditions. Rofecoxib inhibits COX-2 enzyme, that also plays a role in the production of prostaglandin, an anti-clotting agent [4]. Therefore Rofecoxib decreases prostaglandin production, leading to an inefficiency in declumping and vasorelaxation. Rofecoxib was withdrawn from the market in 2004 because it is related with risk of heart attack. Over 80 million people were prescribed rofecoxib before it was withdrawn. Moreover, the possibility that a designed drug binds molecules other than the target, *off-target side effects*, is ignored until in vivo evidence emerge. For instance, the phosphodiesterase (PDE) inhibitor Viagra is designed to target PDE-5 and to promote the relaxation of smooth muscle [5]. The drug also binds PDE-6 in the eye, leading to a documented "blue vision" side effects [6], difficult to discover with tests on animals. These situations need to be identified early in the drug discovery process. Systems modelling can help to improve our knowledge of the effects of pharmacological products. In particular, *systems biology* [7] integrates into *consistent models* different levels of information for understanding and eventually predicting the operation underlying complex biological systems. Therefore it appears as the "right" paradigm to overcome the limits of the reductionist approach adopted during the development of new drugs [8,9,10,11]. An effective model needs to be *extensible*, additional real properties can be added in the same framework, and *compositional*, the behaviour of a complex system is determined by the behaviour of its elementary components. Extensibility assures that information can be added to the model as well as it emerges in the drug discovery process. Compositionality allows to test the effects of a drug in different contexts, "simply" composing the models of the drug and of the context.

Among different proposals, formal methods from concurrency theory and process calculi are promising [12], because extensibility and compositionality are deeply studied in that context [13]. Here we propose Beta-binders$^\mathbb{D}$, a specialization of Beta-binders [14], for integrating pathway information from systems biology and binding prediction from molecular docking. Beta-binders introduces a special class of binders, used to model mobile processes [15,16] encapsulated into *boxes* with interaction capabilities. A molecule M is represented as a box B_M, depicted below:

$$x_1 : \Delta_1 \ \ldots \ x_n : \Delta_n$$

The pairs $x_i : \Delta_i$ indicate the sites through which B_M may interact with other boxes. The *types* Δ_i denote the interaction capabilities at x_i. In [14], types are set of names, and the authors observe that the typing policy could be changed to accommodate more refined kinds of interactions. Here, we specialise binder types to represent information from molecular docking, to drive interactions between boxes. The dynamic behaviour of B_M is specified by the internal pi-process M. A pi-process is a π-calculus process for representing biomolecular interactions [17,18], extended for manipulating the interaction sites of a box. The parallel composition of different boxes abstracts a biological system that evolves relying on the *semantics* of Beta-binders. For instance, consider the enzyme-catalyzed reaction schema [19]:

$$E + S \rightharpoonup ES \rightharpoonup EP \rightharpoonup E + P$$

The substrate S binds the enzyme E to form the complex ES; then the enzyme catalyses the reaction to EP and finally the product P is released without changing the structure of E. Beta-binders models such reactions as:

$$
\begin{array}{ccccc}
x_E : \Delta_E \;\; x_S : \Delta_S & x_E : \Delta_E \;\; x_S : \Delta_S & x_E : \Delta_E \;\; x_P : \Delta_P & x_E : \Delta_E \;\; x_P : \Delta_P \\
\boxed{\;E\;}\;\boxed{\;S\;} \;\rightarrow\; & \boxed{\;E \mid S\;} \;\rightarrow\; & \boxed{\;E \mid P\;} \;\rightarrow\; & \boxed{\;E\;}\;\boxed{\;P\;} \\
(B_E) \quad (B_S) & (B_{ES}) & (B_{EP}) & (B_E) \quad (B_P)
\end{array}
$$

Boxes B_E, for the enzyme, and B_S, for the substrate, can complex into box B_{ES}, if the types Δ_E and Δ_S are *compatible* up to a certain molecular docking algorithm. Then, the internal pi-process $E \mid S$ evolves into $S \mid P$ and Δ_S into Δ_P. Finally the complex unbinds releasing the product B_P.

The paper is organized as follow. Sect. 2 introduces drug discovery and molecular docking. We also propose *DockSpace* as a uniform model to handle different molecular docking algorithms. Sect. 3 shows some enzymatic reaction schemas as running examples. Then, in Sect. 4, we present Beta-binders$^{\mathbb{D}}$ that integrates process calculi formalism with information from molecular docking. The section concludes showing the models and the dynamic evolution of the enzymatic reactions presented. Finally, Sect. 5 closes the paper.

2   Drug Discovery

The normal activity of a specific biological receptor may be altered by different factors causing minor symptoms (e.g. runny eyes due to allergies) or life-threatening events. A *drug* is a small molecule designed to correct the activity of these receptors.

The *Drug discovery pipeline* in Fig. 1 is composed by the processes that allow to discover and design new drugs.[1] The development of a new drug starts with years of study to identify the biochemistry underlying a medical problem. The outcome is a specific receptor, called *target*, that needs to be regulated (i.e. alter its activity) by the drug. *High Throughput Screening* (HTS) allows to compare the target with large libraries of known substances, called *compounds*, to find anything that binds to the receptor in any fashion. Instead, *rational drug design* studies biological and physical properties of the target to predict the structure of possible ligands. Once a set of *hits* has been established they are *validated* and *refined* to obtain a *lead compound*. From this point onward a loop between validation and optimization starts until a lead compound with sufficient target

---

[1] The pipeline may vary depending on the pharmaceutical company, here we summarise main steps.

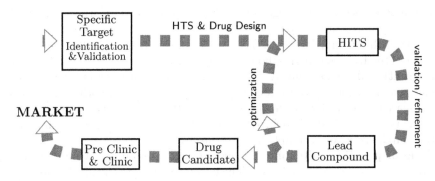

Fig. 1. Drug discovery pipeline

potency and selectivity is reached, obtaining a *drug candidate*. Then, in the *preclinical trial* the drug candidate is tested for safety, toxicity, pharmacokinetics and metabolism impacts in human. If succeed, the drug candidate is then tested in *human clinical trials*. Trials are designed to evaluate the safety and efficacy of an experimental therapy. Finally the new drug is approved for the market.

Even the process of defining new drugs requires years of studies and million of dollars [20], many marketed drugs fail because they are *not sufficiently effective* or because they *cause unwanted side effects*. Also when a drug is approved for marketing, success is not assured. There is an increasing need for a better approach to drug development, and pharmacological companies are moving to new technologies to understand cell responses to pharmacological intervention [21]. A system approach allows to find pathways related with the disease and also predict unwanted on- and off-target side effects, improving the drug discovery pipeline.

2.1   Molecular Docking

Molecular docking is a technique used in rational drug design for predicting whether one molecule will bind to another. Molecular recognition is a central question in biology, because *"life is crucially dependent on molecular binding: to the right target, at the right time, in the right place, with the right affinity, and (sometimes) at the right speed"* [22]. Molecular docking simulates the interaction of two ligand surfaces by arranging molecules in favorable configurations that match complementary features. Current docking methodologies varies considering, e.g., small molecules binding instead of macromolecular interactions, or rigid vs. flexible body [2]. However, there are three common ingredients in docking:

Representation of the molecular structure: The structure of a molecule is first determined in laboratory relying on biophysical techniques as x-ray crystallography or nuclear magnetic resonance (NMR) spectroscopy. Therefore the basic description of a ligand surface is its atomic representation.

The first pioneering work [23] describes only geometric features: a ligand surface is represented as a set of spheres that fill the space occupied by the atoms of the ligand. More advanced systems, e.g. DARWIN [24], represent also electrostatic and hydrophobic properties.

Conformational space search: The search space is all possible orientations and conformations of the interacting ligands. A search algorithm explores the search space to locate the most stable conformation. Each conformation of the paired molecules is referred to as a *pose*. Many strategies for sampling the search space are available in literature [2].

Ranking of possible solution: A *scoring function* computes the affinity between the receptor and the ligand. The idea is to estimate chemical properties with mathematical models. A scoring function must rank poses correctly, i.e. score best the most closely experimental structures, and must be fast to be applied concretely.

Molecular docking algorithm screens large databases of molecules (e.g. Protein DataBank[2]) orienting and scoring them in the binding site of a target. Top-ranked molecules are then tested for binding in vitro. Integrating pathways modelling with molecular docking enables researchers to incorporate experimental data on pathways with information on the structure of the compounds making more confident decisions on the future of new drugs. Unfortunately, algorithms and programs available differ for representation of the molecular structure, accuracy, computational costs, parameters, etc. There is not a standard (or a set of standards) and the available results are presented without uniformity in the literature. To abstract the particular algorithm used we introduce a consistent representation called *DockSpace*.

2.2   The DockSpace

The above short survey on molecular docking outlines the main characteristics needed for estimating binding affinity between molecular ligands: a *space of the structure* $\mathbb{D}$, where an element $D \in \mathbb{D}$ is a representation of the structure of a molecular ligand and an output space $\mathbb{S}$, that abstracts the output of a scoring function. Then, a molecular docking algorithm is a function m that takes the representation of two molecular ligands D_1 and D_2, a set of parameters $P \in \mathbb{P}$ (e.g. the Ph of the system), and returns a value in $\mathbb{S}$. We introduce *DockSpace* to uniformly represent molecular docking algorithms.

Definition 1 (DockSpace). *A DockSpace is a 4-tuple* $(\mathbb{D}, \mathbb{P}, \mathbb{S}, m)$ *where* $\mathbb{D}$ *is the space of the structure,* $\mathbb{P}$ *is the space of the parameters and* $\mathbb{S}$ *is the scoring space. The distance function* $m : \mathbb{D} \times \mathbb{D} \times \mathbb{P} \to \mathbb{S}$, *takes two molecular structures* D_1 *and* D_2, *a set of parameters* P, *and returns the scoring of the bind between* D_1 *and* D_2 *with parameters* P.

As a simple example we define the DockSpace $G = (\mathbb{D}_G, \mathbb{P}_G, \mathbb{S}_G, m_G)$. A molecule is described as a graph $D \in \mathbb{D}_G$, where nodes are labelled by atom types and edges

---

[2] http://www.rcsb.org/pdb/Welcome.do

(a)	$E+S \leftrightarrow ES \rightarrow EP \rightarrow E + P$		
(b) $E + S_1 \leftrightarrow ES_1 \rightarrow E^*P_1 \rightarrow E^* + P_1$ $+$ $E + P_2 \leftarrow EP_2 \leftarrow E^*S_2 \leftrightarrow S_2$		(c) $E + S \leftrightarrow ES \rightarrow EP \rightarrow E + P$ $+$ $I \leftrightarrow EI$	

Fig. 2. Enzymatic reactions

by the corresponding inter-atom distance. The distance function $m_G(D_1, D_2, n)$ is *true* iff it is possible to find a common subgraph between D_1 and D_2 with at least n nodes. A DockSpace where $\mathbb{S} = \{true, false\}$ is a *Boolean DockSpace*.

3   Example: Enzymatic Reactions

Enzymes are molecules that speed up biochemical reactions without themselves being changed, that is, they act as catalysts [19]. Enzymes bind to one or more ligands, called *substrates*, and convert them into one or more chemically modified *products*. The catalysis of organized sets of chemical reactions by enzymes creates and maintains the cell. Enzymes are genetically designed to be specific for a particular molecular target and any error could have dangerous consequences. Many drugs modify or regulate the activity of specific enzymes [25]. In this section we present three enzymatic reaction schema that we will use later as test cases for Beta-binders$^{\mathbb{D}}$.

Simple enzyme-catalyzed reaction Fig. 2(a). In this schema, the enzyme E and the substrate S bind to form the enzyme-substrate complex ES. The reaction takes place in ES to form the enzyme-product complex EP. Finally the product P is released and the enzyme E regenerated.

Multi-substrate systems Fig. 2(b). Multi-substrate systems are enzymatic reactions in which enzyme catalysis involves two or more substrates. For instance, in the *Ping-Pong mechanism* of Fig. 2(b) the substrate S_1 binds the enzyme E resulting in a product P_1 and an enzyme E^*, a modified version of E which often carries a fragment of S_1. Then, a second substrate S_2 binds E^* releasing a second product P_2 and the enzyme E. This process is called Ping-Pong mechanism because of the bouncing between E and E^*.

Enzyme inhibition Fig. 2(c). An *inhibitor* is a molecule that decreases the speed of an enzymatic reaction. The study of enzyme inhibition is crucial for drug discovery because in many cases a drug acts as an inhibitor or has to suppress an inhibitor. For instance, in *Competitive inhibition* of Fig. 2(c) the substrate S and the inhibitor I compete for the same enzyme E. The complex EI cannot interact with S inhibiting the production of P.

4   Docked Bets-Binders

Beta-binders abstracts biological systems as parallel processes that interact through interfaces. For instance, proteins have backbones as internal control structure and motifs for interacting with other entities. Here we introduce Beta-binders$^{\mathbb{D}}$, a refined version of [14], to integrate molecular docking information into biomolecular reactions. In particular, we are interested in representing molecular complexation driven by the shape of the ligands involved and the subsequent molecular changes.

We assume a Boolean DockSpace $G = (\mathbb{D}_G, \mathbb{P}_G, \mathbb{S}_G, m_G)$ and a countably infinite set $\mathcal{N}$ of names (ranged over by lower-case letters). Beta-binders$^{\mathbb{D}}$ represents a molecules M with a *box* B_M depicted as

$$x_1 : \Delta_{M_1} \ \ldots \ x_n : \Delta_{M_n}$$

$$\boxed{M}$$

and written as $\beta(x_1, \Delta_{M_1}) \ldots \beta(x_n, \Delta_{M_n}) [\ M\ ]$.

A box is a π-calculus process for representing biological interactions [17,18] prefixed by specialised binders, named *beta binders*, that represent interaction capabilities. *An elementary beta binder* (also binder for simplicity) is $\beta(x, \Gamma)$ (active) or $\beta^h(x, \Gamma)$ (hidden), where the name x is the *subject* and $\Gamma \in \mathbb{D}_G$ is the *type* of x. Hidden binders cannot be used in interaction. We let $\hat{\beta} \in \{\beta, \beta^h\}$. A beta binder (ranged over by $\boldsymbol{B}, \boldsymbol{B}_1, \boldsymbol{B}', \ldots$) is a non-empty string of elementary beta binders whose subjects are all distinct. The set of the subjects of all the elementary beta binders in $\boldsymbol{B}$ is $\mathsf{sub}(\boldsymbol{B})$, and $\boldsymbol{B}^*$ denotes either a beta binder or the empty string.

The π-calculus syntax is enriched to manipulate beta binders obtaining the set of *pi-processes* $\mathcal{P}$ defined by the following syntax:

$$\pi ::= \overline{x}\langle z \rangle \mid x(y) \mid \tau$$
$$\pi_\beta ::= \mathsf{hid}(x) \mid \mathsf{unh}(x) \mid \mathsf{exp}(x, \Gamma) \mid \mathsf{ch}(x, \Gamma)$$
$$P ::= M \mid P \mid P' \mid \nu \tilde{x}\, P \mid [x = y]\, P \mid A\,(\tilde{y})$$
$$M ::= \mathsf{nil} \mid \pi.\, P \mid \pi_{\boldsymbol{B}}.\, P \mid M {+} M'$$

The process nil is inactive. The prefix $\pi.\, P$ ($\pi_\beta.\, P$) assures that action π (π_β) has to be fired before executing P. The output prefix $\overline{x}\langle z \rangle$ sends name z on link x. The input prefix $x(y)$ receives over x a name y. The name y is a binder for the prefixed process. We will use $\overline{z}$ and x where the objects of the output or of the input are empty. The silent prefix τ abstracts a non visible action. Prefixes $\mathsf{hid}(x)$ and $\mathsf{unh}(x)$ make the elementary beta binder with subject x not available (hidden) and available (unhidden), respectively. The prefix $\mathsf{exp}(x, \Gamma)$ adds to the box the binder $\beta(x, \Gamma)$. Finally prefix $\mathsf{ch}(x, \Gamma)$ changes the current type of x with Γ. The process $P \mid P'$ represents a system composed by two parallel sub-processes P and P'. The process $M + M'$ behaves either as M or as M'. The names $\tilde{x}$ ($\tilde{x}$ denotes the sequence $x_1, \ldots, x_n$) in $\nu \tilde{x}\, P$ are static binders for $\tilde{x}$ in P. Matching $[x = y]\, P$ behaves as P if $x = y$. Finally, the agent identifier $A\,(\tilde{y})$ has a unique defining equation $A\,(\tilde{x}) \overset{def}{=} P$. Each occurrences of an agent identifier $A\,(\tilde{y})$ will be replaced by the process P, with the formal parameter $\tilde{x}$ substituted by the actual parameters $\tilde{y}$.

Table 1. Laws for structural congruence

a. $P_1 \equiv P_2$ if P_1 and P_2 are α-equivalent	i. $B[\ P_1\ ] \equiv B[\ P_2\ ]$ if $P_1 \equiv P_2$
b. $(\mathcal{P}/\equiv, \mid, \mathsf{nil})$ is an abelian monoid	j. $B_1 B_2[\ P\ ] \equiv B_2 B_1[\ P\ ]$
c. $(\mathcal{P}/\equiv, +, \mathsf{nil})$ is an abelian monoid	k. $B^* \hat{\beta}(x : \Gamma)[\ P\ ] \equiv B^* \hat{\beta}(y : \Gamma)[\ P\{y/x\}\ ]$
d. $\nu z\,\nu w\,P \equiv \nu w\,\nu z\,P$ e. $\nu z\,\mathsf{nil} \equiv \mathsf{nil}$	if y fresh in P and $y \notin \mathsf{sub}(B^*)$
f. $\nu y\,(P_1 \mid P_2) \equiv P_1 \mid \nu y\,P_2$ if $y \notin \mathsf{fn}(P_1)$	l. $(\mathcal{B}/\equiv, \parallel, \mathsf{Nil})$ is an abelian monoid
g. $[x = x]\,P \equiv P$	
h. $A(\tilde{y}) \equiv P\{\tilde{y}/\tilde{x}\}$ if $A(\tilde{x}) \overset{def}{=} P$	

The usual definitions of *free* and *bound names* (denoted by $\mathsf{fn}(-)$ and $\mathsf{bn}(-)$, respectively) and of *name substitution* $\{x/y\}$ are extended by stipulating that $\exp(x, \Gamma).P$ is a binder for x in P. The set of names of a pi-process results to be $\mathsf{n}(P) = \mathsf{fn}(P) \cup \mathsf{bn}(P)$. We also define $\{|x/y|\}$, that behaves as $\{x/y\}$ but it does not rename π_β prefixes:

$$(\pi.P)\{|x/y|\} = \pi\{x/y\}.(P\{|x/y|\}) \qquad (\pi_\beta.P)\{|x/y|\} = \pi_\beta.(P\{|x/y|\})$$

A biological system is modelled as the parallel composition of boxes, named *bio-processes*. The set of bio-processes $\mathcal{B}$ (denoted as $B, B_1, B', \ldots$) is defined as:

$$B ::= \mathsf{Nil} \mid B[\ P\ ] \mid B \parallel B$$

A system is the parallel composition $(B \parallel B)$ of boxes $(B[\ P\ ])$, with nullary element Nil. For instance, the bio-process $B = B_{M_1}[\ M_1\ ] \parallel B_{M_2}[\ M_2\ ]$ represents two molecules M_1 and M_2 in the same solution. The set of names of a box $B[\ P\ ]$ is $\mathsf{n}(B[\ P\ ]) = \mathsf{sub}(B) \cup \mathsf{n}(P)$.

The reduction semantics for Beta-binders uses the structural congruence over pi- and bio-processes defined as the smallest relations satisfying the laws of Tab. 1, where we overload the symbol $\equiv$ when unambiguous. Structural congruence allows to manipulate the structure of pi- and bio-processes. Rule a states that $P_1 \equiv P_2$ if P_2 can be obtained by a finite number of changes in the bound names of P_1, and vice versa (i.e. P_1 and P_2 are α-equivalent). Rules b and c say that $\mid$ and $+$ are commutative, associative and have identity element nil. Rules d, e and f are for restriction, in particular f moves the scope of a restriction to include or exclude a process in which the restricted name is not free. Rule g allows to proceed the process $[x = y]\,P$ iff x is equal to y. Rule h instantiates the agent identifier A with parameters $\tilde{y}$ iff $A(\tilde{x})$ is defined as the pi-process P. Rule i lifts to bio-processes structural congruence between pi-processes; rule j lets to write beta binders in any order, and rule k enables α-conversion for the subject of a binder. Finally rule l states that $\parallel$ is commutative, associative with identity element Nil. In the following we shall consider processes up to $\equiv$.

The reduction transition system is $TS = (\mathcal{B}, \rightarrow)$ where $\mathcal{B}$ is the set of states (equivalence classes of bio-processes w.r.t. $\equiv$) and the *reduction relation* $\rightarrow$ is the smallest relation over bio-processes obtained by applying the axioms and rules of Tab. 2.

Table 2. Axioms and rules for the reduction relation

(intra) $\boldsymbol{B}\big[\ \nu\tilde{u}\,(x(w).\,P_1{+}M_1 \mid \overline{x}\langle z\rangle.\,P_2{+}M_2 \mid P_3)\ \big]{\rightarrow}\boldsymbol{B}\big[\ \nu\tilde{u}\,(P_1\{z/w\} \mid P_2 \mid P_3)\ \big]$

(tau) $\boldsymbol{B}\big[\ \nu\tilde{u}\,(\tau.\,P_1{+}M_1 \mid P_2)\ \big]{\rightarrow}\boldsymbol{B}\big[\ \nu\tilde{u}\,(P_1 \mid P_2)\ \big]$

(expose) $\boldsymbol{B}\big[\ \nu\tilde{u}\,(\mathsf{exp}(x,\Gamma).\,P_1{+}M_1 \mid P_2)\ \big]{\rightarrow}\boldsymbol{B}\,\beta(x,\,\Gamma)\big[\ \nu\tilde{u}\,(P_1 \mid P_2)\ \big]$
   provided $x \notin \tilde{u}$, $x \notin \mathsf{sub}(\boldsymbol{B})$ and $x \notin \Gamma$

(change) $\boldsymbol{B}^*\,\hat{\beta}(x,\,\Gamma)\big[\ \nu\tilde{u}\,(\mathsf{ch}(x,\Delta).\,P_1{+}M_1 \mid P_2)\ \big]{\rightarrow}\boldsymbol{B}^*\,\hat{\beta}(x,\,\Delta)\big[\ \nu\tilde{u}\,(P_1 \mid P_2)\ \big]$
   provided $x \notin \tilde{u}$

(hide) $\boldsymbol{B}^*\,\beta(x,\,\Gamma)\big[\ \nu\tilde{u}\,(\mathsf{hid}(x).\,P_1{+}M_1 \mid P_2)\ \big]{\rightarrow}\boldsymbol{B}^*\,\beta^h(x,\,\Gamma)\big[\ \nu\tilde{u}\,(P_1 \mid P_2)\ \big]$
   provided $x \notin \tilde{u}$

(unhide) $\boldsymbol{B}^*\,\beta^h(x,\,\Gamma)\big[\ \nu\tilde{u}\,(\mathsf{unh}(x).\,P_1{+}M_1{'} \mid P_2)\ \big]{\rightarrow}\boldsymbol{B}^*\,\beta(x,\,\Gamma)\big[\ \nu\tilde{u}\,(P_1 \mid P_2)\ \big]$
   provided $x \notin \tilde{u}$

(bind) $\beta(x_1,\,\Delta_1)\,\boldsymbol{B}_1^*\big[\ P_1\ \big] \parallel \beta(x_2,\,\Delta_2)\,\boldsymbol{B}_2^*\big[\ P_2\ \big]{\rightarrow}$

$\qquad\qquad \beta^h(x_1,\,\Delta_1)\,\boldsymbol{B}_1^*\,\beta^h(x_2,\,\Delta_2)\,\boldsymbol{B}_2^*\big[\ P_1\{l/x_1|\} \mid P_2\{l/x_2|\}\ \big]$

   if $m_G(\Delta_1,\Delta_2)$ and $l \notin \mathsf{n}(\beta(x_1,\,\Delta_1)\,\boldsymbol{B}_1^*\big[\ P_1\ \big] \parallel \beta(x_2,\,\Delta_2)\,\boldsymbol{B}_2^*\big[\ P_2\ \big])$

(unbind) $\beta^h(x_1,\,\Delta_1)\,\boldsymbol{B}_1^*\,\beta^h(x_2,\,\Delta_2)\,\boldsymbol{B}_2^*\big[\ P_1\{l/x_1|\} \mid P_2\{l/x_2|\}\ \big]{\rightarrow}$

$\qquad\qquad \beta(x_1,\,\Delta_1)\,\boldsymbol{B}_1^*\big[\ P_1\ \big] \parallel \beta(x_2,\,\Delta_2)\,\boldsymbol{B}_2^*\big[\ P_2\ \big]$

   if $\mathcal{D}(\beta(x_1,\,\Delta_1)\,\boldsymbol{B}_1^*\big[\ P_1\ \big],\beta(x_2,\,\Delta_2)\,\boldsymbol{B}_2^*\big[\ P_2\ \big])$

(redex) $\dfrac{B{\rightarrow}B'}{B \mid B''{\rightarrow}B' \mid B''}$ (struct) $\dfrac{B_1{\equiv}B_1' \qquad B_1'{\rightarrow}B_2}{B_1{\rightarrow}B_2}$

The axiom (intra) concerns communications between pi-processes within the same box. The rule states that given a box B, if its internal pi-process can perform a communication then B can be reduced leading to a box with the same interface as B and with the internal process changed by the communication. The axiom (tau) does the same for prefix τ. The axiom (expose) adds a new binder to a box. The name x declared in the prefix $\mathsf{exp}(x,\Gamma)$ is a placeholder which can be renamed to avoid clashes with the subjects of the other binders of the containing box. The axiom (change) changes the type of a binder. The axiom (hide) forces a binder to become hidden (when made invisible, a binder named x is graphically represented by x^h). The (unhide) axiom, dual to (hide), makes visible a hidden binder.

Wrt [14], the axioms (bind) is an instance of the join axiom schema, as well as the axiom (unbind) is an instance of the split rule. The axiom (bind) is for complexation. Consider the following transition:

$$x : \Delta_E \qquad z : \Delta_S \qquad\qquad x^h : \Delta_E \quad z^h : \Delta_S$$

(1) $\boxed{x.\,E}\ \boxed{\bar{z}.\,\mathsf{ch}(z, \Delta_P).\,P} \ \rightarrow\ \boxed{l.\,E\{|1/x|\} \mid \bar{l}.\,\mathsf{ch}(z, \Delta_P).\,P\{|1/z|\}}$

$$(B_E) \qquad\qquad (B_S) \qquad\qquad\qquad (B_{ES})$$

Boxes B_E and B_S can complex into B_{ES} if $m_G(\Delta_E, \Delta_S)$ is true. The biological mechanism underlying complexation allows two complexed molecules to interact through their binding sites. This is the key mechanism of many biological interactions. Beta-binders$^{\mathbb{D}}$ provides an abstraction of such mechanism relying on names substitution. For instance the internal process of B_{ES} can interact through the new name l. The idea is to introduce a new name for allowing communications between internal pi-processes after the complexation, without inhibiting the ability of changing the interface of the box. That is, only input and output prefixes are renamed, while hide, unhide, change and expose do not modify their links. To formally handle communication after complexation, we introduced $\{|1/x|\}$. For instance, the prefix $\mathsf{ch}(z, \Delta_P)$ remains unaltered after the complexation of B_E and B_S. Finally, after a bind the two beta binders involved, e.g. $\beta(x, \Delta_E)$ and $\beta(y, \Delta_S)$, become hidden and they are not available for further complexation.

The axiom (unbind) reverses the axiom (bind). Theoretically, a complex can be broken at any time if enough energy is available. In practice, only some complexes can be broken depending on their conformation. We abstract energy valuation as a *decomplexation* relation $\mathcal{D} \subseteq \mathcal{B} \times \mathcal{B}$. Once $B_1[\,P_1\,]$ and $B_2[\,P_2\,]$ bind, they can unbind iff $(B_1[\,P_1\,], B_2[\,P_2\,]) \in \mathcal{D}$, written $\mathcal{D}(B_1[\,P_1\,], B_2[\,P_2\,])$. For instance, if $\mathcal{D}(B_E, B_S)$ then we can derive the following transition

$$x^h : \Delta_E \quad z^h : \Delta_S \qquad\qquad\qquad x : \Delta_E \qquad z : \Delta_S$$

$\boxed{l.\,E\{|1/x|\} \mid \bar{l}.\,\mathsf{ch}(z, \Delta_P).\,P\{|1/z|\}}\ \rightarrow\ \boxed{x.\,E}\ \boxed{\bar{z}.\,\mathsf{ch}(z, \Delta_P).\,P}$

$$(B_{ES}) \qquad\qquad\qquad\qquad (B_E) \qquad\qquad (B_S)$$

that reverses the complexation in (1).

The rules redex and struct are standard in reduction semantics. They allow to interpret the reduction of a subcomponent as a reduction of the global system, and to infer a reduction after a proper structural shuffling of the bio-process at hand, respectively.

$$x : \Delta_E \qquad\qquad\qquad y : \Delta_S$$

$$B_E = \boxed{E}\quad E = x.\,E \qquad B_S = \boxed{S} \qquad \begin{aligned} S &= \bar{y}.\,S^* \\ S^* &= \mathsf{ch}(y, \Delta_P).\,P \end{aligned}$$

$$\mathcal{D}_1 = \{(\beta(x, \Delta_E)[\,E\,], \beta(y, \Delta_S)[\,S\,]), (\beta(x, \Delta_E)[\,E\,], \beta(y, \Delta_P)[\,P\,])\}$$

Fig. 3. Beta-binders$^{\mathbb{D}}$ specification of simple enzyme-catalyzed reaction

4.1 Example: Beta-Binders for Enzymatic Reactions

Now we are able of modeling the enzymatic reaction schemas presented in Sec. 3 within Beta-binders$^{\mathbb{D}}$ formalism. We will describe Beta-binders$^{\mathbb{D}}$ specifications of the molecules involved and we will derive dynamic behaviours relying on the semantics presented above. We assume a Boolean DockSpace $G = (\mathbb{D}_G, \mathbb{P}_G, \mathbb{S}_G, m_G)$.

Simple enzyme-catalyzed reaction. Enzyme E and substrate S are represented as boxes B_E and B_S of Fig. 3, respectively. The decomplexation relation $\mathcal{D}_1$, also in Fig. 3, specifies that the complex between the enzyme and the substrate can be broken, as well as the complex between the enzyme and the product. We can derive the path that leads to the production of product P:

$$x : \Delta_E \; y : \Delta_S \quad x^h : \Delta_E \quad y^h : \Delta_S \quad x^h : \Delta_E \quad y^h : \Delta_S$$

$$\boxed{E} \;\; \boxed{S} \;\to\; \boxed{l.\,E\{|^l/_x|\} \mid \bar{l}.\,S^*\{|^l/_y|\}} \;\to\; \boxed{E\{|^l/_x|\} \mid \mathsf{ch}(y, \Delta_P).\,P\{|^l/_y|\}} \;\to\;$$

$$x^h : \Delta_E \quad y^h : \Delta_P \qquad x : \Delta_E \; y : \Delta_P$$

$$\to\; \boxed{E\{|^l/_x|\} \mid P\{|^l/_y|\}} \;\to\; \boxed{E} \;\; \boxed{P}$$

The boxes B_E and B_S can complex if $m_G(\Delta_E, \Delta_S)$, mimicking the specificity of the enzyme-substrate interaction. After the bind, the internal processes E and S can interact on the new name l, due to the substitutions $\{|^l/_x|\}$ and $\{|^l/_y|\}$. The synchronization on l makes the complex active, enabling the prefix $\mathsf{ch}(y, \Delta_P)$. Then an (unbind) occurs, because $\mathcal{D}_1(\beta(x, \Delta_E)[\,E\,], \beta(y, \Delta_P)[\,P\,])$.

We also highlight another computation:

$$x : \Delta_E \; y : \Delta_S \quad x^h : \Delta_E \quad y^h : \Delta_S \quad x : \Delta_E \; y : \Delta_S$$

$$\boxed{E} \;\; \boxed{S} \;\to\; \boxed{l.\,E\{|^l/_x|\} \mid \bar{l}.\,S^*\{|^l/_y|\}} \;\to\; \boxed{E} \;\; \boxed{S}$$

Boxes B_E and B_S complex again, but they also decomplex immediately, because $\mathcal{D}_1(\beta(x, \Delta_E)[\,E\,], \beta(y, \Delta_S)[\,S\,])$.

Multi-substrate systems. In this reaction schema the enzyme catalysis involves an enzyme E' and two substrates S_1 and S_2, specified in Fig 4 by boxes $B_{E'}$ and B_{S_i} ($i \in \{1, 2\}$), respectively. The structure of the two substrates are the same of the substrate B_S above. We need to refine the structure of the enzyme B_E, in order to capture the bouncing between the two states of the enzyme. We also refine the decomplexation relation as $\mathcal{D}_2$.

We derive the computation that lead to the production of P_1 and P_2:

$$x : \Delta_{E'} y : \Delta_{S_1} y : \Delta_{S_2} \quad x^h : \Delta_{E'} \quad y^h : \Delta_{S_1} \qquad\qquad y : \Delta_{S_2}$$

$$\boxed{E'} \;\; \boxed{S_1} \;\; \boxed{S_2} \;\to\; \boxed{l.\,\mathsf{ch}(x, \Delta_{E^*}).\,E^*\{|^l/_x|\} \mid \bar{l}.\,S_1^*\{|^l/_y|\}} \;\; \boxed{S_2} \;\to\;$$

The enzyme E' and the substrate S_1 binds if $m_G(\Delta_{E'}, \Delta_{S_1})$ is true.

$$x^h : \Delta_{E'} \quad y^h : \Delta_{S_1} \qquad\qquad y : \Delta_{S_2} \quad x^h : \Delta_{E^*} \; y^h : \Delta_{S_1} \quad y : \Delta_{S_2}$$

$$\to\; \boxed{\mathsf{ch}(x, \Delta_{E^*}).\,E^*\{|^l/_x|\} \mid S_1^*\{|^l/_y|\}} \;\; \boxed{S_2} \;\to\; \boxed{E^*\{|^l/_x|\} \mid S_1^*\{|^l/_y|\}} \;\; \boxed{S_2} \;\to\;$$

$$x : \Delta_{E'} \qquad\qquad\qquad\qquad\qquad\qquad y : \Delta_{S_i}$$

$$B_{E'} = \boxed{E'} \quad \begin{array}{l} E' = x.\,\mathsf{ch}(x, \Delta_{E^*}).\,E^* \\ E^* = x.\,\mathsf{ch}(x, \Delta_{E'}).\,E' \end{array} \qquad B_{S_i} = \boxed{S_i} \quad \begin{array}{l} S_i = \overline{y}.\,S_i^* \\ S_i^* = \mathsf{ch}(y, \Delta_{P_i}).\,P_i \end{array}$$

$$\mathcal{D}_2 = \{(\beta(x, \Delta_E)\,[\,E\,], \beta(y, \Delta_{S_1})\,[\,S_1\,]), (\beta(x, \Delta_{E^*})\,[\,E^*\,], \beta(y, \Delta_{P_1})\,[\,P_1\,]),$$
$$(\beta(x, \Delta_{E^*})\,[\,E^*\,], \beta(y, \Delta_{S_2})\,[\,S_2\,]), (\beta(x, \Delta_E)\,[\,E\,], \beta(y, \Delta_{P_2})\,[\,P_2\,])\}$$

Fig. 4. Beta-binders$^{\mathbb{D}}$ specification of a multi-substrate system

The complex $\mathsf{E}'\mathsf{S}_1$ is activated by an interaction on the new channel l. Then the type $\Delta_{E'}$ becomes Δ_{E^*} and therefore the enzyme changes its state in E^*.

$$x^h : \Delta_{E^*} \quad y^h : \Delta_{P_1} \qquad y : \Delta_{S_2} \qquad x : \Delta_{E^*}\ y : \Delta_{P_1}\ y : \Delta_{S_2}$$

$$\rightarrow \boxed{E^*\{|^l\!/x|\} \mid P_1^*\{|^l\!/y|\}} \quad \boxed{S_2} \rightarrow \boxed{E^*} \quad \boxed{P_1} \quad \boxed{S_2} \rightarrow$$

The product P_1 is ready and then it is released because the decomplexation relation specify $\mathcal{D}_2(\beta(x, \Delta_{E^*})\,[\,E^*\,], \beta(y, \Delta_{P_1^*})\,[\,P_1\,])$.

$$x^h : \Delta_{E^*} \quad y^h : \Delta_{S_2} \qquad\qquad\qquad y : \Delta_{P_1}$$

$$\rightarrow \boxed{l.\,\mathsf{ch}(x, \Delta_{E'}).\,E'\{|^l\!/x|\} \mid \bar{l}.\,S_2^*\{|^l\!/y|\}} \quad \boxed{P_1} \rightarrow$$

The types Δ_{E^*} and Δ_{S_2} are affine, i.e. $m_G(\Delta_{E^*}, \Delta_{S_2})$ is true, and the second substrate S_2 can bind the modified enzyme E^*.

$$x^h : \Delta_{E^*} \quad y^h : \Delta_{S_2} \qquad y : \Delta_{P_1} \quad x^h : \Delta_{E'} \qquad y^h : \Delta_{S_2}\,y : \Delta_{P_1}$$

$$\rightarrow \boxed{\mathsf{ch}(x, \Delta_{E'}).\,E'\{|^l\!/x|\} \mid S_2^*\{|^l\!/y|\}} \quad \boxed{P_1} \rightarrow \boxed{E'\{|^l\!/x|\} \mid S_2^*\{|^l\!/y|\}} \quad \boxed{P_1} \rightarrow$$

The complex $\mathsf{E}^*\mathsf{S}_1$ is activated by an interaction on l and then the enzyme returns to its initial state E'.

$$x^h : \Delta_{E'} \quad y^h : \Delta_{P_2} \qquad y : \Delta_{P_1} \quad x : \Delta_{E'}\,y : \Delta_{P_1}\,y : \Delta_{P_2}$$

$$\rightarrow \boxed{E'\{|^l\!/x|\} \mid P_2\{|^l\!/y|\}} \quad \boxed{P_1} \rightarrow \boxed{E'} \quad \boxed{P_1} \quad \boxed{P_2}$$

Finally, the product P_2 is ready and released.

This is only one among the different possible evolutions of the system composed by E', S_1 and S_2. However, despite of the complexity of the transition system the model is obtained "simply" *extending* the specification of the previous (and simpler) example.

Enzyme inhibition. The example of a multisubstrate system above outline the extensibility of Beta-binders$^{\mathbb{D}}$, while the enzyme inhibition of the present example is significant for *compositionality*. In fact, to specify the competitive inhibition presented in Sec. 3 we add the inhibitor I, specified as

$$y : \Delta_I$$

$$B_I = \boxed{I} \quad I = \mathsf{nil}$$

to the specification of Fig. 3, without changing the boxes B_E and B_S. We also extend the decomplexation relation as

$$\mathcal{D}_3 = \mathcal{D}_1 \cup \{(\beta(x,\, \Delta_E)\,[\ E\ ], \beta(y,\, \Delta_I)\,[\ I\ ])\}$$

to allow the inhibitor frees the enzyme. Now the system can again evolve to produce the product P, but also we can infer the computation

$$x : \Delta_E \quad y : \Delta_S \quad y : \Delta_I \qquad x^h : \Delta_I \qquad y^h : \Delta_I \qquad y : \Delta_S$$

| E | S | I | $\rightarrow$ | $l.\,E\{|{}^{l}\!/_{x}|\}\mid\text{nil}$ | S |

where the complex EI cannot be activated because I is inactive. Moreover, the substrate S cannot interact with E until the inhibitor frees it (i.e. EI decomplex).

5   Conclusion

Many potential drugs fail to reach the market because of unexpected effects on human metabolism, such as toxicity. There is the need of early elimination of such compounds in the drug discovery pipeline considering the high costs, in time and money, of the production of a new drug. The problem is that scientists have to make decisions on the future of new drugs without a detailed understanding of the mechanisms underlying the disease. A better system would endow researchers to make accurate decisions based on the structure of the new compound, and on the information about the disease. We focused on two areas of knowledge: (i) *computational models of molecular structures* used by the pharmaceutical companies; we outlined molecular docking, a technique for predicting whether one molecule will bind to another; (ii) *systems biology*, that studies physiology and diseases at the level of molecular pathways and regulatory networks.

In this paper we suggested a direction for integrating these two resources, relying on concurrency theory and formal languages. In particular, we introduced Beta-binders$^{\mathbb{D}}$, a process calculus that incorporates molecular docking prediction with dynamic information of the system under investigation. The formal semantics of Beta-binders$^{\mathbb{D}}$ will serve as foundation of tools and methods for qualitative analysis of non-linear flow of information, studying non-trivial effects of perturbing a system. Moreover, Beta-binders$^{\mathbb{D}}$ aims to enhance *compositionality* offered by process calculi to formally organise biological knowledge and eventually predict the behavior of complex systems. We also outline *extensibility* as a key feature that may improve systems biology approach.

We tested Beta-binders$^{\mathbb{D}}$ modelling some enzymatic reaction schemas. Enzymes are proteins essential to sustain life. For instance, metabolic pathways comprises several enzymes that work together with a precise order: the product of an enzyme is the substrate in the next enzymatic reaction. A malfunction of a critical enzyme can lead to severe diseases, therefore many drugs regulate the activity of an enzyme acting as, e.g., an inhibitor. Here we presented qualitative models of three reaction schemas and we highlighted some computations. Quantitative reasoning is also feasible relying on a stochastic extension of Beta-binders [26]: it allows simulation relying on Gillespie's algorithm [27].

References

1. McMartin, C., Bohacek, S.: QXP: Powerful, rapid computer algorithms for structure-based drug design. Journal of Computer-Aided Molecular Design **11** (1997) 333

2. Halperin, I., Ma, B., Wolfson, H., Nussinov, R.: Principles of Docking: An Overview of Search Algorithms and a Guide to Scoring Functions. PROTEINS: Structure, Function, and Genetics **47** (2002) 409

3. Butcher, E., Berg, E., Kunkel, E.: Systems biology in drug discovery. Nature Biotechnology **22** (2004) 1253

4. FitzGerald, G.: Coxibs and Cardiovascular Disease. The New England journal of medicine **351**(17) (2004) 1709

5. Kuthe, A., Montorsi, F., Andersson, K., Stief, C.: Phosphodiesterase inhibitors for the treatment of erectile dysfunction. Current opinion in investigational drugs **10** (2002) 1489

6. McCulley, T., Lam, B., Marmor, M., Hoffman, K., Luu, J., Feuer, W.: Acute effects of sildenafil (viagra) on blue-on-yellow and white-on-white Humphrey perimetry. Journal of neuro-ophthalmology **20** (2000) 227

7. Kitano, H.: Foundations of System Biology. MIT Press (2002)

8. Bugrim, A., Nikolskaya, T., Nikolsky, Y.: Early prediction of drug metabolism and toxicity: systems biology approach and modeling. Drug Discovery Today **9** (2004) 127

9. Rao, B., Lauffenburger, D., Wittrup, K.: Integrating cell-level kinetic modeling into the design of engineered protein therapeutics. Nature Biotechnology **23** (2005) 191

10. Apica, G., Ignjatovicb, T., Boyerb, S., Russellc, R.: Illuminating drug discovery with biological pathways. FEBS Letters **579** (2005) 1872

11. Rajasethupathy, P., Vayttaden, S., Bhalla, U.: Systems modeling: a pathway to drug discovery. Current Opinion in Chemical Biology **9** (2005) 400

12. Regev, A., Shapiro, E.: Cells as computations. Nature **419** (2002) 343

13. Bergstra, J.A.: Handbook of Process Algebra. Elsevier Science Inc. (2001)

14. Priami, C., Quaglia, P.: Beta Binders for Biological Interactions. In: CMSB '04. Volume 3082 of LNBI., Springer (2005)

15. Sangiorgi, D., Walker, D.: The π-calculus: a Theory of Mobile Processes. Cambridge Universtity Press (2001)

16. Milner, R.: Communicating and mobile systems: the π-calculus. Cambridge Universtity Press (1999)

17. Priami, C., Regev, A., Silverman, W., Shapiro, E.: Application of a stochastic name-passing calculus to representation and simulation of molecular processes. Information Processing Letters **80**(1) (2001) 25–31

18. Phillips, A., Cardelli, L.: A Correct Abstract Machine for the Stochastic Picalculus. In: BioConcur '04, Workshop on Concurrent Models in Molecular Biology. (2004)

19. Alberts, B., Johnson, A., Lewis, J., Raff, M., Roberts, K., Walter, P.: Molecular biology of the cell (IV ed.). Garland science (2002)

20. DiMasi, J., Hansen, R., Grabowski, H.: The price of innovation: new estimates of drug development costs. Journal of Health Economics **22** (2003) 151

21. Mack, G.: Can complexity be commercialized? Nature Biotechnology **22** (2004) 1223

22. Tame, J.: Scoring Functions the First 100 Years. Journal of Computer-Aided Molecular Design **19** (2005) 445

23. Kuntz, I., Blaney, J., Oatley, S., Langridge, R., Ferrin, T.: A geometric approach to macromolecule-ligand interactions. Journal of molecular biology **161** (1982) 269

24. Taylor, J., Burnett, R.: DARWIN: A program for docking flexible molecules. PRO-TEINS: Structure, Function, and Genetics **41** (2000) 173

25. Copeland, R.: Evaluation of Enzyme Inhibitors in Drug Discovery : A Guide for Medicinal Chemists and Pharmacologists (Methods of Biochemical Analysis). Wiley-Interscience (2005)

26. Degano, P., Prandi, D., Priami, C., Quaglia, P.: Beta-binders for biological quantitative experiments. In: 4rd Int. Workshop on Quantitative Aspects of Programming Languages (QAPL 06). (2006) to appear.

27. Gillespie, D.: Exact stochastic simulation of coupled chemical reactions. Journal of Physical Chemistry **81**(25) (1977) 2340–2361

Feedbacks and Oscillations in the Virtual Cell VICE

D. Chiarugi[1], M. Chinellato[2], P. Degano[2,3], G. Lo Brutto[2], and R. Marangoni[2]

[1]Dipartimento di Scienze Matematiche e Informatiche, Università di Siena
[2]Dipartimento di Informatica, Università di Pisa
[3]The Microsoft Research - University of Trento Centre for Computational and Systems Biology
chiarugi3@unisi.it, {chinella, degano, lobrutto, marangon}@di.unipi.it

Abstract. We analyse an enhanced specification of VICE, a hypothetical prokaryote with a genome as basic as possible. Besides the most common metabolic pathways of prokaryotes in interphase, VICE also posseses a regulatory feedback circuit based on the enzyme phosphofructokinase. We use as formal description language a fragment of the stochastic π-calculus. Simulations are run on BEAST, an abstract machine specially tailored to run *in silico* experimentations. Two kinds of virtual experiments have been carried out, depending on the way nutrients are supplied to VICE. The result of our experimentations *in silico* confirm that our virtual cell "survives" in an optimal environment, as it exhibits the homeostatic property similary to real living cells. Additionally, oscillatory patterns in the concentration of fructose-6-phosphate and fructose-1,6-bisphosphate show up, similar to the real ones.

1 Introduction

One of the major challenges of contemporary biology is addressing the complexity underlying the dynamics of the various molecules inside the cellular machinery, when they give rise to a living organism [19]. Even though many details of the single "building blocks" are nowadays known, the way they interact is still unclear.

Additionally, the so-called high-throughput techniques have provided us a with large collection of biological data in a relatively short period of time. In this way, the catalogue of the components of many living organisms is rapidly growing up. Nevertheless, it is absolutely not trivial to fill in the gap between the description at the molecular level and the behaviour exhibited by the system as a whole: complex properties of biological systems only emerge when the various "building block" interact.

Experimental approaches result are not adequate to cope with these systemic problems: even the so-called -*omic* techniques seem able to provide only snapshots of the complete movie. A promising approach is to represent all the known relationship between the elements of a metabolome *in silico*, so building up a sort of *virtual cell* [11,21,5]. This method consists in defining a formal model of

C. Priami (Ed.): CMSB 2006, LNBI 4210, pp. 93–107, 2006.

the studied phenomenon and in investigating its properties through computer-based simulation tools. The various models proposed so far differ mainly for the formalism they rely upon.

The most widespread one is based on ordinary differential equation (ODE for short). They describe the known relationship between the elements of the modelled organism *via* a set of reaction rate equations, possibly constrained, as in the case of Flux Balance Analysis [27]. Computer based differential models are characterized by a high descriptive power and have been successfully used to investigate features of crucial metabolic pathways. Building up ODE-based models, however, requires to set up a lot of details. For example, representing a metabolic pathway requires the knowledge of the involved differential kinetic equation for each reaction of the pathway. Unfortunately, the necessary mechanicistic details and kinetic parameters are often unavailable. Moreover, ODEs seem not flexible enough, in that they are hard to compose, update and solve. Moreover it is difficult to express alternative behaviour resulting, e.g. from dynamic changes in the topological relations between the objects modelled.

A recent alternative consists in specifying the living matter through so-called calculi for concurrent systems and run the correspondent simulation "programs" [26,29]. The paradigm of concurrency results particularly suitable to describe biological organism, from the molecular to (multi)cellular level. Indeed, biological components and organisms can be seen as processes and as networks of processes, respectively, while cell interactions are represented as communications between processes. A relevant feature is that process calculi are compositional and thus each biological component is specified independently of (most of) the others. Specifications are then put aside and run, with no *a priori* constraint on temporal or causal relations of the computations.

Our goal is studying a whole cell, in an holistic fashion typical of Systems Biology. A successfull example of this goal is given by Virtual *E. coli* [1]. However, the choice of the organism to model is crucial, because even simple biological entities, like bacteria, have a complexity extremely high; their simulation thus requires huge computational resources. We faced the problem of specifying a cell using process calculi. To decrease complexity of modelled organisms, in [5] some of us proposed VICE, a hypothetical cell with a genome as basic as possible, derived from the Minimal Gene Set of [23]. The basic genome of VICE was obtained by eliminating duplicated genes and other redundancies from MGS, and further modified to obtain a very basic prokaryote-like genome, which only contains 187 different genes. We then specified VICE in (enhanced) π-calculus [22,7] and run several virtual experiments. The results show that VICE exhibits *in silico* a behaviour typical of real prokaryotes in the same experimental conditions, e.g. the time course distribution of metabolites concentration along the glycolitic pathway significantly resembles the real one. A major point of our approach is that we were able to study the interplay of *all* metabolic pathways in VICE, because we modelled a *whole* cell.

We report on our recent work on a deeper specification, simulation and analysis of VICE. Since the timing of physiological events is a chief feature of living organisms, we are interested in investigating whether VICE possesses some sort of capability for autonomously regulating its own internal "clock." In other words, we look for those components driving the physiological pace maker of the cell. As a very first step in this direction we are interested in detecting some rhythmic behaviours that eventually involve physiological parameters. In the literature there are proposals of oscillatory patterns exhibited by metabolites along pathways, particularly along the glycolysis. This substained oscillations appear to emerge from the presence of feedback control circuits, involving some enzymes active in the glycolitic pathway. The main responsible for oscillatory behaviour has been proposed in [9] to be a specific enzyme, namely phosphofructokinase, the rate of which increases as the concentration of the metabolite fructose-1,6-bisphosphate grows. We enhanced the specification of VICE with this positive feedback regulation (see Section 3 and the Appendix).

To perform our *in silico* experiments, we designed a specially tailored version of the stochastic π-calculus, described in Section 2. We implemented its abstract machine, very similar to SPIM [4]. This abstract machine, called BEAST (Biological Environment Analysis and Simulation Tool), includes the stochastic simulation algorithm SSA by Gillespie [14]. To our surprise, a small fragment of the stochastic π-calculus proved sufficient to specify our virtual prokaryote, which, as any real ones, has no intracellular structure, e.g. it is not compartmentalized. We studied VICE in interphase, under two different feeding conditions. In the first, we supplied a large reservoire of glucose, while in the second this nutrient was fed at a constant rate — we recall from [5] that VICE only has carriers for glucose, to compare our results with those in the literature obtained for prokaryotes cultivated in a glucose-limited medium [2].

The results of our simulations are in Section 4. They confirm those in [5] in that VICE has homeostatic properties, when glucose is given in a single big supply. Indeed, our model reaches a steady state that is resistent to non shocking changes in the external environment, in particular when glucose is instead fed continuously. The simulations of VICE under this second experimental condition show that oscillations emerge, are substained with constant period and amplitude, and enjoy the typical properties of prokaryotes.

Our model behaves then in surprising agreement with the living prokaryotes, when exerted under similar experimental conditions. In particular, our results add a little token to the hypothesis that a phosphofructokinase based feedback circuit provides a cell with an internal pace maker. These results further asses the validity of our model and the feasability of our approach, hopefully also for a future development of predictive tools.

2   The Calculus and the Abstract Machine

We assume the reader is familiar with process calculi, in particular with the π-calculus and its usage for specifying bio-chemical reactions; more on this topic can be found in [29].

Here, we use the stochastic version of the π-calculus proposed in [25], that enables its users to specify both qualitative and quantitative aspects of distributed systems. The main difference with the standard π-calculus is that prefixes become pairs (μ, r) where μ is an action and r is a random variable with an exponential distribution, called *activity rate*. Also, nondeterminism is replaced by stochastic choice, defined by the random variable: a *race condition* will actually choose the fastest among two or more actions enabled at the same time. The semantics of the stochastic π-calculus is a transition system, whose transitions are labelled with a rate r. As a matter of fact, in building up our model of the basic cell VICE, we found it sufficient a small subset of the whole calculus, actually a subset of CCS. In particular neither message passing was needed, as synchronization suffices, nor restriction (ν). Additionally we adopted only stochastic guarded choices, and we used constant definition in place of replication. Moreover, a channel is only used for communication between two processes, and the same channel can not be used by a process both for output and for input. These restrictions sometimes make our specifications less natural, but hel in simplifying the calculus of channel activities. As a new features we imposed an upper bound to the rate of channels, called top-rate. In this way, we describe *saturation*, a typical feature of reactions catalysed by enzymes, like those occurring in metabolic pathways. Actually, the capability of an enzyme to catalyse a reaction grows up until it reaches its maximum value. To define the rate associated with actions, we followed the line used in SPIM.

We associate each channel x with a corresponding reaction rate, written $rate(x)$. These rates model kinetic constants, so the actions involving a specific channel will always and coherently be associated with its rate. The actual rate of a communication, i.e. the apparent rate of the corresponding bio-chemical reaction, is computed using the StochasticSimulation Algorithm (SSA for short) by Gillespie [14].

More formally, let $Chan = \{a, b, \ldots\}$ be a set of communication channels and $Hid = \{\tau_1, \tau_2, \ldots\}$ be a set of hidden, internal channels, with $Chan \cap Hid = \emptyset$. Let $rate, top\_rate : Chan \cup Hid \rightarrow \Re^+$ be the functions associating channels with their basal and top rate respectively, with the condition:

$$\forall x \in Chan \cup Hid \qquad 0 < rate(x) \leq top\_rate(x)$$

Finally, let $A = \{A, B, \ldots\}$ be the set of *constant names*. The set of *processes*, $P = \{P, Q, \ldots\}$, is defined by the following BNF-like grammar:

$$P ::= Nil \mid \pi.A \mid P|Q \mid \sum_{i \in I} \pi_i.A$$

where (i) π is a prefix of the form a, $\bar{a}$ for an input or output on a, and τ for a silent move; and (ii) constrant A has a unique defining equation $A \stackrel{\triangle}{=} P$.

As usual the operational semantic comprises the standard rules for the structural congruence $\equiv$, i.e. $(P_{/\equiv}, +, Nil)$ and $(P_{/\equiv}, |, Nil)$ are abelian monoids and $P + P \equiv P$.

The inference rules defining the dynamics of our tiny calculus are layered: the final step only computes the apparent rate $g(P, Q, r)$ of the transition from P to Q according to Gillespie's algorithm. We have designed and implemented a modular abstract machine, called BEAST, for a significant super-set of the calculus presented in Table 1. For the present investigation, we however switched

Table 1. Inference rules

$$(a.A + \textstyle\sum_i P_i)|(\bar{a}.B + \sum_j Q_j) \overset{rate(a)}{\rightarrow} A|B$$

$$\frac{P \xrightarrow{r}_R Q}{P|R \xrightarrow{r}_R Q|R} \qquad\qquad \frac{P \xrightarrow{r} Q}{A \xrightarrow{r} Q} A \overset{\triangle}{=} P$$

$$\frac{P \xrightarrow{r}_R Q}{P \overset{q}{\Rightarrow} Q} \qquad\qquad \text{where } q = (P, Q, r)$$

off most of its features that, surprisingly enough, resulted unecessary in modelling VICE.

In designing BEAST, we have been greatly inspired bu SPIM [4]. In particular we used a data structure for storing channels and their rates and for linking channels to the process definitions where they occur. Crucial to our study is the possibility of inspecting internal states, that store the intermediate concentrations of metabolites during the simulation. This information is collected in an output text files, used later on to infer causality relationships and to perform statistical analysis.

At the present time, we naively implemented Gillespie's algorithm for computing the actual rates of transitions. As a matter of fact BEAST spends most of its time in running the SSA, despite of the restricted usage of channels that reduce the model complexity. This is one of the crucial points as far as efficiency of simulations is concerned. We feel it necessary to find new algorithms for computing the apparent rate of transitions, e.g. [13,3], or at least implement in a smarter way the Gillespie's SSA.

Just to give a rough idea of the computational burden of simulating VICE, consider a system made of about 2,000,000 processes, each with 10 stochastic choices in average. On an AMD Athlon 1.5 GHz duo with 1 Gb of RAM, the simulation of about 30,000 transitions took about one night.

3   The Model

In building up our model, we strictly followed in the first phase the line of [5]. We chose the virtual cell VICE, because its genome is extremely reduced, while related to that of a real prokaryote, in particular because simulations show VICE "surviving" *in silico*.

The genome of VICE has been obtained from the hypothetical Minimal-Gene-Set (MGS) proposed in [20,23], through a functional screening. According to this analysis, some genes were further eliminated. Also, some critical steps missing in pathways were found, and so two other genes not originally contained in MGS were introduced. The working hypothesis were that VICE is placed in an optimal environment containing enough essential nutrients, and shaped to dilute or remove all the potentially toxic catabolites. Also, competition and other stressing factor are completely banned. As done in [5], to keep small the specification of VICE in the π-calculus, a few further slight simplification were made. Typically, we grouped in a single entity all the multi-enzymatic complexes, when acting like a single cluster.

A further simplification is that our specification only has carriers for glucose, just as it was for [5]. This is because, the results obtained *in silico* are to be compared with those in the literature obtained for prokaryotes cultivated in a glucose-limited medium [2].

Our virtual cell possesses the following main features:

1. The cell relies on a complete glycolytic pathway for the oxydation of glucose to pyruvate and reduced-NAD. Pyruvate is then converted to acetate which, being catabolite, can diffuse out of the cell. A transmembrane reduced-NAD dehydrogenase complex catalyzes the oxydation of reduced-NAD; this reaction is coupled with the synthesis of ATP through the ATP synthase/ATPase transmembrane system. This set of reactions enables the cell to manage its energetic metabolism.

2. The cell has a Pentose Phosphate Pathway, coposed by enzymes leading to the synthesis of ribose phosphate and 2-deoxyribose phosphate.

3. For lipid metabolism, the cell has enzymes for glycerol-fatty acids condensation, but no pathways for fatty acids synthesis. So, these metabolites must be taken from the outside.

4. The cell has no pathways for amino acid synthesis and, therefore, we assume all amino acids be present in the environment.

5. Thymine is the only nucleotide the cell is able to synthesize *de novo*; the other nucleotides are provided by "salvage pathways".

6. The cell possesses a proper set of carriers for metabolites uptake:
   (a) a Glycerol Uptake Facilitator Protein;
   (b) a PtsG System for sugar uptake;
   (c) an ACP carrier protein for fatty acids uptake;
   (d) a broad specificity amino acids uptake ATPase;
   (e) broad specificity permeases for other essential metabolites uptake;

7. The cell possesses the necessary enzymes for protein synthesis, including DNA-transcription and translation. The cell possesses also the whole machinery necessary for DNA synthesis.

8. All the nucleotide biosynthetic pathways are present in our model, so the cell is equipped with the means for the cell reproduction; however, at the present stage we have not designed these activity.

Some metabolites are considered to be ubiquitary, among which water, inorganic phosphate, some metals ions, and Nicotinammide. Their concentration in external or internal environment is assumed to be constant and not to be significantly affected by cellular metabolism.

Summing up, the cell can take metabolites from external environment using the set of permeases and ATPases specified above. Among the pathways of our virtual cell, there is Glycolysis: glucose and fructose taken from the outside are oxidized helding energy in the form of ATP and reduced-NAD. Pyruvate, the last metabolite of conventional Glycolysis, becomes then acetate which, in turn, diffuses out of the cell. The cell "imports" fatty acids, glycerole and some other metabolites, e.g. Choline, and uses them for the synthesis of tryglicerides and phospholipids; these are essential components of the plasma membrane. Our virtual cell is also able to synthesize DNA, RNA and proteins; the needed metabolites are mostly taken form the external environment or synthesized along its own pathways (e.g. Thymine and Ribose).

Additionally, as summarized in Section 1, we enhanced this model with a regulatory feedback circuit on the enzyme phosphofructokinase whose scheme is shown in Figure 1.

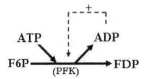

Fig. 1. The scheme of a positive feedback control circuit. The involved metabolites are fructose-6-phosphate (f6p), fructose-1,6-bisphosphate (fdp), phosphofructokinase (pfk), ATP and ADP.

The considered enzyme catalyses the phosphorilation of fructose-6-phosphate using ATP as the phosphate donor. In turn ATP becomes ADP. The catalytic rate of phosphofructokinase depends on the concentration of ADP. This metabolite acts on the enzyme enhancing its catalytical capability and accelerates the correspondent reaction. This kind of loop is known as positive feedback. We described it according to [17], where this control circuit is proposed under the name of "mechanism for glycolytical oscillation." Its formal specification is detailed in the Appendix.

We follow the standard way of using process calculi for modelling biological organism [29]. However, we slightly deviate in that channels represent enzymes, to easy tracking the occurrences of certain reactions and the usage of catalysers along our virtual experiments. Briefly, we have built up our model according to the following correspondences, as done in [5]:

- A *metabolite* is rendered as a *process*
- An *enzyme* is rendered as a *channel* (or as a *set of channels*, one for each reaction the enzyme catalyzes)
- The occurrence of a *bio-chemical reaction*, with apparent rate r, is rendered as a *synchronization*, labelled by r

As mentioned in Section 2, each channel is associated with a rate r, that takes values according to an exponential distribution. In our model r is related with a biological parameter of the described enzyme, i.e. with the Michaelis-Menten constant of the reaction catalysed by the considered enzyme.

Furthermore we assume that the occurrence of a transition corresponds to the production of a fixed quantity of a specific metabolite. For example the following transition, modelling a step in the glycolysis:

β-D-fructose_1-6bP $\xrightarrow{r}$ D-Glyceraldehyd_3-phosphate | Dihydroxyacetone_phosphate

will not describe the behaviour of a *single* molecule of β-D-fructose-1-6bP, but it models the production of a certain quantity of D-Glyceraldehyde-3-phosphate and Dihydroxyacetone-phosphate from β-D-fructose-1-6bP, with the stoichiometric ratio of 1 : 1.

It is worth noting that each molecule is specified independently of the others, except for channel sharing. Then, we built up the whole VICE by making the wanted number of copies, from thousand to millions, of the specified processes, and by putting them in parallel. The resulting system is finally run and interpreted as a virtual experiment.

4 *In Silico* Experimentation and Results

We briefly discuss now the adequacy and the results of our proposal. First, we describe the experiments made *in silico*, and we then interpret their outcome. The results suggest us that VICE can "live" in an optimal environment and that it exhibits some biological behaviour in accordance with real prokaryotes.

4.1 *In Silico* Experimentations

We performed two classes of experiments, depending on the way glucose has been supplied to VICE. In the first class, we provided VICE with a large reservoire of glucose, while in the second this nutrient was given at a constant rate. The first feeding regimen is intended to check whether the new implementation of VICE still has the homeostatic properties shown by [5]. The second regimen is used to detect the emergence of oscillatory patterns in presence of the feedback control circuit discussed above.

All our tests have been carried on assuming that the virtual cell acts in the ideal environment discussed in Section 3. These experiments *in silico* have been performed under the following initial conditions:

- *Concentration of essential nutrients in the environment*: we assume that there are 1,000,000 copies of processes for extra-cellular glucose, made available in the two ways discussed above; instead there are 1,000 copies of the processes for the other essential nutrients (external amynoacids, nitrous basis, complex lipids precursors, etc).
- *Membrane carriers*: we assume that there is one process for each of the three carriers for glucose, ATP and reduced-NAD.
- *Metabolites*: we assume that there are 1,000 copies of processes for each metabolite inside the cell.

We ran about 50 simulations for a time period of about 12 hours on the AMD Athlon 1.5 GHz duo with 1 Gb of RAM.

The performed computations resulted to be composed by approximately 30,000 transitions each. Recall that our virtual unit of time is represented by the occurrence of a transition, that produces a fixed amount of molecules, rather than a single one. It is therefore meaningful to plot the quantity of each metabolites versus the number of transitions.

The BEAST abstract machine gives an output file (`.csv`) displaying the amount of monitored metabolites. Then we sampled this output using a routine that inspects the `.csv` file, picks up the relevant data, and produces a smaller `.csv` file. This step helps in making more evident the form of the curve.

4.2   Results

We first report on the outcomes of the simulations, when VICE is fed with a large amount of glucose. In this case, the cell can uptake this nutrient upon need, with an increasing rate, till the top rate is reached.

We are interested in observing whether the time course distribution of metabolites reaches a plateau, i.e. whether the virtual cell reaches its steady state after a certain initial period of time/number of transitions. This property mimics the homeostatic capability, typical of real cells that require it, in order to regulate their internal medium. Homeostatic biological systems oppose external environment change to maintain internal equilibrium and succeed in reestablishing their balance, while non homeostatic ones eventually stop functioning.

It turns out that VICE reaches its steady state, as shown in Figure 2, that depicts the time course distribution of the concentrations of three selected metabolites. These metabolites represents critical nodes in the entire metabolic network, so their behaviour gives a sketch of the overall trend of VICE. The distributions displayed in Figure 2 are affected by white gaussian noise. This is because BEAST turns out to be stochastic, due to Gillespie's SSA it embodies.

We also compared some aspects of the behaviour of the cell with that of real prokaryotes acting *in vivo* in similar circumstances [15]. To do that, we examined the time course distribution of certain metabolites concentration that are representative of the modelled pathways. As our unit of measurement is arbitrary, we took the ratio between metabolite quantities. In particular, we investigated the glycolytic pathway, on which the literature has a relatively large quantity of biological data. Within this pathway, we selected three ratios between its most significant metabolites: ATP vs. ADP and NAD vs. ATP, that roughly measure the cellular energy content in two different ways; and glucose-6-phosphate vs. fructose-1,6-bisphosphate, giving the trend of metabolic flux along glycolisis. The following table shows that the selected ratios significantly match those of real organisms, we computed from the values of [2], obtained *in vivo*. Virtual ratios are computed from metabolite concentrations at the steady state; because this quantity fluctuates due to white gaussian noise, we sampled 20 steady state concentration values for each metabolite and we took their average — at this stage we limited ourselves to this shallow statistics, as standard deviation is quite small, especially for the virtual values.

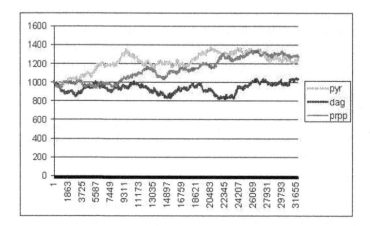

Fig. 2. Time course distribution of pyruvate (pyr), diacilglycerol (dag), phosphoribo-sylpyrophosphate (prpp). The concentration of metabolites is plotted vs the number of transitions.

Table 2. Metabolites Ratio Comparison

	ATP/ADP	glu6p/fru16bp	NAD/ATP
Real	0.775	0.067	11.452
Virtual	0.697	0.053	10.348

These first results confirm that VICE exhibits some capability of "living" *in silico*, just as it was the case for the simulations reported in [5]. In particular, its homeostatic property guarantees that we can change the feeding regimen and still it reestablishes its internal balance. So we modified the rate of the reaction for uptaking glucose. More precisely, we set equal the basal and the top rate of the channel representing the enzymatic complex catalysing the uptake. In this way, the apparent rate is kept constant, and VICE assumes glucose in a constant manner. This kind of supplying a prokaryote with glucose makes it detectable *in vivo* the effect of the regulatory feedback circuit based on phosphofructokinase. The same happens with VICE. Figure 3 displays the oscillations of fructose-6-phosphate and of fructose-1,6-bisphosphate. In spite of white gaussian noise, the two plots have a clear constant period and amplitude, showing that an oscillatory pattern is emerging. Compare Figure 3 with Figure 4, that shows the oscillations of the same metabolites in experiments carried on *in vivo*.

5   Related Work

As previously mentioned in the introduction, most of the recent biological models describing metabolic networks rely either on ODE formalisms or on process algebras. The first approach dates back to the '60s, while the other started with

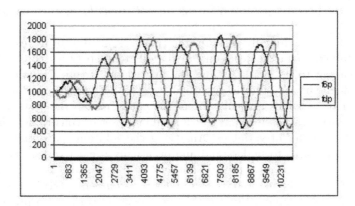

Fig. 3. *In silico* oscillations of fructose-6-phosphate (f6p) and fructose-1,6-bisphosphate (fdp) in the glycolysis

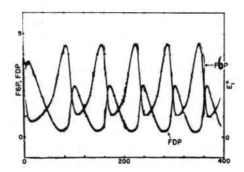

Fig. 4. *In vivo* oscillations of fructose-6-phosphate (f6p) and fructose-1,6-bisphosphate (fdp) in the glycolysis. Adapted from [17]

the pioneeristic paper by Regev and Shapiro [28,26]. The literature has a huge number of papers, but we only consider below a few, admittedly far from giving a comprehensive survey.

The first kind of models are used in two different kind of approaches, namely Metabolic Control Analysis (MCA) and Flux Balance Analysis (FBA). In both of them, metabolic networks are analyzed under the steady-state approximation. The MCA grounds on a set of theoremes formally presented in [16]. The central problem to solve is how the steady state variables change when the steady-state itself changes in response to a perturbation in one or more parameters. In order to solve this problem it is necessary to differentiate a set of steady-state equations with respect to the parameters. Metabolic control analysis has been applied to many types of systems e.g. modular systems [30], signal transduction pathways [18], time-dependent phenomena [16] and oscillating systems [8]. These papers contributed to shed some light on the overall organization of the investigated networks. In particular, the virtual simulations shown that the various enzymes composing the network possess different *control strength*, i.e. they have not the same weight in regulating metabolic fluxes.

The FBA approach analyses the target network in term of fluxes. The flux of the metabolite j in the reaction i is the difference between the rate of the reaction leading to its synthesis and the one leading to its degradation. The all set of fluxes in a system is described by a set of *balance equations*. Often the number of variables exceeds that of the equations, so the system has not a unique solution; a possible way out is defining a set of constraints, that help reducing the number of variables. Analyzing the solution space under certain conditions is possible to understand and eventually predict certain behaviours of the object of study. This is the way followed by Palsson and co-workers [10,12], who reconstructed the metabolic map of *Escherichia coli* MG1655 from its sequenced genome. Applying FBA to this metabolic network and a proper set of constraints, they were able to qualitatively predict the extent of utilization of the whole metabolic network, and showed their predictions consistent with experimental data collected *in vivo*. Within this line, an important project is Virtual *E. coli* [1]. This virtual prokaryote possesses most of the metabolic pathways of the real *Escherichia coli*. It has been successfully used to investigate some basic properties of the behaviour of this bacterium, in particular of its genetic control network and of time course of its metabolic fluxes.

Among the modelling approaches grounded on process algebras, we already cited the pioneeristic paper by Regev and Shapiro and by Priami et al. Along the same line, but using an enhanced versions of the π-calculus is [6], which specifies simple pathways, such as some signalling pathways and the glycolysis, offering stochastic and causality-based representations of them. Particularly related to our work is Cardelli and Phillips' [4] in which the abstract machine SPIM is presented. While, SPIM seems very effective when modelling genetic regulatory circuits, it seems less efficient when applied to modelling other metabolic networks, e.g those related with energetic metabolism.

A main problem affecting both SPIM and our BEAST is the huge computational cost of Gillespie's SSA. Some refinements have been recently proposed by different authors. The *Next Reaction Method* (NRM) [13] is based on the idea that the reaction with the shortest firing time must be chosen at each step among all possible ones. The *Logarithmic Direct Method* (LDM) [24], dynamically sorts the channel list reducing the computational cost of channel scanning, and to the best of our knowledge, offers the fastest way for stochastically computing the apparent rates of transitions.

6   Conclusion

We considered the virtual prokaryote VICE, proposed by [5]. We gave a more detailed specification of its metabolic pahtways, and we analysed its behaviour *in silico*. In particular, we enhanced VICE with a positive feedback control circuit involving phosphofructokinase, an enzyme active in the glycolysis.

The results of our experimentations show that VICE has homeostatic proper-ties, and that an oscillating behaviour emerges. Indeed, VICE reaches a steady state, balancing the quantities of its internal metabolites. This first result makes us confident that VICE can "survive" in an optimal environment. Also, we com-puted the ratio between the quantities of relevant metabolites of VICE. It turns out that these values are rather close to those the literature reports about real prokariotes under similar experimental conditions. The validation of our model received further support by mimicking real experiments that detect oscillatory patterns of important metabolites in the glycolytic pathway. The outcomes of this simulation show that oscillations emerge with constant period and ampli-tude, with a shape comparable with the real ones.

To carry on our experimentations *in silico*, we designed and implemented BEAST, an abstract machine for a variant of the stochastic π-calculus, similar to SPIM [4]. Our simulation tool is considerably more efficient than the one used in the previous studies on VICE [5]. The simulation time was significantly reduced so enabling us to extensively experiment on our virtual cell. This has also been possible because a small fragment of the π-calculus suffices for specifing a simple cell with "no compartments/membranes" like our prokariote.

Currently, we are going to complete the specification of the more complex bac-teria *Escherichia coli*. Very preliminary results show it feasible to have a model closer to this real organism. Also here, our fragment of pure CCS was enough to specify the metabolome of *E. coli*. This came a real surprise to us, and we are still wondering how far we can go without message passing or more sophisticated linguistic features. However, the time required by the first significant simulations grows very high. We feel that the real bottleneck is the current implementation of Gillespie's SSA. A smarter implementation of this algorithm is thus in order. Also, we plan to study a parallel version of the abstract machine BEAST in order to improve its performance in time. The parallelization of SSA seems to be quite hard, possibly requiring to design a new version of the stochastic simulation algorithm.

Acknowledgements. This work has been partially supported by EU-IST project 016004 SENSORIA, and PRIN project SYBILLA.

References

1. E. Almaas and B. Kowács *et al.* Global organization of metabolic fluxes in the bacterium *Escherichia Coli*. *Nature*, 427:839–843, 2004.
2. A. Buchholz, R. Takors, and C. Wandrey. Quantification of intracellular metabo-lites in escherichia coli k12 using liquid chromatographic-electrospray ionization tandem mass spectrometric techniques. *Analytical Biochemistry*, 295:129–137, 2001.
3. Y. Cao, H. Li, and L. Petzold. Efficient formulation of the stochastic simulation algorithm for chemically reacting system. *Journal of Chemical Physics*, 121:4059–4067, 2004.
4. L. Cardelli and A. Phillips. A correct abstract machine for stochastic *pi*-calculus. In *Procs. BioConcur*, 2004.

5. D. Chiarugi, M. Curti, P. Degano, and R. Marangoni. ViCe: a VIrtual CEll. In *Procs. of the 2st Int. W/S Computational Methods in Systems Biology*, volume 3082 of *LNCS*. Springer, 2004.
6. M. Curti, P. Degano, C. Priami, and C.T. Baldari. Modelling biochemical pathways through enhanced π-calculus. *Theoretical Computer Science*, To appear, 2004.
7. P. Degano and C. Priami. Enhanced operational semantics. *ACM Computing Surveys*, 28(2):352–354, 1996.
8. O. V. Demin, B. N. Kholodenko, and H. V. Westerhoff. Control analysis of stationary forced oscillations. *Journal of Physical Chemistry*, 103:10696–10710, 1999.
9. J.C. Diaz Ricci. Adp modulates the dynamic behavior of the glycolytic pathway of escherichia coli. *Biochemical and Biophysical Research Communications*, 271:244–249, 2000.
10. J. Edwards, R. Ibarra, and Palsson B. In silico prediction of escherichia coli metabolic capabilities are consistent with experimental data. *Nature Biotecnology*, 19:125–130, 2001.
11. Tomita Masaru *et al.* E–CELL: software environment for whole–cell simulation. *Bioinformatics*, 15:72–84, 1998.
12. S. Fong, J. Marciniak, and B. Palsson. Description and interpretation on adaptive evolution of escherichia coli k-12 mg1655 by using a genome-scale in silico metabolic model. *Journal of Bacteriology*, 185:6400–6408, 2003.
13. M. Gibson and J. Bruck. Efficient exact stochastic simulation of hcemical systems with many species and many channels. *Journal of Physical Chemistry*, 104:1876–1889, 2005.
14. D.T. Gillespie. Exact stochastic simulation of coupled chemical reactions. *Journal of Physical Chemistry*, 81 (25):2340–2361, 1977.
15. G.G. Hammes and P.R. Shimmel. *The Enzymes, vol. 2*. P.D. Boyer (New York Academic Press), 1970.
16. R. Heinrich and S. Schuster. *The regulation of cellular systems*. Chapmann and Hall, 1996.
17. J. Higgins. A chemical mechanism for oscillation of glycolytic intermediates in yeast cell. *Proceeding of National Academy Science USA*, 51:989–994, 1964.
18. B. N. Kholodenko, J. B. Hoek, H. V. Westerhoff, and Brown G.C. Quantification of information tranfer via cellular transduction pathways. *FEBS letters*, 414:430–434, 1997.
19. H. Kitano. *Foundations of System Biology*. MIT Press, 2002.
20. E.V. Koonin. How many genes can make a cell:the minimal-gene-set concept. *Annual Review Genomics and Human Genetics*, 01:99–116, 2000.
21. L.M. Loew and J.C. Schaff. The virtual cell: a software environment for computational cell biology. *Trends Biotechnology*, 19(10):401–406, Oct. 2001.
22. R. Milner. *Communicating and Mobile Systems: the π-calculus*. Cambridge Univ. Press, 1999.
23. A.R. Mushegian and E.V. Koonin. A minimal gene set fir cellular life derived by comparison of complete bacterial genome. *Proceedings of National Academy of Science USA*, 93:10268–10273, 1996.
24. L. Petzold and H. Li. Logarithmic direct method for discrete stochastic simulation of chemically reacting systems. *http://www.engineering.ucsb.edu/cse/Files/ldm0513.pdf*, 2006.
25. C. Priami. Stochastic π-calculus. *The Computer Journal*, 38, 6:578–589, 1995.
26. C. Priami, A. Regev, W. Silverman, and E. Shapiro. Application of a stochastic passing-name calculus to representation and simulation of molecular processes. *Information Processing Letters*, 80:25–31, 2001.

27. J. Reed and Palsson B. Thirteen years of building contraint-based in silico models of escherichia coli. *Journal of Bacteriology*, 185:2692–2699, 2003.

28. A. Regev and E. Shapiro. Cells as computations. *Nature*, 419:343, 2002.

29. A. Regev, W. Silverman, and E. Shapiro. Representation and simulation of biochemical processes using the π-calculus process algebra. In *Pacific Symposium of Biocomputing (PSB2001)*, pages 459–470, 2001.

30. S. Schuster, D. Kahn, and H. V. Westerhoff. Modular analysis of the control of complex metabolic pathways. *Biophysical Chemistry*, 48:1–17, 1993.

Appendix

We briefly summarize here the main features of the "mechanism for glycolytical oscilations" upon which we build up the formal specification of the feedback control circuit mentioned in Section 3. It essentially consists in the following chemical reactions:

$$ATP + F6P + E_1^* \longrightarrow FDP + ADP + E_1^* \tag{1}$$

$$ADP + E_1 \longleftrightarrow E_1^* \tag{2}$$

The rate of (1) depends on the concentration of the auxiliary reactant E_1^* (see below). Reaction (2) produces E_1^* starting from ADP and the enzyme phosphofructokinase, represented here as E_1. The more the ADP the more E_1^* making reaction (1) faster

We formalize this system of chemical reactions as follows:

$$ATP = \bar{a}.ADP \tag{3}$$

$$E_1^* = a.E_1^* + b.E_1^* + \tau.(ADP \mid E_1) \tag{4}$$

$$F6P = \bar{b}.FDP \tag{5}$$

$$ADP = \bar{c}.0 \tag{6}$$

$$E_1 = c.E_1^* \tag{7}$$

As done before, we abbreviate fructose-6-phosphate with F6P and fructose-1,6-bisphosphate with FDP. In the above specification, the process E_1^* is only used for conciseness; in fact there is no biological counterpart for this auxiliary process. Note also that its quantity is kept constant, and thus it does not affect the behaviour of the overall system, especially as far as the SSA algorithm is concerned.

Recall that the oscillatory effects of this regulatory circuit become detectable only when glucose is continuously fed, i.e. when the basal and the top rate of glucose carrier, namely PtsG, are set equal.

Modelling Cellular Processes Using Membrane Systems with Peripheral and Integral Proteins

Matteo Cavaliere and Sean Sedwards

Microsoft Research – University of Trento
Centre for Computational and Systems Biology, Trento, Italy
{matteo.cavaliere, sean.sedwards}@msr-unitn.unitn.it

Abstract. Membrane systems were introduced as models of computation inspired by the structure and functioning of biological cells. Recently, membrane systems have also been shown to be suitable to model cellular processes. We introduce a new model called *Membrane Systems with Peripheral and Integral Proteins*. The model has compartments enclosed by membranes, floating objects, objects associated to the internal and external surfaces of the membranes and also objects integral to the membranes. The floating objects can be processed within the compartments and can interact with the objects associated to the membranes. The model can be used to represent cellular processes that involve compartments, surface and integral membrane proteins, transport and processing of chemical substances. As examples we model a circadian clock and the G-protein cycle in yeast *saccharomyces cerevisiae* and present a quantitative analysis using an implemented simulator.

1 Introduction

Membrane systems are models of computation inspired by the structure and the function of biological cells. The model was introduced in 1998 by Gh. Păun and since then many results have been obtained, mostly concerning computational power. A short introductory guide to the field can be found in [12], while an updated bibliography is available via the web-page [18]. Recently (see, e.g., [10]), membrane systems have been successfully applied to systems biology and several models have been proposed for simulating biological processes (e.g., see the monograph dedicated to membrane computing applications [5]).

By the original definition, membrane systems are composed of an hierarchical nesting of membranes that enclose regions (*the cellular structure*), in which free-floating objects (*molecules*) exist. Each region can have associated rules, called *evolution rules*, for evolving the free-floating objects and modelling the biochemical reactions present in cell regions. Rules also exist for moving objects across membranes, called *symport* and *antiport* rules, modelling cellular transport. Recently, inspired by *brane calculus* [3], a model of a membrane system, having free-floating objects and objects attached to the membranes, was introduced in [2]. The attached objects model the proteins that are embedded in lipid bilayer cell membranes. In [2], however, objects are associated to an *indivisible* membrane which has no concept of inner or outer surface, while in [4] objects

C. Priami (Ed.): CMSB 2006, LNBI 4210, pp. 108–126, 2006.

(peripheral proteins) are attached to either side of a membrane. In reality, many biological processes are driven and controlled by the presence of specific proteins on the appropriate side of and *integral* to the membrane: there is a constant interaction between floating chemicals and embedded proteins and between peripheral and integral proteins (see, e.g., [1]). Receptor-mediated processes, such as endocytosis (illustrated in Figure 1) and signalling, are crucial to cell function and by definition are critically dependent on the presence of peripheral and integral membrane proteins.

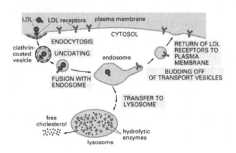

Fig. 1. Endocytosis of LDL (*Essential Cell Biology*, 2/e, ©2004 Garland Science)

One model of the cell is that of compartments and sub-compartments in constant communication, with molecules being passed from donor compartments to target compartments by interaction with membrane proteins. Once transported to the correct compartment, the substances are then *processed* by means of local biochemical reactions.

Motivated by these ideas we extend the model presented in [4], introducing a model having peripheral as well as integral proteins.

In each region of the system there are floating objects (the floating chemicals) and, in addition, objects can be associated to each side of a membrane or integral to the membrane (the peripheral and integral membrane proteins). Moreover, the system can perform the following operations: (*i*) the floating objects can be processed/changed inside the regions of the system (emulating biochemical rules) and (*ii*) the floating and attached objects can be processed/changed when they interact (modelling the interactions of the floating molecules with membrane proteins).

The proposed model can be used to represent cellular processes that involve floating molecules, surface and integral membrane proteins, transport of molecules across membranes and processing of molecules inside the compartments. As examples, we model a circadian clock and the G-protein cycle in *saccharomyces cerevisiae*, where the possibility to use, in *an explicit way*, compartments, membrane proteins and transport rules is very useful. A *quantitative analysis* of the models is also presented, performed using an extended version of the simulator presented in [4] (downladable at [19]). The simulator employs a stochastic algorithm and uses intuitive syntax based on chemical equations (described in appendix B).

2   Formal Language Preliminaries

Membrane systems are based on *formal language theory* and *multiset rewriting*. We now briefly recall the basic theoretical notions used in this paper. For more details the reader can consult standard books, such as [8], [15], [6] and handbook [14].

Given the set A we denote by $|A|$ its cardinality and by $\emptyset$ the empty set. We denote by $\mathbb{N}$ and by $\mathbb{R}$ the set of natural and real numbers, respectively.

As usual, an *alphabet* V is a finite set of symbols. By V^* we denote the set of all strings over V. By V^+ we denote the set of all strings over V excluding the empty string. The empty string is denoted by λ. The *length* of a string v is denoted by $|v|$. The concatenation of two strings $u, v \in V^*$ is written uv.

The number of occurrences of the symbol a in the string w is denoted by $|w|_a$.

A *multiset* is a set where each element may have a multiplicity. Formally, a multiset over a set V is a map $M : V \rightarrow \mathbb{N}$, where $M(a)$ denotes the multiplicity of the symbol $a \in V$ in the multiset M.

For multisets M and M' over V, we say that M is *included in* M' if $M(a) \leq M'(a)$ for all $a \in V$. Every multiset includes the *empty multiset*, defined as M where $M(a) = 0$ for all $a \in V$.

The *sum* of multisets M and M' over V is written as the multiset $(M + M')$, defined by $(M + M')(a) = M(a) + M'(a)$ for all $a \in V$. The *difference* between M and M' is written as $(M - M')$ and defined by $(M - M')(a) = max\{0, M(a) - M'(a)\}$ for all $a \in V$. We also say that $(M + M')$ is obtained by *adding* M to M' (or viceversa) while $(M - M')$ is obtained by *removing* M' from M. For example, given the multisets $M = \{a, b, b, b\}$ and $M' = \{b, b\}$, we can say that M' is included in M, that $(M+M') = \{a, b, b, b, b, b\}$ and that $(M-M') = \{a, b\}$.

If the set V is finite, e.g. $V = \{a_1, \ldots, a_n\}$, then the multiset M can be explicitly described as $\{(a_1, M(a_1)), (a_2, M(a_2)), \ldots, (a_n, M(a_n))\}$. The *support* of a multiset M is defined as the set $supp(M) = \{a \in V \mid M(a) > 0\}$. A multiset is empty (hence finite) when its support is empty (also finite).

A compact notation can be used for finite multisets: if $M = \{(a_1, M(a_1)), (a_2, M(a_2)), \ldots, (a_n, M(a_n))\}$ is a multiset of finite support, then the string $w = a_1^{M(a_1)} a_2^{M(a_2)} \ldots a_n^{M(a_n)}$ (and all its permutations) precisely identify the symbols in M and their multiplicities. Hence, given a string $w \in V^*$, we can say that it identifies a finite multiset over V, written as $M(w)$, where $M(w) = \{a \in V \mid (a, |w|_a)\}$. For instance, the string bab represents the multiset $M(w) = \{(a, 1), (b, 2)\}$, that is the multiset $\{a, b, b\}$. The empty multiset is represented by the empty string λ.

3   Operations with Peripheral and Integral Proteins

Let V denote a finite alphabet of *objects* and Lab a finite set of labels.

As is usual in the membrane systems field, a membrane is represented by a pair of square brackets, [ ]. A *membrane structure* is an *hierarchical nesting* of membranes enclosed by a main membrane called the *root membrane*. To each membrane is associated a label that is written as a superscript of the membrane, e.g. $[\,]^1$. If a membrane has the label i we call it membrane i.

A membrane structure is essentially that of a tree, where the nodes are the membranes and the arcs represent the containment relation. In this paper we avoid a formal mapping in the interest of the intuitiveness of the description, however, being a tree, a membrane structure can be represented by a string of matching square brackets, e.g., $[\,[\,[\,]^2\,]^1\,[\,]^3\,]^0$.

To each membrane there are associated three multisets, u, v and x over V, denoted by $[\,]_{u|v|x}$. We say that the membrane is *marked* by u, v and x; x is called the *external marking*, u the *internal marking* and v the *integral marking* of the membrane. In general, we refer to them as *markings* of the membrane.

The internal, external and integral markings of a membrane model the proteins attached to the internal surface, attached to the external surface and integral to the membrane, respectively.

In a membrane structure, the region between membrane i and any enclosed membranes is called region i. To each region is associated a multiset of objects w called the *free objects* of the region. The free objects are written between the brackets enclosing the regions, e.g., $[\,aa\,[\,bb\,]^1\,]^0$.

The free objects of a membrane model the floating chemicals within the regions of a cell.

We denote by $int(i)$, $ext(i)$ and $itgl(i)$ the internal, external and integral markings of membrane i, respectively. By $free(i)$ we denote the free objects of region i. For any membrane i, distinct from a root membrane, we denote by $out(i)$ the label of the membrane enclosing membrane i.

For example, the string

$$[\,ab\,[\,cc\,]^2_{a|\,|}\,[\,abb\,]^1_{bba|ab|c}\,]^0$$

represents a membrane structure, where to each membrane are associated markings and to each region are associated free objects. Membrane 1 is internally marked by bba (i.e., $int(1) = bba$), has integral marking ab (i.e., $itgl(1) = ab$) and is externally marked by c (i.e., $ext(1) = c$). To region 1 are associated the free objects abb (i.e., $free(1) = abb$). To region 0 are associated the free objects ab. Finally, $out(1) = out(2) = 0$. Membrane 0 is the root membrane. The string can also be depicted diagrammatically, as in Figure 2.

When a marking is omitted it is intended that the membrane is marked by the empty string λ, i.e., the empty multiset. For instance, in $[\,ab\,]_{u|v|}$ the external marking is missing, while in the case of $[\,ab\,]_{|v|x}$ the internal marking is missing.

3.1 Operations

We introduce rules that describe bidirectional interactions of floating objects with the membrane markings which we call *membrane rules*. These rules are motivated by the behaviour of cell membrane proteins (e.g., see [1]) and therefore

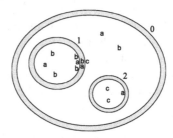

Fig. 2. Graphical representation of $[\ ab\ [\ cc\ ]^2_{a|\ |}\ [\ abb\ ]^1_{bba|ab|c}\ ]^0$

permit a level of abstraction based on the behaviour of real molecules. We denote the rules as $attach_{in}$, $attach_{out}$, $de-attach_{in}$ and $de-attach_{out}$, defined:

$$attach_{in} : [\ \alpha\ ]^i_{u|v|} \rightarrow [\ ]^i_{u'|v'|}\ ,\quad \alpha \in V^+, u, v, u', v' \in V^*, i \in Lab$$

$$attach_{out} : [\ ]^i_{|v|x}\ \alpha \rightarrow [\ ]^i_{|v'|x'}\ ,\quad \alpha \in V^+, v, x, v', x' \in V^*, i \in Lab$$

$$de-attach_{in} : [\ ]^i_{u|v|} \rightarrow [\ \alpha\ ]^i_{u'|v'|}\ ,\quad \alpha, u', v', u, v \in V^*, |uv| > 0, i \in Lab$$

$$de-attach_{out} : [\ ]^i_{|v|x} \rightarrow [\ ]^i_{|v'|x'}\alpha,\quad \alpha, v', x', v, x \in V^*, |vx| > 0, i \in Lab$$

The semantics of these rules is as follows.

The $attach_{in}$ rule is *applicable to membrane* i if $free(i)$ includes α, $int(i)$ includes u and $itgl(i)$ includes v. When the rule *is applied to membrane* i, α is removed from $free(i)$, u is removed from $int(i)$, v is removed from $itgl(i)$, u' is added to $int(i)$ and v' is added to $itgl(i)$. The objects not involved in the application of the rule are left unchanged in their original positions.

The $attach_{out}$ rule is applicable to membrane i if $free(out(i))$ includes α, $itgl(i)$ includes v, $ext(i)$ includes x. When the rule is applied to membrane i, α is removed from $free(out(i))$, v is removed from $itgl(i)$, x is removed from $ext(i)$, v' is added to $itgl(i)$ and x' is added to $ext(i)$. The objects not involved in the application of the rule are left unchanged in their original positions.

The $de-attach_{in}$ rule is applicable to membrane i if $int(i)$ includes u and $itgl(i)$ includes v. When the rule is applied to membrane i, u is removed from $int(i)$, v is removed from $itgl(i)$, u' is added to $int(i)$, v' is added to $itgl(i)$ and α is added to $free(i)$. The objects not involved in the application of the rule are left unchanged in their original positions.

The $de-attach_{out}$ rule is applicable to membrane i if $itgl(i)$ includes v and $ext(i)$ includes x. When the rule is applied to membrane i, v is removed from $itgl(i)$, x is removed from $ext(i)$, v' is added to $itgl(i)$, x' is added to $ext(i)$ and α is added to $free(out(i))$. The objects not involved in the application of the rule are left unchanged in their original positions.

We denote by $\mathcal{R}^{att}_{V,Lab}$ the set of all possible *attach* and *de−attach* rules over the alphabet V and set of labels Lab. Instances of $attach_{in}$, $attach_{out}$, $de-attach_{in}$ and $de-attach_{out}$ rules are depicted in Figure 3.

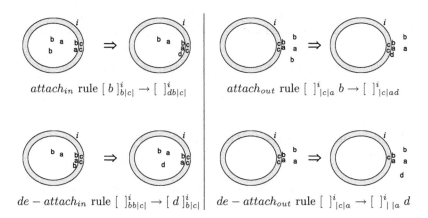

$attach_{in}$ rule $[\, b\,]^i_{b|c|} \to [\ ]^i_{db|c|}$ $\qquad$ $attach_{out}$ rule $[\ ]^i_{|c|a}\, b \to [\ ]^i_{|c|ad}$

$de-attach_{in}$ rule $[\ ]^i_{bb|c|} \to [\, d\,]^i_{b|c|}$ $\quad$ $de-attach_{out}$ rule $[\ ]^i_{|c|a} \to [\ ]^i_{|\ |a}\, d$

Fig. 3. Examples of $attach_{in}$, $attach_{out}$, $de-attach_{in}$ and $de-attach_{out}$ rules, showing how free and attached objects may be rewritten. E.g., in the $attach_{in}$ rule one of the two free instances of b is rewritten to d and added to the membrane's internal marking.

We next introduce *evolution rules* that rewrite the free objects contained in a region conditional on the markings of the enclosing membrane. These rules can be considered to model the biochemical reactions that take place within the cytoplasm of a cell. We define an evolution rule:

$$evol : \ [\, \alpha \to \beta\,]^i_{u|v|}$$

where $u, v, \beta \in V^*$, $\alpha \in V^+$, and $i \in Lab$.

The semantics of the rule is as follows. The rule is applicable to region i if $free(i)$ includes α, $int(i)$ includes u and $itgl(i)$ includes v. When the rule is applied to region i, α is removed from $free(i)$ and β is added to $free(i)$. The membrane markings and the objects not involved in the application of the rule are left unchanged in their original positions.

We denote by $\mathcal{R}^{ev}_{V,Lab}$ the set of all evolution rules over the alphabet V and set of labels Lab. An instance of an evolution rule is represented in Figure 4.

In general, when a rule has label i we say that a rule is associated to *membrane* i (in the case of *attach* and *de − attach* rules) or is associated to *region* i (in the case of *evol* rules). For instance, in Figure 3 the $attach_{in}$ is associated to membrane i.

The objects of α, u and v for $attach_{in}/evol$ rules, of α, v and x for $attach_{out}$ rules, of u and v for $de-attach_{in}$ rules and of v and x for $de-attach_{out}$ rules are the *reactants* of the corresponding rules. E.g., in the attach rule $[\, b\,]_{a|c|} \to [\ ]_{d|c|}$, the reactants are a, b and c.

Fig. 4. *evol* rule $[\, a \rightarrow b \,]^{i}_{b|c|}$. Free objects can be rewritten inside the region and the rewriting can depend on the integral and internal markings of the enclosing membrane.

We note that a single application of an evol rule may be simulated by an application of an attach$_{in}$ rule followed by an application of an de $-$ attach$_{in}$ rule. This may be biologically realistic in some cases, but not in all. Hence the need for evolution rules.

4   Membrane Systems with Peripheral and Integral Proteins

In this section we define membrane systems having membranes marked with peripheral proteins, integral proteins, free objects and using the operations introduced in Section 3.

Definition 1. *A membrane system with peripheral and integral proteins and n membranes (in short, a P_{pi} system), is a construct*

$$\mathcal{P} = (V_{\mathcal{P}}, \mu_{\mathcal{P}}, (u_0, v_0, x_0)_{\mathcal{P}}, \ldots, (u_{n-1}, v_{n-1}, x_{n-1})_{\mathcal{P}}, w_{0,\mathcal{P}}, \ldots, w_{n-1,\mathcal{P}}, R_{\mathcal{P}},$$
$$t_{in,\mathcal{P}}, t_{fin,\mathcal{P}}, \; rate_{\mathcal{P}})$$

- $V_{\mathcal{P}}$ *is a finite, non-empty alphabet of objects.*
- $\mu_{\mathcal{P}}$ *is a membrane structure with $n \geq 1$ membranes injectively labelled by labels in $Lab_{\mathcal{P}} = \{0, 1, \cdots, n-1\}$, where 0 is the label of the root membrane.*
- $(u_0, v_0, x_0)_{\mathcal{P}} = (\lambda, \lambda, \lambda), (u_1, v_1, x_1)_{\mathcal{P}}, \cdots, (u_{n-1}, v_{n-1}, x_{n-1})_{\mathcal{P}} \in V^* \times V^* \times V^*$ *are called* initial markings *of the membranes.*
- $w_{0,\mathcal{P}}, w_{1,\mathcal{P}}, \cdots, w_{n-1,\mathcal{P}} \in V^*$ *are called* initial free objects *of the regions.*
- $R_{\mathcal{P}} \subseteq \mathcal{R}^{att}_{V,Lab_{\mathcal{P}}-\{0\}} \cup \mathcal{R}^{ev}_{V,Lab_{\mathcal{P}}}$ *is a finite set of evolution rules, attach/de-attach rules.*[1]
- $t_{in,\mathcal{P}}, t_{fin,\mathcal{P}} \in \mathbb{R}$ *are called the* initial time *and the* final time, *respectively.*
- $rate_{\mathcal{P}} : R_{\mathcal{P}} \longmapsto \mathbb{R}$ *is the* rate mapping. *It associates to each rule a rate.*

Let Π be an arbitrary P_{pi} system. An *instantaneous description* I of Π consists of the membrane structure μ_{Π} with markings associated to the membranes and free objects associated to the regions. We denote by $\mathbb{I}(\Pi)$ the set of *all instantaneous descriptions* of Π. We say in short *membrane (region) i of I* to denote the membrane (region, respectively) i present in I.

---

[1] *The root membrane may contain objects and evolution rules but not attach or de $-$ attach rules, since it has no enclosing region. It may therefore be viewed as an extended version of a membrane systems environment (as defined in [12]), with objects and evol rules. Alternatively, it can be seen as a membrane systems skin membrane, where the environment contains nothing and is not accessible.*

Let I be an arbitrary instantaneous description from $\mathbb{I}(\Pi)$ and r an arbitrary rule from R_Π. Suppose that r is associated to membrane $i \in Lab_\Pi$ if $r \in \mathcal{R}^{att}_{V, Lab_\Pi - \{0\}}$ (or to region $i \in Lab_\Pi$ if $r \in \mathcal{R}^{ev}_{V, Lab_\Pi}$).

Then, if r is applicable to membrane i (or to region i, accordingly) of I, in short we say that r *is applicable to I*. We denote by $r(I) \in \mathbb{I}(\Pi)$ the instantaneous description of Π obtained when the rule r is applied to membrane i (or to region i, accordingly) of I (in short, we say r *is applied to I*).

The *initial instantaneous description* of Π, $I_{in,\Pi} \in \mathbb{I}(\Pi)$, consists of the membrane structure μ_Π with membrane i marked by $(u_i, v_i, x_i)_\Pi$ for all $i \in Lab_\Pi - \{0\}$ and free objects $w_{i,\Pi}$ associated to region i for all $i \in Lab_\Pi$.

A *configuration* of Π is a pair (I, t) where $I \in \mathbb{I}(\Pi)$ and $t \in \mathbb{R}$; t is called the *time* of the configuration. We denote by $\mathcal{C}(\Pi)$ the set of all configurations of Π. The *initial configuration* of Π is $C_{in,\Pi} = (I_{in,\Pi}, t_{in,\Pi})$.

Suppose that $R_\Pi = \{rule^1, rule^2, \ldots, rule^m\}$ and let S be an arbitrary sequence of configurations $\langle C_0, C_1, \cdots, C_j, C_{j+1}, \cdots, C_h \rangle$, where $C_j = (I_j, t_j) \in \mathcal{C}(\Pi)$ for $0 \leq j \leq h$. Let $a_j = \sum_{i=1}^{m} p_j^i$, $0 \leq j \leq h$, where p_j^i is the product of $rate(rule^i)$ and the *mass action* combinatorial factor for $rule^i$ and I_j (see Appendix A).

The sequence S is an *evolution* of Π if

- for $j = 0$, $C_j = C_{in,\Pi}$
- for $0 \leq j \leq h - 1$, $a_j > 0$, $C_{j+1} = (r_j(I_j), t_j + dt_j)$ with r_j, dt_j as in [7]:
  - $r_j = rule^k$, $k \in \{1, \cdots, m\}$ and k satisfies $\sum_{i=1}^{k-1} p_j^i < ran'_j \cdot a_j \leq \sum_{i=1}^{k} p_j^i$
  - $dt_j = (-1/a_j) ln(ran''_j)$
  where ran'_j, ran''_j are two random variables over the sample space $(0, 1]$, uniformly distributed.
- for $j = h$, $a_j = 0$ or $t_j \geq t_{fin,\Pi}$.

In other words, an evolution of Π is a sequence of configurations, starting from the initial configuration of Π, where, given the current configuration $C_j = (I_j, t_j)$, the next one, $C_{j+1} = (I_{j+1}, t_{j+1})$, is obtained by applying the rule r_j to the current instantaneous description I_j and adding dt_j to the current time t_j. The rule r_j is applied as described in Section 3. Rule r_j and dt_j are obtained using the Gillespie algorithm [7] over the current instantaneous description I_j. The evolution halts when all rules have zero probability of being applied ($a_j = 0$) or when the current time is greater or equal to the specified final time.

5   Modelling and Simulation of Cellular Processes

Having established a theoretical basis, we now wish to demonstrate the quantitative behaviour of the presented model. To this end we have extended the simulator presented in [4] to produce evolutions of an arbitrary P_{pi} system. In Sections 5.2 and 5.3 we demonstrate the model and the simulator using two examples from the literature.

5.1   The Stochastic Algorithm

We use a discrete stochastic algorithm based on Gillespie's which can more accurately represent the dynamical behaviour of small quantities of reactants, in comparison, say, to a deterministic approach based on *ordinary differential equations* [11]. Moreover, Gillespie has shown that the algorithm is fully equivalent to the chemical master equation.

The Gillespie algorithm is specifically designed to model the interaction of chemical species and imposes a restriction of a maximum of three reacting molecules. This is on the basis that the likelihood of more than three molecules colliding is vanishingly small. Hence the simulator is similarly restricted. Note that in the evolution of a P_{pi} system, the stochastic algorithm does not distinguish between floating objects and objects attached or integral to the membrane. That is, the algorithm is applied to the objects irrespective of where they are in the compartment on the assumption that the interaction between floating and attached molecules can be considered the same as between floating molecules. Our application of the Gillespie algorithm to membranes is further described in Appendix A.

5.2   Modelling a Noise-Resistant Circadian Clock

Many organisms use circadian clocks to synchronise their metabolisms to a daily rhythm, however the precise mechanisms of implementation vary from species to species. One common requirement is the need to maintain a measure of stability of timing in the face of perturbations of the system: *the clock must continue to tick and keep good time.* A general model which captures the essence of such stability, based on common elements of several real biological clocks, is presented in [16]. We choose this as an interesting, non-trivial example to model and simulate with a P_{pi} system using evolution rules alone. Moreover, we choose this example because it has been modelled in other formalisms, such as in stochastic Π calculus (see, e.g., [17], [13]).

The model is described diagrammatically in Figure 5. The system consists of two different genes (gA and gR) which produce two different proteins (pA and pR, respectively) via two different mRNA species (mA and mR, respectively). Protein pA up-regulates the transcription of its own gene and also the transcription of the gene that produces pR. The proteins are removed from the system by simple degradation to nothing (dashed lines) and by the formation of a complex AR. In this latter way the production of pR reduces the concentration of pA and has the consequence of down-regulating pR's own production. Thus, in turn, pA is able to increase, increasing the production of pR and causing the cycle to repeat. Key elements of the stable dynamics are the rapid production of pA, by virtue of positive feedback, and the relative rate of growth of the complexation reaction.

A description of the P_{pi} system used to model the circadian clock is given in Figure 6, together with the corresponding simulator script for comparison. The alphabet, V_{clock}, is specified to contain all the reacting species. This corresponds to the object statement of the simulator script. The sixteen chemical reactions of Figure 5 are simply transcribed into corresponding rules mapped to reaction rates. In the simulator script they are grouped under one identifier, clock. The membrane structure, μ_{clock}, comprises just the *root* membrane. The root region initially contains one copy each of the two genes as free objects. These facts are reflected in the system statement of the simulator script, which also associates to the contents the set of rules clock.

The results of running the script are shown in Figure 5: the two proteins exhibit anti-phase periodicity of approximately 24 hours, as expected.

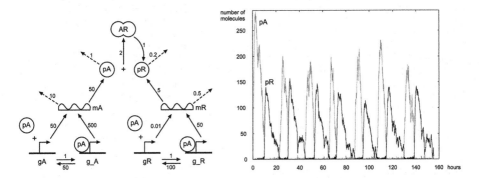

Fig. 5. Reaction scheme and simulation results of noise-resistant circadian clock of [16]

The simulator has the capability to add or subtract reactants from the simulation in runtime. We use this facility to discover the effect of switching off gR in the circadian clock by making the following addition to the system statement:

$$-1 \text{ gR } @50000, \ -1 \text{ g\_R } @50000$$

These *instructions* request a subtraction from the system at time step 50000 of one gR and one g_R. Note that to switch off the gene it is necessary to remove both versions (i.e., with and without pA bound), since it is not possible to know in what state it will exist at a particular time step. Negative quantities are not allowed in the simulator, so only the existent specie will be deleted. In general, the number subtracted is the minimum of the existent quantity and the requested amount. The same syntax, without the negative sign, is used to add reactants.

The effect of switching off gR, shown in Figure 7, is to reduce the amount of pR to near zero and to thus allow pA to reach a maximum governed by its relative rates of production and decay. Note that a small amount of pR continues to exist long after its gene has been switched off. This is the result of a so-called *hidden* pathway from the AR complex, which decays at a much slower rate than pR (second graph of Figure 7). Although this model is a generalisation of biological circadian clocks and may not represent the behaviour of a specific example, the existence of an unexpected pathway exemplifies an important problem encountered when attempting to predict the behaviour of biological systems.

P_{pi} system *clock*   Simulator script

$V_{clock} = \{gA, g\_A, gR, g\_R, mA, mR, pA, pR, RA\}$

```
object gA,g_A,gR,g_R,mA,mR,pA,pR,RA
```

$rate_{clock} =$

```
rule clock
```

$$
\begin{aligned}
&\{ \\
&[\, gA \to gA\, mA \,]^0_{\ |\,|} &&\mapsto 50 \\
&[\, pA\, gA \to g\_A \,]^0_{\ |\,|} &&\mapsto 1 \\
&[\, g\_A \to g\_A\, mA \,]^0_{\ |\,|} &&\mapsto 500 \\
&[\, gR \to gR\, mR \,]^0_{\ |\,|} &&\mapsto 0.01 \\
&[\, g\_R \to g\_R\, mR \,]^0_{\ |\,|} &&\mapsto 50 \\
&[\, mA \to pA \,]^0_{\ |\,|} &&\mapsto 50 \\
&[\, mR \to pR \,]^0_{\ |\,|} &&\mapsto 5 \\
&[\, pA\, pR \to AR \,]^0_{\ |\,|} &&\mapsto 2 \\
&[\, AR \to pR \,]^0_{\ |\,|} &&\mapsto 1 \\
&[\, pA \to \lambda \,]^0_{\ |\,|} &&\mapsto 1 \\
&[\, pR \to \lambda \,]^0_{\ |\,|} &&\mapsto 1 \\
&[\, mA \to \lambda \,]^0_{\ |\,|} &&\mapsto 10 \\
&[\, mR \to \lambda \,]^0_{\ |\,|} &&\mapsto 0.5 \\
&[\, g\_R \to pA\, gR \,]^0_{\ |\,|} &&\mapsto 100 \\
&[\, pA\, gR \to g\_R \,]^0_{\ |\,|} &&\mapsto 1 \\
&[\, g\_A \to pA\, gA \,]^0_{\ |\,|} &&\mapsto 50 \\
&\}
\end{aligned}
$$

```
{
gA 50-> gA + mA
pA+gA 1-> g_A
g_A 500-> g_A + mA
gR 0.01-> gR + mR
g_R 50-> g_R + mR
mA 50-> pA
mR 5-> pR
pA+pR 2-> AR
AR 1-> pR
pA 1-> 0A
pR 0.2-> 0R
mA 10-> 0mA
mR 0.5-> 0mR
g_R 100-> pA+gR
pA+gR 1-> g_R
g_A 50-> pA+gA
}
```

$w_{0,clock} = gA\, gR$
$\mu_{clock} = [\ ]^0$

```
system 1 gA, 1 gR, clock
```

$t_{in,clock} = 0$
$t_{fin,clock} = 155\ hours$

```
evolve 0-150000

plot pA, pR
```

Fig. 6. P_{pi} system model of circadian clock of [16] with corresponding simulator script. Note the similarities between the definitions of V_{clock} and `object` and between the definitions of the elements of $rate_{clock}$ and of `rule clock`.

5.3 Modelling Saccharomyces Cerevisiae Mating Response

To demonstrate the ability of P_{pi} systems to represent compartments and membranes we model and simulate the G-protein mating response in yeast *saccharomyces cerevisiae*, based on experimental rates provided by [9]. The G-protein transduction pathway involves membrane proteins and the transport of substances between regions and is a mechanism by which organisms detect and respond to environmental signals. It is extensively studied and many pharmaceutical agents are aimed at components of the G-protein cycle in humans. The diagram in Figure 8 shows the relationships between the various reactants and regions modelled and simulated.

A description of the biological process is that the yeast cell receives a signal ligand (pL) which binds to a receptor pR, integral to the cell membrane. The receptor-ligand dimer then catalyses (dotted line in the diagram of Figure 8) the reaction that converts the inactive G-protein Gabg to the active GA. A competing sequence of reactions, which dominate in the absence of RL, converts GA to Gabg via Gd in combination with Gbg. The bound and unbound receptor (RL and pR, respectively) are degraded by transport into a vacuole via the cytoplasm. Figure 9 contains the P_{pi} system model and corresponding simulator script. Note that while additional quantities of the receptor pR are created in

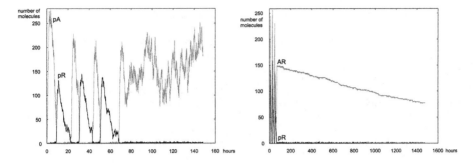

Fig. 7. Simulated effect of switching off **gA** in circadian clock of [16]

runtime, no species is *deleted* from the system; the dynamics are created by transport alone.

Figure 8 shows the results of the stochastic simulation plotted with experimental results from [16] equivalent to simulated GA. There is an apparent correspondence between the simulated and experimental data, in line with the deterministic simulation presented in the original paper. The stochastic noise evident in Figure 8 may explain why some measured points do not lie exactly on the deterministic curve, however further analysis of the original model is beyond the scope of this paper.

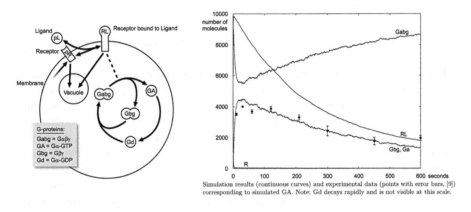

Simulation results (continuous curves) and experimental data (points with error bars, [9]) corresponding to simulated GA. Note: Gd decays rapidly and is not visible at this scale.

Fig. 8. Model and simulation results of saccharomyces cerevisiae mating response

6   Perspectives

We have introduced a model of membrane systems (called a P_{pi} system) with objects integral to the membrane and objects attached to either side of the membrane. We have also introduced operations that can rewrite floating objects conditional on the existence of integral and attached objects and operations that facilitate the interaction of floating objects with integral and attached objects.

P_{pi} system *gprot*   Simulator script

$V_{gprot} = \{pL, pr, pR, RL, Gd, Gbg, Gabg, GA\}$ object pL,pr,pR,RL,Gd,Gbg,Gabg,GA
$rate_{gprot} =$ rule g_cycle
$\{$ {
$[\ ]^1_{|pr|} \rightarrow [\ ]^1_{|pR\ pr|}$ $\mapsto 4.0$ |pr| 4-> |pR,pr|
$[\ ]^1_{|pR|}\ pL \rightarrow [\ ]^1_{|RL|}$ $\mapsto 3.32e^{-18}$ |pR| + pL 3.32e-18-> |RL|
$[\ ]^1_{|RL|} \rightarrow [\ ]^1_{|pR|}\ pL$ $\mapsto 0.011$ |RL| 0.011-> |pR| + pL
$[\ ]^1_{|RL|} \rightarrow [\ RL\ ]^1_{|\ |}$ $\mapsto 4.1e^{-3}$ |RL| 4.1e-3-> RL + ||
$[\ ]^1_{|pR|} \rightarrow [\ pR\ ]^1_{|\ |}$ $\mapsto 4.1e^{-4}$ |pR| 4.1e-4-> pR + ||
$[\ Gabg \rightarrow GA\ Gbg\ ]^1_{|RL|}$ $\mapsto 1.0e^{-5}$ Gabg + |RL| 1.0e-5-> GA, Gbg + |RL|
$[\ Gd\ Gbg \rightarrow Gabg\ ]^1_{|\ |}$ $\mapsto 1.0$ Gd + Gbg 1-> Gabg
$[\ GA \rightarrow Gd\ ]^1_{|\ |}$ $\mapsto 0.11$ GA 0.11-> Gd
                                                         }
                                                         rule vac_rule
                                                         {
$[\ ]^2_{|\ |}\ pR \rightarrow [\ pR\ ]^2_{|\ |}$ $\mapsto 4.1e^{-4}$ || + pR 4.1e-4-> pR + ||
$[\ ]^2_{|\ |}\ RL \rightarrow [\ RL\ ]^2_{|\ |}$ $\mapsto 4.1e^{-3}$ || + RL 4.1e-3-> RL + ||
$\}$ }

$w_{2,gprot} = \lambda$ compartment vacuole [vac_rule]
$(u_2, v_2, x_2)_{gprot} = (\lambda, \lambda, \lambda)$
$w_{1,gprot} = Gd^{3000}\ Gbg^{3000}\ Gabg^{7000}$ compartment cell [vacuole,3000 Gd,...
$(u_1, v_1, x_1)_{gprot} = (\lambda, pR^{10000}\ pr, \lambda)$ ... 3000 Gbg,7000 Gabg,g_cycle : |10000 pR,pr|]
$w_{0,gprot} = pL^{6.022e17}$ system cell, 6.022e17 pL
$\mu_{gprot} = [\ [\ [\ ]^2\ ]^1\ ]^0$
$t_{in,gprot} = 0$ evolve 0-600000
$t_{fin,gprot} = 600\ seconds$
                                                         plot cell[Gd,Gbg,Gabg,GA:|pR,RL|]

Fig. 9. P_{pi} system model of G-protein cycle and corresponding simulator script

With these we are able to model in detail many real biochemical processes occurring in the cytoplasm and in the cell membrane.

Evolutions of a P_{pi} system are obtained using an algorithm based on Gillespie [7] and in the second part of the paper we have presented a simulator which can produce evolutions of an arbitrary P_{pi} system, using syntax based on chemical equations. To demonstrate the utility of P_{pi} systems and of the simulator we have modelled and simulated a circadian clock and the G-protein cycle mating response of saccharomyces cerevisiae. The latter makes extensive use of membrane operations.

Several different research directions are now proposed. The primary direction is the application of P_{pi} systems and of the simulator to real biological systems, with the aim of *prediction by in-silico experimentation*. Such application is likely to lead to the need for new bio-inspired features and these constitute another direction of research. The features will be implemented in the model and simulator as necessary, however it is already envisaged that operations of *fission* and *fusion* will be required to permit the modification of a membrane structure in runtime.

A further direction of research is the investigation of the theoretical properties of the model. Reachability of configurations and of markings have already been proved to be decidable for the more restricted model presented in [4] and these proofs should be extended accordingly for the model presented here. Other work in this area might include the modification of the way a P_{pi} system evolves, for example, to allow other semantics (such as that of *maximal parallel* [12]) or to use algorithms that more accurately model the behaviour of biological membranes. In this way we will be able to explore the limits of the model and perhaps discover a more useful level of abstraction.

References

1. B. Alberts, A. Johnson, J. Lewis, M. Raff, K. Roberts, P. Walter, *Molecular Biology of the Cell, 4th* Ed., Garland Science, 2002, p. 593.
2. R. Brijder, M. Cavaliere, A. Riscos-Núñez, G. Rozenberg, D. Sburlan, Membrane Systems with Marked Membranes. *Electronic Notes in Theoretical Computer Science.* To appear.
3. L. Cardelli, Brane Calculi. Interactions of Biological Membranes. *Proceedings Computational Methods in System Biology 2004* (V. Danos, V. Schächter, eds.), Lecture Notes in Computer Science, 3082, Springer-Verlag, Berlin, 2005.
4. M. Cavaliere, S. Sedwards, Membrane Systems with Peripheral Proteins: Transport and Evolution. *Electronic Notes in Theoretical Computer Science.* To appear.
5. G. Ciobanu, Gh. Păun, M.J. Pérez-Jiménez, eds., *Applications of Membrane Computing.* Springer-Verlag, Berlin, 2006.
6. J. Dassow, Gh. Păun, *Regulated Rewriting in Formal Language Theory.* Springer-Verlag, Berlin, 1989.
7. D. T. Gillespie, A General Method for Numerically Simulating the Stochastic Time Evolution of Coupled Chemical Reactions. *Journal of Computational Physics*, 22, 1976.
8. J.E. Hopcroft, J.D. Ullman, *Introduction to Automata Theory, Languages, and Computation.* Addison-Wesley, 1979.
9. T.-M. Yi, H. Kitano, M. I. Simon, A quantitative characterization of the yeast heterotrimeric G protein cycle. *Proceedings of the National Academy of Science*, 100, 19, 2003.
10. M. J. Pérez-Jiménez, F. J. Romero-Campero, Modelling EGFR signalling network using continuous membrane systems. *Proceedings of the Third Workshop on Computational Method in Systems Biology*, Edinburgh, 2005.
11. H. McAdams, A. Arkin, Stochastic mechanisms in gene expression. *Proceedings of the National Academy of Science*, 94, 1997.
12. Gh. Păun, G. Rozenberg, A Guide to Membrane Computing. *Theoretical Computer Science*, 287-1, 2002.
13. A. Regev, W. Silverman, N. Barkai, E. Shapiro, Computer Simulation of Biomolecular Processes using Stochastic Process Algebra. *Poster at 8th International Conference on Intelligent Systems for Molecular Biology, ISMB*, 2000.
14. G. Rozenberg, A. Salomaa, eds., *Handbook of Formal Languages.* Springer-Verlag, Berlin, 1997.
15. A. Salomaa, *Formal Languages.* Academic Press, New York, 1973.
16. M. G. Vilar, H. Y. Kueh, N. Barkai, S. Leibler, Mechanisms of noise-resistance in genetic oscillators. *Proceedings of the National Academy of Science*, 99, 9, 2002.
17. http://www.wisdom.weizmann.ac.il/ biospi/
18. http://psystems.disco.unimib.it
19. http://www.msr-unitn.unitn.it/downloads.php

Appendices

A The Gillespie Algorithm Applied to Membranes

The Gillespie algorithm is an exact stochastic simulation of a 'spatially homogeneous mixture of molecular species which inter-react through a specified set
of coupled chemical reaction channels' [7]. It is unclear whether a biological cell
contains a *spatially homogeneous mixture of molecular species* and less clear still
whether integral and peripheral proteins can be described in this way, however
for the purposes of the P_{pi} system model we choose to regard them as such.
Hence we treat the objects attached to the membrane as homogeneously mixed
with the floating objects, however objects of the same type (i.e. having the same
name) but existing in different regions are considered to be of different types in
the stochastic algorithm.

The mass action combinatorial factors of the Gillespie algorithm, defined by
equations (14a...g) in [7], are calculated over the set of chemical reactions given
in equations (2a...g) of [7], using standard stoichiometric syntax of the general
form

$$S_1 + S_2 + S_3 \rightarrow P_1 + P_2 + \ldots + P_n$$

S_1, S_2 and S_3 are the reactants and $P_1, \ldots, P_n$ are the products of the reaction.
Since the order of the reactants and products is unimportant they may be repre-
sented as multisets $S_1 S_2 S_3$ and $P_1 P_2 \cdots P_n$, respectively, over the set of objects
V. Hence a chemical reaction may be expressed using the notation

$$S_1 S_2 S_3 \rightarrow P_1 P_2 \cdots P_n$$

In the definition of the evolution of a P_{pi} system, the mass action combinatorial factor is calculated using equations (14a...g)[7] after transforming the
membrane and evolution rules into chemical reactions and the objects of the
current instantaneous description, using the following procedure.

Let $V_i = \{a_i | a \in V\}$, $V_{i,int} = \{a_{i,int} | a \in V\}$, $V_{i,itgl} = \{a_{i,itgl} | a \in V\}$ and
$V_{i,out} = \{a_{i,out} | a \in V\}$. We then define morphisms $free^i : V \rightarrow V_i$, $int^i :$
$V \rightarrow V_{i,int}$, $itgl^i : V \rightarrow V_{i,itgl}$ and $out^i : V \rightarrow V_{i,out}$ such that $free^i(a) = a_i$,
$int^i(a) = a_{i,int}$, $itgl^i(a) = a_{i,itgl}$ and $out^i(a) = a_{i,out}$ for $a \in V$. Hence we map
an evolution rule of the type

$$[\, \alpha \rightarrow \beta \,]^i_{u|v|}$$

with $u, v, \alpha, \beta \in V^*$ and $i \in Lab$, to the chemical reaction

$$free^i(\alpha) \cdot int^i(u) \cdot itgl^i(v) \rightarrow free^i(\beta) \cdot int^i(u) \cdot itgl^i(v)$$

We map membrane rules, generally described by

$$[\, \alpha \,]^i_{u|v|x}\, \beta \rightarrow [\, \alpha' \,]^i_{u'|v'|x'}\, \beta'$$

with $u, v, x, \alpha, \beta, u', v', x', \alpha', \beta' \in V^*$ and $i \in Lab$, to the chemical equation

$$free^i(\alpha) \cdot int^i(u) \cdot itgl^i(v) \cdot out^i(x) \cdot free^j(\beta) \rightarrow$$
$$free^i(\alpha') \cdot int^i(u') \cdot itgl^i(v') \cdot out^i(x') \cdot free^j(\beta')$$

where $j \in Lab$ is the marking of the membrane surrounding the region enclosing membrane i.

The objects of the current instantaneous description are similarly transformed, using the morphisms defined above, in order to correspond with the transformed membrane and evolution rules.

B   The Simulator Syntax

The simulator syntax aims to be an intuitive interpretation of the P_{pi} system model. A simulator script conforms to the following grammar:

$$
\begin{aligned}
SimulatorScript = &\{Object\ Declaration,\ NewLine\}^+ \\
&\{Rule\ Definition,\ NewLine\}^+ \\
&\{Compartment\ Definition,\ NewLine\} \\
&System\ Statement,\ NewLine \\
&Evolve\ Statement,\ NewLine \\
&Plot\ Statement,\ [NewLine]
\end{aligned}
$$

where $NewLine$ is an appropriate sequence of characters to generate a new line.

An example of a simple simulator script is shown below, together with its P_{pi} system counterpart.

Simulator script P_{pi} system $lotka$

```
// Lotka reactions
object X,Y1,Y2,Z
```
$V_{lotka} = \{X, Y1, Y2, Z\}$

$rate_{lotka} = \{$

```
rule r1 X + Y1 0.0002-> 2Y1 + X
rule r2 Y1 + Y2 0.01-> 2Y2
rule r3 Y2 10-> Z
```
$[\,XY1 \rightarrow Y1Y1X\,]^0_{|\,|} \mapsto 0.0002$
$[\,Y1Y2 \rightarrow Y2Y2\,]^0_{|\,|} \mapsto 0.01$
$[\,Y2 \rightarrow Z\,]^0_{|\,|} \mapsto 10\ \}$

```
system 100000 X,1000 Y1,1000 Y2,r1,r2,r3
```
$w_{0,lotka} = X^{100000} Y1^{1000} Y2^{1000}$

$\mu_{lotka} = [\,]^0$

```
evolve 0-1000000
```
$t_{in,lotka} = 0$

```
plot Y1,Y2
```

The syntax of the sections of a simulator script are described below.

B.1   Comments

Comments begin with a double forward slash (`//`) and include all subsequent text on a single line. They may appear anywhere in the script.

B.2   Object Declaration

The reacting objects are defined in one or more statements beginning with the keyword `object` followed by a comma separated list of unique reactant names.

E.g.:

```
object X,Y1,Y2,Z
```

The names are case-sensitive and must start with a letter but may include digits and the underscore character (_). This corresponds to defining the alphabet V of the P_{pi} system.

B.3   Rule Definition

The reaction rules are defined using rule definitions comprising the keyword `rule` followed by a unique name and the rewriting rule itself. E.g.:

```
rule r1 X + Y1 0.0002-> 2Y1 + X
```

These correspond to the attach / de-attach and evolution rules of the P_{pi} system model. Note, however, that simulator rules are user-defined types which may be instantiated in more than one region. The value preceding the implication symbol (`->`) is the average reaction rate and corresponds to an element of the range of the mapping *rate* given in Definition 1. In the simulator it is also possible to define a reaction rate as the product of a constant and the rate of a previously defined rule, using the name of the previous rule in the following way:

```
rule r2 Y1 + Y2 50 r1-> 2Y2
```

This has the meaning that rule `r2` has a rate 50 times that of `r1`. In addition, in the simulator it is possible to define a group of rules using a single identifier and braces. E.g.,

```
rule lotka {
      X + Y1 0.0002-> 2Y1 + X
      Y1 + Y2 0.01-> 2Y2
      Y2 10-> Z  }
```

To include membrane operations the simulator rule syntax is extended with the `||` symbol. Objects listed on the left hand side of the `||` represent the internal markings, objects listed on the right hand side represent the external markings and objects listed between the vertical bars are the integral markings of the membrane. E.g.:

```
rule r4 X + |Y2| 0.1-> |X,Y2|
```

means that if one `X` exists within the compartment and one `Y2` exists integral to the membrane, then the `X` will be added to the integral marking of the membrane. The P_{pi} system equivalent is the following *attach*$_{in}$ rule:

$$[\,X\,]_{|Y2|} \rightarrow [\;]_{|XY2|}$$

To represent an *attach*$_{out}$ rule in the simulator the following syntax is used:

```
rule r4 |Y2| + X 0.1-> |X,Y2|
```

Here the `X` appears to the right of the `||` symbol following a `+`, meaning that it must exist in the region surrounding the membrane for the rule to be applied. Hence the `+` used in simulator membrane rules is non-commutative.

B.4   Compartment Definition

Compartments may be defined using the keyword `compartment` followed by a unique name and a list of contents and rules, all enclosed by square brackets. For example,

    compartment c1 [100 X, 100 Y1, r1, r2]

instantiates a compartment having the label `c1` containing 100 `X`, 100 `Y1` and rules `r1` and `r2`. In a P_{pi} system such a compartment would have a P_{pi} system (partial) initial instantaneous description

$$[\, X^{100}Y1^{100}\,]^1$$

Note that a P_{pi} system requires a numerical membrane label and that any rules associated to the region or membrane must be defined separately.

Compartments may contain other pre-defined compartments, so the following simulator statement

    compartment c2 [100 Y2, c1]

corresponds to the P_{pi} system (partial) initial instantaneous description

$$[\, Y2^{100}[\, X^{100}Y1^{100}\,]^1\,]^2$$

Membrane markings in the simulator are added to compartment definitions using the symbol | |, to the right of and separated from the floating contents by a colon. E.g.,

    compartment c3 [100 X, c2 : 10 Y2||10 Y1]

has the meaning that the compartment `c3` contains compartment `c2`, 100 `X`, and the membrane surrounding `c3` has 10 `Y2` attached to its inner surface and 10 `Y1` attached to its outer surface. This corresponds to the P_{pi} system (partial) initial instantaneous description

$$[\, X^{100}[\, Y2^{100}[\, X^{100}Y1^{100}\,]^1\,]^2\,]^3_{Y2^{10}|\,|Y1^{10}}$$

B.5   System Statement

The system is instantiated using the keyword `system` followed by a comma-separated list of constituents. E.g.:

    system 100000 X,1000 Y1,1000 Y2,r1,r2,r3

This statement corresponds to the definition of $u_0 \ldots u_n$, $v_0 \ldots v_n$, $w_0 \ldots w_n$, $x_0 \ldots x_n$ and μ of the P_{pi} system.

The system statement may be extended to multiple lines by enclosing the list of constituents between braces. E.g.:

    system {
        100000 X,
        1000 Y1,
        1000 Y2,
        r1,r2,r3 }

It is also possible to add or subtract reactants from the simulation in runtime using the following syntax in the `system` statement:

```
-10 X @50000, 10 Y1 @50000
```

These *instructions* request a subtraction of ten X from the system and an addition of ten Y1 to the system at time step 50000. Negative quantities are not allowed in the simulator, so if a subtraction requests a greater amount than exists, only the existing amount will be deleted.

B.6   Evolve Statement

The simulator requires a directive to specify the total number of evolution steps to perform and also the number of the evolution step at which to start recording data. This is achieved using the keyword `evolve` followed by the minimum and maximum evolution steps to record. E.g.,

```
evolve 0-1000000
```

Note that the minimum evolution step does not correspond to t_{in} of the P_{pi} system, since the simulation always starts from the 0th step. By convention, the simulator sets the initial time of the simulation to 0, hence $t_{in} = 0$ for all simulations. Note that although t_{fin} of a P_{pi} system evolution *corresponds* to the maximum evolution step, the units are different and there is no explicit conversion.

B.7   Plot Statement

To specify which objects are to be observed during the evolution the `plot` keyword is used followed by a list of reactants. To plot the contents of a specific compartment the `plot` statement uses syntax similar to that used in the compartment definition. E.g.,

```
plot X, c3[X,Y1 : Y1|Y2|]
```

plots the number of free-floating X in the environment and the specified contents of compartment c3 and its membrane.

Modelling and Analysing Genetic Networks: From Boolean Networks to Petri Nets

L.J. Steggles, Richard Banks, and Anil Wipat

School of Computing Science, University of Newcastle, Newcastle upon Tyne, UK
{L.J.Steggles, Richard.Banks, Anil.Wipat}@ncl.ac.uk

Abstract. In order to understand complex genetic regulatory networks researchers require automated formal modelling techniques that provide appropriate analysis tools. In this paper we propose a new qualitative model for genetic regulatory networks based on Petri nets and detail a process for automatically constructing these models using logic minimization. We take as our starting point the Boolean network approach in which regulatory entities are viewed abstractly as binary switches. The idea is to extract terms representing a Boolean network using logic minimization and to then directly translate these terms into appropriate Petri net control structures. The resulting compact Petri net model addresses a number of shortcomings associated with Boolean networks and is particularly suited to analysis using the wide range of Petri net tools. We demonstrate our approach by presenting a detailed case study in which the genetic regulatory network underlying the nutritional stress response in *Escherichia coli* is modelled and analysed.

1 Introduction

The development and function of cellular systems is regulated by complex networks of interacting genes, proteins and metabolites known as *genetic regulatory networks* [3]. With the advent of improved post–genomic technology the data is now available to allow researchers to study genetic regulatory networks at a holistic level [26]. However, interpreting and analysing this data is still problematic and further work is needed to develop automated formal techniques that provide appropriate tools for modelling and analysing genetic regulatory networks.

In this paper, we present a new technique for qualitatively modelling and analysing genetic regulatory networks. We take as our starting point *Boolean networks* [1,3], an existing modelling approach for regulatory networks in which regulatory entities (i.e. genes, proteins, and external signals) are viewed abstractly as binary switches. While Boolean networks have proved successful in modelling real world regulatory networks [14,27], they suffer from a number of shortcomings: analysis can be problematic due to the exponential growth in Boolean states and the lack of tool support; and they do not cope well with the inconsistent and incomplete data that often occurs in practice. To address these problems, we propose a new model for genetic regulatory networks based on *Petri nets* [21,18], a well developed formal framework for modelling and analysing

C. Priami (Ed.): CMSB 2006, LNBI 4210, pp. 127–141, 2006.

complex concurrent systems [22,28]. A range of initial investigations into using Petri nets to model biological systems have appeared in the literature to date, including: Place/Transition nets [19,5,25,12]; stochastic nets [9,24]; high–level nets [11,6]; and hybrid nets [17]. The results we present significantly extend the related ideas presented in [5], both semantically and in the provision of automated tool support for model construction and analysis.

The Petri net model we propose is based on using an intuitive Petri net structure to represent the Boolean relationships between regulatory entities. We start by defining each entities individual behaviour as a *truth table* [10] from which we extract Boolean terms by applying *logic minimization techniques* [10,4]. These Boolean terms compactly represent the fundamental relationships between regulatory entities and we directly translate them into appropriate Petri net control structures. The result is a compact Petri net model that completely captures the original Boolean behaviour of a genetic regulatory network. Both the *synchronous* and *asynchronous* semantic interpretation of Boolean networks [8] can be modelled using our approach. We choose to focus on the synchronous semantics here and develop a simple two phase commit protocol to allow synchronized state updates within the asynchronous Petri net framework. To support the modelling process a prototype tool has been developed which is able to automatically construct Petri net models of genetic networks from their truth table definitions. The resulting models can then be analysed using the wide range of available Petri net techniques and tools [22,7,28].

We illustrate our approach by presenting a detailed case study in which the genetic regulatory network for the carbon starvation stress response in the bacterium *E. coli* [20] is modelled and analysed. Using the detailed data provided in [20] we define the Boolean behaviour of the key regulatory entities involved using truth tables. We then apply our prototype tool to automatically construct a qualitative Petri net model capturing the behaviour of the given genetic regulatory network. This Petri net is then validated and analysed using PEP [29] and in particular, we illustrate the application of model checking techniques [7,15] for detailed model analysis.

This paper is organised as follows. In Section 2 we give a brief introduction to Boolean networks and Petri nets. In Section 3 we describe a new approach to modelling the Boolean behaviour of genetic regulatory networks using Petri nets. In Section 4 we consider a case study in which we apply our techniques to modelling and analysing the genetic regulatory network for the carbon starvation stress response in *E. coli*. Finally, in Section 5, we present some concluding remarks on our work.

2   Background

In this section we give a brief overview of the modelling formalisms discussed in this paper: *Boolean networks* [1,3] and *Petri nets* [21,18]. In the sequel we assume the reader is familiar with the basic Boolean operators *not*, *or* and *and* (for example, see [10]).

2.1   Boolean Networks

In a Boolean network [1,3] the state of each regulatory entity g_i is represented as a Boolean value, either 1 representing the entity is active (e.g. a gene is expressed or a protein is present) or 0 representing the entity is inactive (e.g. a gene is not expressed or a protein is absent). The state of a gene regulatory network containing n entities is then naturally represented as a Boolean vector $[g_1, \ldots, g_n]$ and this gives us a state space containing 2^n states [4]. The behaviour of each g_i is described using a Boolean function f_i which, given the current states of the entities in its neighbourhood (i.e. those entities which directly affect it), defines the next state for g_i. As an example consider the Boolean network in Figure 1.(a) [1] which contains three entities g_1, g_2 and g_3, where the next state g_i' of each entity is defined by the following Boolean functions:

$$g_1' = g_2, \qquad g_2' = g_1\, g_3, \qquad g_3' = \overline{g_1}$$

where the notation $\overline{x}$, $x+y$ and $x\,y$ is used to represent the Boolean operators *not*, *or* and *and* [10] respectively. The dynamic behaviour of a Boolean network can be semantically interpreted in two distinct ways [8]: *asynchronously* where genes update their state independently; and *synchronously* where all genes update their state together. We focus on the synchronous semantics in this paper which appears to be widely used in the literature [3,8]. The synchronous behaviour for our example Boolean network is shown as a *truth table* in Figure 1.(b) and a *state transition graph* [4] in Figure 1.(c).

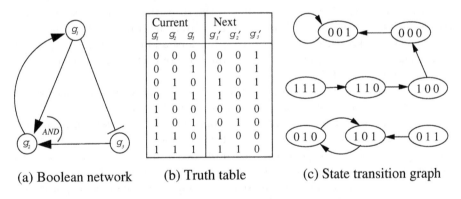

(a) Boolean network   (b) Truth table   (c) State transition graph

Current			Next		
g_1	g_2	g_3	g_1'	g_2'	g_3'
0	0	0	0	0	1
0	0	1	0	0	1
0	1	0	1	0	1
0	1	1	1	0	1
1	0	0	0	0	0
1	0	1	0	1	0
1	1	0	1	0	0
1	1	1	1	1	0

Fig. 1. An example of a Boolean network for three entities g_1, g_2 and g_3

Boolean networks have proved successful in modelling real world regulatory networks [14,27]. However, their application in practice is hindered by a number of shortcomings. In particular, analysis can be problematic due to the exponential growth in Boolean states and the lack of tool support in this area. They are also unable to cope with the inconsistent and incomplete regulatory network data that often occurs in practice. For this reason we consider extending the Boolean network approach by developing a Petri net based Boolean model.

2.2   Petri Nets

The theory of Petri nets [21,18] provides a graphical notation with a formal mathematical semantics for modelling and reasoning about concurrent, distributed systems. A Petri net is a directed bipartite graph and consists of four basic components: *places* which are denoted by circles; *transitions* denoted by black rectangles; *arcs* denoted by arrows; and *tokens* denoted by black dots. A simple example of a Petri net is given in Figure 2. The places, transitions and arcs

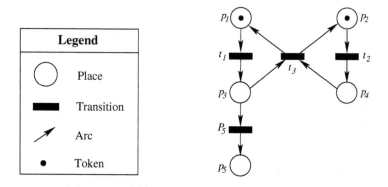

Fig. 2. A simple example of a Petri net

describe the static structure of the Petri net. Each transition has a number of *input places* (places with an arc leading to the transition) and a number of *output places* (places with an arc leading to them from the transition). We normally view places as representing resources or conditions and transitions as representing actions or events [21]. Note arcs that directly connect two transitions or two places are not allowed.

The state of a Petri net is given by the distribution of tokens on places within it, referred to as a *marking*. The *state space* of a Petri net is therefore the set of all possible markings. The dynamic properties of the system are modelled by transitions which can fire to move tokens around the places in a Petri net. Transitions are said to be *enabled* if each of their input places contain at least one token. An enabled transition can *fire* by consuming one token from each of its input places and then depositing one token on each of its output places. For example, in Figure 2 both transitions t_1 and t_2 are enabled. Firing transition t_1 would result in a token being taken from place p_1 and a new token being deposited on place p_3. Often, more than one transition is enabled to fire at any one time (as in the example above). In such a case, a transition is chosen non–deterministically to fire. A marking m_2 is said to be *reachable* from a marking m_1 if there is a sequence of transitions that can be fired starting from $m1$ which results in the marking $m2$. A Petri net is said to be *k–bounded* if in all reachable markings no place has more than k tokens. A Petri net which is 1–bounded is said to be *safe*. Safeness is an important property since any safe Petri net has a restricted state space which is well–suited to automatic analysis [22].

An important advantage of Petri nets is that they are supported by a wide range of techniques and tools for simulation and analysis [22,28]. For example, Petri nets can be automatically checked for boundedness and the presence of deadlocks (markings in which no transitions are enabled to fire) [28]. A Petri net can also be analysed by constructing its *reachability graph* [18] which captures the possible firing sequences that can occur from a given initial marking. A range of techniques based on *model checking* [7,15] have been developed for analysing reachability properties of a Petri net and these provide a means of coping with the potentially large state space of a Petri net model.

3   Modelling Genetic Networks Using Petri Nets

In this section we present a new qualitative model for gene regulatory networks based on *Petri nets* [18] and detail a process for automatically constructing these models using *logic minimization* [4].

3.1   Deriving Regulatory Relationships Using Logic Minimization

Given a set of truth tables defining the Boolean behaviour of all the entities in a genetic network we would like to extract a compact representation of the regulatory relationships between entities. We address this using well–known techniques from Boolean logic [4,10] which allow us to derive Boolean terms describing the functional behaviour of each entity. The idea is to consider the truth table for each entity and to list all the states which result in a next state in which the entity is active (i.e. in state 1). For example, consider the truth table given in Figure 1.(b) for a simple Boolean network (see Section 2.1). Then by considering the truth table for g_1 we can see that states 010, 011, 110, and 111 result in g_1 being 1 in its next state (where xyz denotes the state $g_1 = x$, $g_2 = y$, and $g_3 = z$). We can represent each state as a product term, called a *minterm* [10], using the *and* Boolean operator, where the variable g_i represents that an entity g_i is in state 1, and the negated variable $\overline{g_i}$ represents that an entity g_i is 0. So the state 010 for g_1 is represented by the minterm $\overline{g_1}\,g_2\,\overline{g_3}$. Applying this approach and then summing the derived minterms using the *or* Boolean operator allows us to derive a Boolean term in *disjunctive normal form* [10] that defines the functional behaviour of an entity. Continuing with our example, we derive the following Boolean term for gene g_1:

$$\overline{g_1}\,g_2\,\overline{g_3} \;+\; \overline{g_1}\,g_2\,g_3 \;+\; g_1\,g_2\,\overline{g_3} \;+\; g_1\,g_2\,g_3$$

Note that this term completely defines the functional behaviour of g_1, i.e. whenever the term above evaluates to 1 in a state we know g_1 will be active in the next state, and whenever the term is 0 we know g_1 will be inactive. Using this technique we can construct a Boolean network that completely specifies the functional behaviour of a genetic network. In our example, we derive the following terms defining the behaviour of g_1, g_2 and g_3:

$$g_1' = \overline{g_1}\, g_2\, \overline{g_3} \;+\; \overline{g_1}\, g_2\, g_3 \;+\; g_1\, g_2\, \overline{g_3} \;+\; g_1\, g_2\, g_3,$$
$$g_2' = g_1\, \overline{g_2}\, g_3 \;+\; g_1\, g_2\, g_3,$$
$$g_3' = \overline{g_1}\, \overline{g_2}\, \overline{g_3} \;+\; \overline{g_1}\, \overline{g_2}\, g_3 \;+\; \overline{g_1}\, g_2\, \overline{g_3} \;+\; \overline{g_1}\, g_2\, g_3.$$

The Boolean terms derived above are often unnecessarily complex and can normally be simplified using *logic minimization* [4,10]. From a biological point of view, this simplification process is important as it helps to identify the underlying regulatory relationships that exist between entities in a genetic network. The idea behind logic minimization is to simplify Boolean terms by merging minterms that differ by only one variable. As an example, consider the term $\overline{g_1}\, g_2\, \overline{g_3} \;+\; \overline{g_1}\, g_2\, g_3$ which contains two minterms that differ by only one variable g_3. This term can be simplified by merging the two minterms to produce a simpler term $\overline{g_1}\, g_2$ which is logically equivalent [4,10]. For brevity we omit the full details of Boolean logic minimization here (we refer the interested reader to [4]) and instead illustrate the idea behind the algorithm using our running example:

$$\overline{g_1}\, g_2\, \overline{g_3} \;+\; \overline{g_1}\, g_2\, g_3 \;+\; g_1\, g_2\, \overline{g_3} \;+\; g_1\, g_2\, g_3 \;\Longrightarrow\; \overline{g_1}\, g_2 \;+\; g_1\, g_2 \;\Longrightarrow\; g_2,$$
$$g_1\, \overline{g_2}\, g_3 \;+\; g_1\, g_2\, g_3 \;\Longrightarrow\; g_1\, g_3,$$
$$\overline{g_1}\, \overline{g_2}\, \overline{g_3} \;+\; \overline{g_1}\, \overline{g_2}\, g_3 \;+\; \overline{g_1}\, g_2\, \overline{g_3} \;+\; \overline{g_1}\, g_2\, g_3 \;\Longrightarrow\; \overline{g_1}\, \overline{g_2} \;+\; \overline{g_1}\, g_2 \;\Longrightarrow\; \overline{g_1}.$$

Note that the final minimized Boolean terms presented above correctly correspond to the Boolean network definitions given in Section 2.1.

3.2   Modelling Boolean Networks Using Petri Nets

While the Boolean terms derived in Section 3.1 compactly capture the behaviour of a Boolean network they are not amenable to analysis in their current form. We address this by translating these terms directly into appropriate Petri net control structures. The resulting Petri net model can then be simulated and analysed using the wide range of available tool support [28].

The approach we take is to represent the Boolean state of each entity g_i in a Petri net by the well–known approach (see for example [21,5]) of using two complementary places Pi and $\overline{Pi}$, where a token on place Pi indicates the entity is active, $g_i = 1$, and a token on place $\overline{Pi}$ that it is not, $g_i = 0$. Note the total number of combined tokens on places Pi and $\overline{Pi}$ will therefore always be equal to 1. Since Petri nets fire transitions asynchronously it is straightforward to model the asynchronous behaviour of a Boolean network in this setting (see [5] for a related approach). We focus on modelling the synchronous behaviour of a Boolean network [8] here and make use of a two phase commit protocol to synchronise updates in our model. In the first phase of the protocol each entity g_i in the model decides whether it should be active or not in the next state. This decision is recorded using two places, Pi_On and Pi_Off, where a token on Pi_On indicates g_i is active in the next state and a token on Pi_Off that it is not. When all the entities have made a decision about their next state the

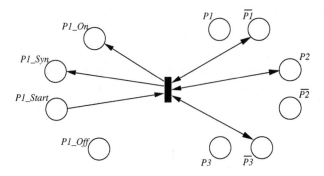

Fig. 3. A transition for gene g_1 modelling the minterm $\overline{g_1}\, g_2\, \overline{g_3}$

second phase of the protocol begins and the state of each entity is synchronously updated according to the recorded decision.

Let us consider how we construct the appropriate Petri net structure to model the decision process for an entity g_i in the first phase of the protocol. We begin by considering under what conditions the entity will be active (i.e. in state 1) and use the process detailed in Section 3.1 to derive a minimized Boolean term which compactly captures these conditions. We model this minimized Boolean term in our Petri net by adding a separate transition to represent each minterm it contains. The idea is that each transition will fire, placing a token on Pi_On, precisely when the corresponding minterm is true. As an example, consider the Boolean term

$$\overline{g_1}\, g_2\, \overline{g_3} \; + \; \overline{g_1}\, g_2\, g_3 \; + \; g_1\, g_2\, \overline{g_3} \; + \; g_1\, g_2\, g_3$$

derived for gene g_1 in our running example (see Section 3.1). Then the first minterm $\overline{g_1}\, g_2\, \overline{g_3}$ tells us that gene g_1 should be expressed, $g_1 = 1$, in the next state when genes $g_1 = 0$, $g_2 = 1$, and $g_3 = 0$ in the current state. We model this minterm using the transition depicted in Figure 3. This transition fires when places $\overline{P1}$, $P2$, and $\overline{P3}$ contain a token (i.e. when $g_1 = 0$, $g_2 = 1$, and $g_3 = 0$) and results in a token being placed on $P1\_On$ (indicating g_1 is expressed in the next state). Note the use of *read arcs* [18] here, i.e. bidirectional arcs which do not consume tokens but just check they are present. This ensures the tokens on places $\overline{P1}$, $P2$, and $\overline{P3}$ are not removed at this stage (doing so would corrupt the current state of genes g_1, g_2 and g_3). The start place $P1\_Start$ is used as a control input to the transition to ensure only one decision is made for gene g_1 during a single protocol step (update transitions can only fire if a token is present on $P1\_Start$). The place $P1\_Syn$ is used to indicate when an update decision has been made for gene g_1, information needed by the protocol to determine when the first phase is complete. This process is then repeated to add transitions to model the remaining three minterms in the Boolean term for g_1.

It remains to model the complementary decision procedure for deciding when an entity is inactive in the next state, that is when Pi_Off should be marked. To do this we simply apply the process detailed in Section 3.1 again amended to derive a Boolean term which compactly captures the conditions under which the entity becomes inactive. We then repeat the procedure detailed above for modelling a minterm as a transition except this time we mark place Pi_Off to record the decision for the next state instead of Pi_On. Note the resulting Petri net structure will contain at most $n2^k$ transitions where n is the number of entities and k is the maximum neighbourhood size. Since k is usually small in practice [8] the size of the model is normally linear with respect to n.

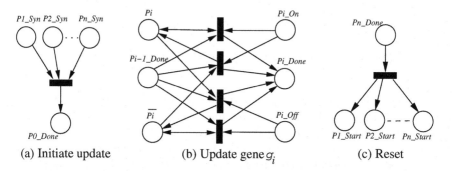

(a) Initiate update   (b) Update gene g_i (c) Reset

Fig. 4. Petri net fragments for controlling synchronous updates

After all the entities have made their update decisions all the synchronisation places will be marked and this allows the control transition depicted in Figure 4.(a) to fire, initiating the second phase of the protocol. This phase performs a synchronised update step in which the state of each entity g_i is updated in turn by placing a token on Pi if place Pi_On is marked or on $\overline{Pi}$ if place Pi_Off is marked. An example fragment of the Petri net structure used for this update is given in Figure 4.(b) for an arbitrary gene g_i. The fragment contains four transitions which represent the four possible update situations that can occur: move token from place $\overline{Pi}$ to Pi; leave token on Pi; move token from place Pi to $\overline{Pi}$; leave token on $\overline{Pi}$. Only one of these transitions will be enabled to fire. Once the gene g_i has updated its state a token is placed on place Pi_Done to indicate that the next entity can be updated. When the last entity g_n has been updated place Pn_Done will be marked and the control transition depicted in Figure 4.(c) initiates a reset step which re-marks the start places, allowing the whole synchronisation protocol to begin again.

So far we have assumed that we are always able to derive complete and consistent truth tables which correctly capture the behaviour of each entity in a regulatory network. However, in practice it is rarely the case that a regulatory network is fully understood and indeed, this is one important reason for modelling such networks. The data provided may be *incomplete* in the sense that information is missing about what happens in certain states, or it may be *inconsistent* in that we have conflicting information about states. The result is that the behaviour of some entities under certain conditions may be unknown. Such incomplete and/or inconsistent information is problematic for the standard

Boolean network model which is unable to represent the possibility of more than one next state. However, Petri nets are a non-deterministic modelling language [21] and so are able to represent unknown behaviour by incorporating all possible next state transitions. The idea is to simply allow the states with unknown behaviour to be used when deriving both the active and inactive Boolean formulas for an entity. The resulting non-deterministic choices within the Petri net model can then be meaningfully taken into account when analysing its behaviour.

The Petri net modelling approach presented above, while theoretically well–founded, is not practical by hand for all but the smallest of models. To support our modelling approach we have developed a prototype tool to automate the model construction process detailed in Sections 3.1 and 3.2. The tool takes as input a series of truth tables describing the behaviour of the entities in a Boolean network. These input tables are allowed to contain inconsistent and incomplete data as discussed above. From these tables the tool is able to automatically construct a Petri net model which is based on either the synchronous or asynchronous Boolean network semantics [8]. This prototype tool is freely available for academic use and can be obtained from the project's website[1].

4   Case Study: Nutritional Stress Response in *E. coli*

In this section we present a detailed case study to demonstrate the modelling techniques we have introduced and the practical application of Petri net analysis techniques. We consider the bacterium *E. coli* which under normal environmental conditions, when nutrients are freely available, is able to grow rapidly entering an *exponential phase* of growth [13]. However, as important nutrients become depleted and scarce the bacteria experiences nutritional stress and responds by slowing down growth, eventually resulting in a *stationary phase* of growth. In this case study we model a simplified version of the genetic regulatory network responsible for the carbon starvation nutritional stress response in *E. coli* based on the comprehensive data collated in [20]. We validate and analyse the resulting Petri net model using *PEP* [29], a leading Petri net support tool.

4.1   Constructing the Petri Net Model

The genetic regulatory network underlying the stress response in *E. coli* to carbon starvation is shown abstractly in Figure 5 (adapted from [20]). The network has a single input signal which indicates the presence or absence of carbon starvation and uses the level of stable RNA (ribosomal RNA and transfer RNA) as indicative of the current phase of *E. coli*, i.e. during the exponential phase the level of stable RNA is high to support rapid growth, while under the stationary phase the level drops, since only a maintenance metabolism is required [20]. The carbon starvation signal is transduced by the activation of adenylate cyclase (Cya), an enzyme which results in the production of the metabolite cAMP.

---

[1] http://bioinf.ncl.ac.uk/gnapn

This metabolite immediately binds with and activates the global regulator protein CRP, and the resulting cAMP.CRP complex is responsible for controlling the expression of key global regulators including Fis and CRP itself. The global regulatory protein Fis is central to the stress response and is responsible for promoting the expression of stable RNA from the *rrn operon* [13,20]. Thus, during the exponential phase high levels of Fis are normally observed and the mutual repression that occurs between Fis and cAMP.CRP is thought to play a key role in the regulatory network [20]. The expression of *fis* is also promoted by high levels of *negative supercoiling* being present in the DNA. The level of DNA supercoiling is tightly regulated by two topoisomerases [13,20]: GyrAB (composed of the products of genes *gyrA* and *gyrB*) which promotes supercoiling; and TopA which removes supercoils. An increase in DNA supercoiling results in increased expression of TopA and thus prevents excessive supercoiling. A decrease in supercoiling results in increased expression of *gyrA* and *gyrB*, and the resulting high level of GyrAB acts to increase supercoiling.

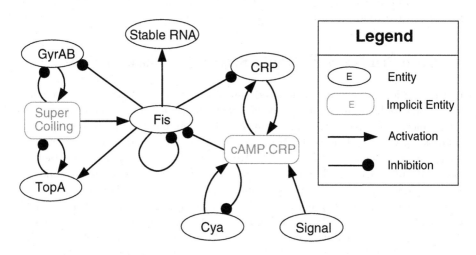

Fig. 5. Genetic network for carbon starvation stress response in *E. coli*

Using the data provided in [20] we are able to derive truth tables defining the Boolean behaviour of each regulatory entity in the nutritional stress response network for carbon starvation. As an example, consider the truth table defining the behaviour of Cya shown in Figure 6. Note following the approach in [20], the level of cAMP.CRP and DNA supercoiling are not explicitly modelled as entities in our model.

The next step is to apply logic minimization to the truth tables we have derived to extract Boolean expressions which compactly define the qualitative behaviour of each regulatory entity. This process is automated by our prototype tool and the result is the following set of Boolean equations:

CRP	Cya	Signal	Cya
0	0	0	1
0	0	1	1
0	1	0	1
0	1	1	1
1	0	0	1
1	0	1	1
1	1	0	1
1	1	1	0

Fig. 6. Truth table defining the Boolean behaviour of Cya

$$\text{Cya} = \overline{\text{Signal}} \; + \; \overline{\text{Cya}} \; + \; \overline{\text{CRP}}, \quad \overline{\text{Cya}} = \text{Signal} \; \text{Cya} \; \text{CRP},$$

$$\text{CRP} = \overline{\text{Fis}}, \quad \overline{\text{CRP}} = \text{Fis},$$

$$\text{GyrAB} = (\overline{\text{GyrAB} \; \overline{\text{Fis}}}) \; + \; (\text{TopA} \; \overline{\text{Fis}}),$$
$$\overline{\text{GyrAB}} = (\text{GyrAB} \; \overline{\text{TopA}}) \; + \; \text{Fis},$$

$$\text{TopA} = \text{GyrAB} \; \overline{\text{TopA}} \; \text{Fis}, \quad \overline{\text{TopA}} = \overline{\text{GyrAB}} \; + \; \text{TopA} \; + \; \overline{\text{Fis}},$$

$$\text{Fis} = (\overline{\text{Fis}} \; \overline{\text{Signal}} \; \text{GyrAB} \; \overline{\text{TopA}}) \; + \; (\overline{\text{Fis}} \; \overline{\text{Cya}} \; \text{GyrAB} \; \overline{\text{TopA}})$$
$$+ \; (\overline{\text{Fis}} \; \overline{\text{CRP}} \; \text{GyrAB} \; \overline{\text{TopA}}),$$
$$\overline{\text{Fis}} = (\text{CRP} \; \text{Cya} \; \text{Signal}) \; + \; \text{Fis} \; + \; \overline{\text{GyrAB}} \; + \; \text{TopA},$$

$$\text{SRNA} = \text{Fis}, \quad \overline{\text{SRNA}} = \overline{\text{Fis}}.$$

The above equations can then be used to construct a Petri net model of the nutritional stress response regulatory network for carbon starvation by applying the approach detailed in Section 3.2. The result is a safe Petri net model that contains 45 places and 49 transitions (based on the synchronous update semantics). The above process can be automated using our prototype tool and the resulting Petri net can then be exported to a wide range of Petri net tools [28].

4.2   Analysing the Petri Net Model

We now consider analysing the Petri net model which results above using the PEP tool [29] and in particular, make use of model checking techniques [7,15]. Our aim is to illustrate the range of analysis possible using available tools, from simple validation tests to more in–depth gene 'knockout' analysis.

We begin our analysis by performing a series of simple validation tests to check the model is able to correctly switch between the exponential and stationary phases of growth. The idea is to initialise the Petri net to a given state and then simulate it, observing the states that occur after each application of the two phase commit protocol. The results of these simulations can then be compared with the expected behaviour [13,20,2] to validate the model. As an example, we consider validating that the model correctly switches from the exponential to the stationary phase of growth. We initialise the Petri net to a state representing the exponential phase but activate Signal to represent the presence of carbon starvation. The resulting simulation run is presented in Figure 7, where the first row represents the models initial state and each subsequent row the next state observed. It shows that the model correctly switches to the stationary phase by entering an attractor cycle of period two (see last three rows in table) in which stable RNA is not present in significant levels (i.e. SRNA remains inactive).

Signal	CRP	Cya	GyrAB	TopA	Fis	SRNA
1	0	1	1	0	1	1
1	0	1	0	1	0	1
1	1	1	1	0	0	0
1	1	0	0	0	0	0
1	1	1	1	0	0	0

Fig. 7. Simulating the switch from exponential to stationary phase

To investigate the behaviour of the model in more detail we make use of the extended reachability analysis provided by the model checking tools of PEP [7,15]. For example, it appears from the literature that the entities GyrAB and TopA should be mutually exclusive, i.e. whenever GyrAB is significantly expressed then TopA shouldn't be and vice a versa. We can verify this in our model by formulating the following constraint on places:

$$\text{GyrAB} + \text{TopA} > 1, \quad \text{GyrAB\_Done} = 1$$

which characterises a state in which the mutual exclusion property does not hold (where the condition GyrAB_Done = 1 is used to ensure we only consider states reached after a complete pass of the two phase commit protocol). The model checking tool is able to confirm that no state satisfying this constraint is reachable from any reasonable initial state and this proves that GyrAB and TopA must be mutually exclusive. We can attempt to prove a similar mutual exclusion property for CRP and Fis using the same approach. However, this time the model checking tool confirms that it is able to reach a state satisfying the constraint, proving that CRP and Fis are not mutually exclusive in our model. In fact, the tool returns a witness firing sequence which leads to such a state to validate the result and we are able to automatically simulate this to gain important insight into how this behaviour occurs.

We can extend our analysis further by experimenting with the underlying structure of the Petri net model, adding or removing regulatory relationships to test possible experimental hypotheses. To illustrate this we can consider investigating the effect of fixing the level of the global regulator CRP which is the target of the carbon starvation signal–transduction pathway [20]. We do this by simply omitting the truth table for CRP from the construction process, resulting in CRP being treated as an input entity (i.e. like the entity Signal) whose state becomes fixed once initialised. We start by 'knocking out' *crp* so that it cannot be expressed and then simulate the amended model to investigate the impact of this change. As expected the results show that the transition from exponential to stationary phase is blocked; the lack of CRP prevents the formation of cAMP.CRP which is needed to initiate the phase transition. Next we fix *crp* to be permanently expressed and again simulate the model. Interestingly the results show that the behaviour of the network is largely unaffected by this change; both the transition from exponential to stationary phase and vice a versa are able to occur as normal.

5   Conclusions

The standard approach of using Boolean networks [1,3] to model genetic regulatory networks has a number of shortcomings: Boolean networks lack effective analysis tools; and have problems coping with incomplete or inconsistent data. In this paper we addressed the shortcomings of Boolean networks by presenting a new approach for qualitatively modelling genetic regulatory networks based on Petri nets [18]. The idea was to use logic minimization [4] to extract Boolean terms representing the genetic network's behaviour and to then directly translate these into Petri net control structures. The result is a compact Petri net model that correctly captures the dynamic behaviour of the original regulatory network and which is amenable to detailed analysis via existing Petri net tools [28].

We illustrated our approach by modelling and analysing the genetic regulatory network underlying the carbon starvation stress response in *E. coli* [20,2]. This case study demonstrated how the PEP tool [29] can be used to validate and analyse our Petri net models. In particular, we considered using simulation tests to validate the correctness of our model and model checking tools [29,15] to investigate the detailed behaviour of the genetic regulatory network.

The results we have presented significantly extend existing work on using Boolean models to analyse genetic regulatory networks (e. g. [5]). In particular, we see the key contributions of this paper as follows: i) A new compact approach to qualitatively modelling genetic regulatory networks based on using logic minimization and Petri nets; ii) Both synchronous and asynchronous semantics of Boolean networks [8] are catered for; iii) Provision of tool support to automate model construction; iv) A detailed case study exploring the application of existing Petri net tools to analyse a Boolean model of a genetic regulatory network.

One drawback of Boolean models is that the high level of abstraction used means behaviour crucial to the operation of a regulatory network may be lost. In future work we intend to address this problem by extending our modelling approach to multi–valued network models [16]. We intend to incorporate our qualitative modelling tools into related work on Stochastic Petri net modelling [23,24] and so provide much needed support in this important area.

Acknowledgments. We are very grateful to O. J. Shaw, M. Koutny and V. Khomenko for many useful discussions concerning this work. We would also like to thank the EPSRC for supporting R. Banks and the BBSRC for supporting this work via the Centre for Integrated Systems Biology of Ageing and Nutrition (CISBAN). Finally we acknowledge the support of the Newcastle Systems Biology Resource Centre.

References

1. T. Akutsu, S. Miyano and S. Kuhara. Identification of genetic networks from small number of gene expression patterns under the Boolean network model. *Proceedings of Pacific Symp. on Biocomputing*, 4:17-28, 1999.
2. G. Batt, D. Ropers, H. de Jong, J. Geiselmann, R. Mateescu, M. Page and D. Schneider. Validation of qualitative models of genetic regulatory networks by model checking: analysis of the nutritional stress response in *Escherichia coli*. *Bioinformatics*, 21:i19–i28, 2005.
3. J. M. Bower and H. Bolouri. *Computational Modelling of Genetic and Biochemical Networks*. MIT Press, 2001.
4. K. J. Breeding. *Digital Design Fundamentals*. Prentice Hall, 1992.
5. C. Chaouiya, E. Remy, P. Ruet, and D. Thieffry. Qualitative modelling of genetic networks: From logical regulatory graphs to standard Petri nets. In: J. Cortadella and W. Reisig (Eds), *Proc. of the Int. Conf. on the Application and Theory of Petri Nets*, Lecture Notes in Computer Science 3099, pages 137-156, Springer–Verlag, 2004.
6. J.-P. Comet, H. Klaudel and S. Liauzu. Modeling Multi-valued Genetic Regulatory Networks Using High-Level Petri Nets. In: G. Ciardo and P. Darondeau (eds), *Proc. of the Int. Conf. on the Application and Theory of Petri Nets*, Lecture Notes in Computer Science 3536, pages 208-227, Springer–Verlag, 2005.
7. J. Esparza. Model checking using net unfoldings. *Science of Computer Programming*, 23(2-3):151-195, 1994.
8. C. Gershenson. Classification of random boolean networks. In: R. K. Standish et al (eds), *Proc. of the 8th Int. Conf. on Artificial Life*, p.1–8, MIT Press, 2002.
9. P. J. E. Goss and J. Peccoud. Quantitative modelling of stochastic systems in molecular biology by using stochastic Petri nets. *Proceedings of the National Academy of Sciences of the United States of America*, 95(12):6750-6755, 1998.
10. P. Grossman. *Discrete Mathematics for Computing*. Palgrave MacMillan, Second Edition, 2002.
11. M. Heiner, I. Koch, and K. Voss. Analysis and simulation of steady states in metabolic pathways with Petri nets. In K. Jensen (ed), *Workshop and Tutorial on Practical Use of Coloured Petri Nets and the CPN Tools (CPN'01)*, pages 15-34, Aarhus University, 2001.
12. M. Heiner, I. Koch, and J. Will. Model validation of biological pathways using Petri nets - demonstrated for apoptosis. *Biosystems*, 75(1-3):15-28, 2004.
13. R. Hengge-Aronis. The general stress response in *Escherichia coli*. In: G. Storz and R. Hengge-Aronis (eds), *Bacterial Stress Responses*, pages 161-178, ASM Press, 2000.
14. S. Huang. Gene expression profiling, genetic networks, and cellular states: an integrating concept for tumorigenesis and drug discovery. *Journal of Molecular Medicine*, Vol. 77:469-480, 1999.

15. V. Khomenko. *Model Checking Based on Prefixes of Petri Net Unfoldings.* Ph. D. Thesis, School of Computing Science, University of Newcastle upon Tyne, 2003.
16. B. Luque and F. J. Ballesteros. Random Walk Networks. *Physica A: Statistical Mechanics and its Applications,* 342(1-2):207-213, 2004.
17. H. Matsuno, A. Doi, M. Nagasaki, and S. Miyano. Hybrid Petri net representation of gene regulatory network. *Pacific Symposium on Biocomputing,* 5:338-349, 2000.
18. T. Murata. Petri nets: properties, analysis and applications. *Proceedings of the IEEE,* 77(4):541–580, 1989.
19. V. N. Reddy, M. N. Liebman, M. L. Mavrovouniotis. Qualitative analysis of biochemical reaction systems. *Computers in Biology and Medicine,* 26(1):9-24, 1996.
20. D. Ropers, H. de Jong, M. Page, D. Schneider, and J. Geiselmann. Qualitative Simulation of the Nutritional Stress Response in *Escherichia coli.* INRIA, Rapport de Reacherche no. 5412, December 2004.
21. W. Reisig. *Petri Nets, An Introduction.* EATCS Monographs on Theoretical Computer Science, W.Brauer *et al* (Eds.), Springer–Verlag, Berlin, 1985.
22. W. Reisig and G. Rozenberg. *Lectures on Petri Nets I: Basic Models.* Advances in Petri Nets, Lecture Notes in Computer Science 1491, Springer-Verlag, 1998.
23. O. J. Shaw, C. Harwood, A. Wipat, and L. J. Steggles. SARGE: A tool for creation of putative genetic networks. *Bioinformatics,* 20(18):3638-3640, 2004.
24. O. J. Shaw, L. J. Steggles, and A. Wipat. Automatic Parameterisation of Stochastic Petri Net Models of Biological Networks. *Electronic Notes in Theoretical Computing Science,* 151(3):111-129, June 2006.
25. E. Simão, E. Remy, D. Thieffry, C. Chaouiya. Qualitative Modelling of Regulated Metabolic Pathways: Application to the Tryptophan Biosynthesis in *E. Coli. Bioinformatics* 21:190-196, 2005.
26. P. T. Spellman, G. Sherlock, M. Q. Zhang, V. R. Iyer, K. Anders, M. B. Eisen, P. O. Brown, D. Botstein, and B. Futcher. Comprehensive identification of cell cycle-regulated genes of the yeast *Saccharomyces cerevisiae* by microarray hybridization. *Molecular Biology of the Cell,* 9(12):3273-3297, 1998.
27. Z. Szallasi and S. Liang. Modeling the Normal and Neoplastic Cell Cycle with "Realistic Boolean Genetic Networks": Their Application for Understanding Carcinogenesis and Assessing Therapeutic Strategies. *Pacific Symposium on Biocomputing* Vol. 3: 66–76, 1998.
28. *Petri nets World,* http://www.informatik.uni-hamburg.de/TGI/PetriNets/, 2006.
29. *PEP Home Page,* http://parsys.informatik.uni-oldenburg.de/~pep/, 2006.

Regulatory Network Reconstruction Using Stochastic Logical Networks

Bartek Wilczyński[1,2] and Jerzy Tiuryn[2]

[1] Institute of Mathematics, Polish Academy of Sciences,
ul. Śniadeckich 8, 00-956 Warszawa, Poland
[2] Institute of Informatics, Warsaw University,
ul. Banacha 2, 00-089 Warszawa, Poland
{bartek, tiuryn}@mimuw.edu.pl,
http://bioputer.mimuw.edu.pl/

Abstract. This paper presents a method for regulatory network reconstruction from experimental data. We propose a mathematical model for regulatory interactions, based on the work of Thomas et al. [25] extended with a stochastic element and provide an algorithm for reconstruction of such models from gene expression time series. We examine mathematical properties of the model and the reconstruction algorithm and test it on expression profiles obtained from numerical simulation of known regulatory networks. We compare the reconstructed networks with the ones reconstructed from the same data using Dynamic Bayesian Networks and show that in these cases our method provides the same or better results. The supplemental materials to this article are available from the website http://bioputer.mimuw.edu.pl/papers/cmsb06

1 Introduction

Understanding the regulatory mechanisms of gene expression is one of the key problems in molecular biology. Since such mechanisms are extremely hard to study *in vivo*, many mathematical models were proposed to help understanding the principles of regulatory network operation. The pioneering work in the field of regulatory network modelling was done in the 1960s by S. Kauffman [9] who showed that such fundamental phenomena of gene regulation as epigenesis and stable convergence can be modelled with a very simple mathematical framework of Boolean networks. This model was extended by Rene Thomas and co-workers [24,25] leading to formulation of generalized logical description of regulatory networks. It allowed to verify important properties of homeostatic networks by examination of negative and positive feedback loops. Also some theorems concerning the correspondence between generalized Boolean models and dynamical systems were proved [20].

Generalized logical modeling approach was successfully applied to many experimentally studied biological regulatory circuits (e.g. [18,23,16,12,17,10,22]) showing that this formalism is well suited for representation of real biological networks. However, it is difficult to reconstruct such networks from experimental data. The problem with reconstruction of such networks lies in the lack of

C. Priami (Ed.): CMSB 2006, LNBI 4210, pp. 142–154, 2006.

a natural scoring function for different models for a given dataset. Even though there are computational ways to effectively simulate such models (i.e. GINSim software package [10]), it doesn't help us with choosing the right model from many possibilities.

On the other hand, one can approach the problem of gene network reconstruction on the basis of statistical data analysis. Particularly, there were many successful applications of Bayesian networks [6] to recover regulatory dependencies from expression data. For the particular case of reconstruction from expression time profiles, the formalism of dynamic Bayesian networks is more appropriate, however currently available results [8,5] indicate that it is difficult to obtain reliable predictions of network topology from unperturbed expression data.

Another approach to this problem is using a stochastic dynamical system to model the dependencies of expression levels of genes. Chen et al. [3] managed to reconstruct parameters for a system of stochastic differential equations from Yeast cell cycle expression data [21]. This was possible with the assumption that the expression of genes depends linearly on the expression of its regulators preventing the method to predict correct dependencies in cases where the regulation is non-linear. Beal et al.[2] proposed a similar approach using Bayesian methods to recover the State Space Model with hidden factors. Despite the fact that it is based on the assumed linear regulatory dependence the possibility of modelling networks with non-observable factors is promising.

Both mentioned approaches have a very interesting feature of their respective scoring functions: a penalizing factor for inclusion of too many edges. This is essential for these methods to prevent over-fitting, since it is known that the more dependencies we include in the model of such kind, the better we can fit it to the data. However it is well known that parsimony is not the correct criterion for selection of biological regulatory networks since the evolution selects the networks that are robust and redundant rather than parsimonious.

In this work we present a novel modelling framework for regulatory networks, called Stochastic Logical Networks (for brevity referred to as SLNs). It is based on the formalism of generalized logical description of networks as introduced by Thomas [24] extended with the stochastic factor leading to a simple and natural scoring function based on calculating the likelihood of the model given the observed data. We also provide an algorithm to find the most likely model for any given dataset and evaluate its performance on simple examples of data simulated from artificial feedback circuits (of size 2,3 and 4) modelling the homeostasis [25]. We compare the networks obtained from these data using our algorithm with the ones obtained using Dynamic Bayesian Networks. In all three cases, the presented method is able to reconstruct the topology of the original network, while using Dynamic Bayesian Networks it is possible only in two simpler cases.

2   Genetic Network Modelling

2.1   Gene Networks as Dynamical Systems

In order to describe the formalism of SLNs, we provide a brief introduction into the way of modelling introduced by R. Thomas [24]. It is based on the assumption that a regulatory system can be accurately described as a dynamical system of ordinary differential equations. We treat the state of the cell, i.e. concentrations of all interesting gene products, as a vector of non-negative real values $v = \langle v_1, v_2, \ldots, v_n \rangle \in R_+^n$, dependent on time t, so the equations have the following form:

$$\frac{\partial v_i}{\partial t} = -v_i \cdot \lambda_i + \mathcal{F}_i(v), \tag{1}$$

where $\mathcal{F}_i(v)$ denotes the production rate of gene v_i depending on the state of all genes, whereas λ_i represents the decay constant responsible for degrading the gene product proportionally to its current concentration. To account for non-linearity and combinatorial nature of the dependence of the production rate on the state of regulators, the production rate of gene i is defined by Snoussi [20] as a linear combination of products of sigmoid functions of expression of regulators:

$$\mathcal{F}_i(v) = \sum_{G \subseteq \{1 \ldots n\}} I_{G,i} \cdot \prod_{j \in G} S_{i,j}(v_j, \theta_{i,j}), \tag{2}$$

where $I_{G,i} \in \mathbb{R}^+$ is the regulatory influence of the set of regulators G on gene i, $\theta_{i,j} \in \mathbb{R}$ are the activation thresholds and $S_{i,j}$ is the sigmoid activation function of gene i by gene j being one of the following:

$$S^+(x, \theta) = sigm_k(x - \theta) \tag{3}$$
$$S^-(x, \theta) = 1 - sigm_k(x - \theta), \tag{4}$$

where $sigm_k(x) = (1 + e^{-kx})^{-1}$ with a notable case of $sigm_\infty$ equal to the Heaviside step function. Different forms of $S_{i,j}$ represent different possible regulatory interactions. If $S_{i,j} = S^-$ we say that j is a repressor of i, otherwise j is an enhancer of i.

2.2 Qualitative Approach

Rene Thomas observed, that qualitative behaviour of such systems can be modelled as a non deterministic discrete process whose states correspond to discretized states of the original dynamical system. This is due to the fact that the production rates of genes change substantially only around the threshold values $\theta_{i,j}$. If we consider the case with $sigm_\infty$ step function, the hyperplanes $v_j = \theta_{i,j}$ divide the phase space of the dynamical system into a finite set of disjoint parts on which the production rates of all genes are constant. In such case, the behavior of the system is determined by the choice of production rates of all genes in all discretized states. A simplistic example of a 2-gene negative feedback loop with its phase space, dependency graph and discrete state graph is depicted in Figure 1.

We use the notion of a discretization mapping $\delta(v) = \langle \delta_1(v_1) \ldots \delta_n \rangle$, where each δ_i is a mapping of the i-th variable into its discrete values using the thresholds $\{\theta_{i,j}\}_{j=1..n}$. We also denote the space of all discrete states by $\Sigma_n = \{0 \ldots n\}^n$

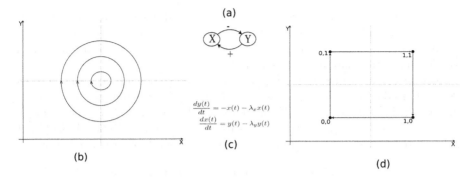

Fig. 1. An example of a dynamical system consisting of two genes: X and Y (a). Its phase space (b), ODEs (c) and state graph (d).

Evolution of the discretized system can be derived from equations governing the dynamical system. If we assume that $S_{i,j}$ are indeed Heaviside step functions, the value of the regulation functions $\mathcal{F}_i(v)$ is constant between the thresholds $\theta_{i,j}$. Therefore for each discrete state σ, corresponding to the discretization domain $\delta^{-1}(\sigma)$, there exist constant production rates $F_i(\sigma)$ of all genes, such that:

$$\forall_{v \in \delta^{-1}(\sigma)} \mathcal{F}_i(v) = F_i(\sigma).$$

However, the Heaviside regulatory functions introduce discontinuity in the right-hand-side of ordinary differential equations for points in the state-space where $v_j = \theta_{i,j}$. We call such points *singular*, and exclude them from further analysis. This can be done without loss of generality since the measure of the set of singular points is 0.

After Thomas[24], we use the notion of an image function $\mathcal{R}$ of a discrete state σ:

$$\mathcal{R}(\sigma) = \langle \delta_1(\frac{F_1}{\lambda_1}), \dots, \delta_n(\frac{F_n}{\lambda_n}) \rangle.$$

If a state is the image of itself, we call it *stable*. Otherwise, since there may be different trajectories of the dynamical system traversing this discretization domain, the state succession is non-deterministic. For each discrete non-stable state σ we define a set of *successor states* $succ_\sigma$ containing all neighbouring discrete states σ' such that there exists a trajectory in the dynamical system going from $\delta^{-1}(\sigma)$ directly to $\delta^{-1}(\sigma')$. We use the notation of $\sigma \to \sigma'$ to denote the fact that $\sigma' \in succ_\sigma$. The generalization of a successor state is its transitive closure: the *reachability* relation $\sigma \to^+ \sigma'$.

2.3   Network Reconstruction

Snoussi [20] showed that the qualitative approach is in strict correspondence with the dynamical system with respect to non-singular steady states i.e. the existence of a non-singular steady state in the dynamical system is equivalent to the existence of a steady state in the discretized system. We are more interested in the analysis of expression time-series data, so we need to focus on the more complex relationship between the deterministic dynamical system and non-deterministic qualitative approach.

Since we are interested in the topology of the dependencies between the variables, we need to be able to reconstruct the dependency graph given the image function $\mathcal{R}$. For this reason need a link between the function $\mathcal{R}$ and the topology of the network $G = \langle V, E \rangle$, where vertices correspond to variables $V = \{1..n\}$ and directed edges $\langle u, v \rangle \in E$ represent regulatory dependencies as follows:

$$\langle u, v \rangle \in E \text{ If and only if there exist states } \sigma_1, \sigma_2$$
$$\text{such that } \sigma_1(u) \neq \sigma_2(u) \text{ and for all genes } u' \neq u \tag{5}$$
$$\sigma_1(u') = \sigma_2(u') \text{ and } \mathcal{R}(\sigma_1)(v) \neq \mathcal{R}(\sigma_2)(v)$$

i.e. variable v depends on u ($\langle u, v \rangle \in E$) if and only if there is a state σ such that we can change its image at v just by changing the concentration of the gene u.

Our task is to reconstruct network topology, given the expression time-series. We assume that the data consists of one or more series of slides, each of which contains the discretized expression levels of all genes in the system at certain times. This can be directly mapped to our discrete model as a set of pairs of observed states $\mathcal{D} = \{\langle o_1, o_2 \rangle, \ldots, \langle o_{m-1}, o_m \rangle\}$. In order to reconstruct the model from such dataset, we need a measure of compatibility between the model and the dataset. We can say that model is a *realization* of a dataset if all the consecutive observed states are reachable in the model:

$$\forall_{\langle \sigma, \sigma' \rangle \in \mathcal{D}} \quad \sigma \rightarrow^+ \sigma'. \tag{6}$$

Such condition is unfortunately not sufficient for reconstruction of the correct network from data. There are just too many models for any dataset that satisfy the equation (6). What's even worse, we cannot apply the parsimony criterion choosing the simplest model for a given dataset because there always exist a very simple "chaotic" network realizing all possible trajectories.

2.4   Stochastic Logical Networks

We propose a solution to the problem of choosing the right realization for a given dataset based on the introduction of a stochastic factor to the differential equations. Doing this at the level of a dynamical system and not at the level of a discrete model, as in other approaches such as Probabilistic Boolean Networks [19], seems more natural since the randomness in biological processes comes from small fluctuations in continuous quantities. To introduce the stochasticity into our system, we follow the common practice of adding white noise in the form of an independent Wiener processes $W_i(t)$ scaled for each variable by ϵ_i to the right-hand side of differential equations (1) obtaining the following

$$dv_i = (\mathcal{F}_i(\boldsymbol{v}) - v_i \cdot \lambda_i)dt + \epsilon_i dW_i(t). \tag{7}$$

If we use the regulation functions $\mathcal{F}_i$ as defined in equation (2) and apply the same discretization mapping δ we obtain a stochastic system consisting of a finite set of disjoint domains. In each of this domains we observe a multivariate Brownian motion with linear drift – a very well studied mathematical model. What we are interested in, is a discrete stochastic process representing the movements of the Brownian motion between domains through time that correspond to the changes of qualitative behaviour of the whole system. For our considerations it is important to find the relationship between the parameters of the discrete process and the dynamical system. Since it is not tractable to analyze the dynamics of such systems in general, we make a simplifying assumption that our process is Markovian i.e. that the probability of moving from one domain to another depends solely on the current discrete state. It is important to note that it is exactly the same assumption that gives raise to the analysis of the successor states in the qualitative approach by Thomas. Given parameters of the dynamical model, probability distribution on the neighbouring states for all states of the discrete Markov process can be calculated. Once we have this distribution, we can consider the obtained Markov chain as a reference model for the regulatory network. This allows us to reformulate our problem into finding the most likely Markov model given the observed trajectories.

2.5   SLN Models and Experimental Data Discretization

As we have noted in our previous work on Dynamic Bayesian Networks reconstruction [5] the parameters of the discretization procedure have a strong impact on the results of the network reconstruction. Since we assume in this approach that we are given already discretized data it follows that we should not rely on the correctness of the discretization process itself. The choice of the discretization thresholds heavily influence the behavior of our model. For this reason, we treat the discretized data as the observations of the trajectories of a Hidden Markov Model whose states correspond to qualitative states of the network.

3   Network Topology Reconstruction from Expression Time-Series

After explaining the rationale behind our methodology we can present the proposed algorithm for network topology reconstruction. We assume that we are given time-series of discretized expression profiles and we try to find the topology of the most likely SLN model. Our method consists of the following three steps:

1. Estimation of the HMM parameters using modified Baum-Welch [1] algorithm,
2. Reduction of the observation probability matrix to the most likely matching between states and observations,
3. Finding the topology of the SLN, given the transition probabilities

which are described in detail in the following sections.

3.1 HMM Reconstruction

The problem of HMM parameter estimation can be stated as follows: Given a set of states $S = \{s_1, \ldots, s_l\}$, a set of possible observations $\Sigma = \{\sigma_1, \ldots, \sigma_m\}$, and a set of observed trajectories (encoded as pairs of consecutive states) $\mathcal{O} = \{\langle o_1, o_2 \rangle, \ldots \langle o_{k-1}, o_k \rangle\}$, estimate the most probable HMM consisting of the transition probability matrix $T = \langle t_{ij} \rangle_{1 \leq i, j \leq l}$ and the observation matrix $O = \langle p_{ij} \rangle_{1 \leq i \leq l, 1 \leq j \leq m}$ where t_{ij} is the probability of a transition from state s_i to s_j and p_{ij} denotes the probability of observing the symbol σ_j while in state s_i. This problem has been thoroughly studied [15] and there is no known way of solving it analytically. However a solution can be approximated with the well-known Baum-Welch algorithm [1] which belongs to the class of Expectation-Maximization (EM) heuristic algorithms.

Since we are dealing with n genes, both the observations O and states S correspond to discretized states of the network, so they can be represented as vectors of length n of discrete variable states. We recall, that in a SLN consisting of n genes, each variable can have at most $n + 1$ discrete states (induced by at most n thresholds), we can use the set $\{0, 1, \ldots, n\}$ for encoding these states. However our case is different from the classical one in an important aspect. Since the model explicitly can change the state of only one variable at a time we cannot assume (as it is often done with Dynamic Bayesian Networks) that we observe all consecutive states on a trajectory. Instead, we have to take into account the possibility that some consecutive observations $\langle o_i, o_{i+1} \rangle \in O$ are not adjacent in the discrete state space. In such cases we decided to remove such pairs from our dataset and replace them with observations of consecutive states on every possible shortest path from o_i to o_{i+1}. This method is very simple and leads to a natural distribution of transition probabilities as shown by the example in Figure 2.

It is clear, that after such pre-processing step, the Baum-Welch algorithm always converges with probability of "jumping" from observed states directly to a non-adjacent state close[1] to 0, which is consistent with the definition of SLNs.

3.2 Reducing the Observation Matrix

Since the procedure of HMM parameter estimation is an EM algorithm, it converges to a local minimum of the likelihood function. The problem, however, with the interpretation of the result is the fact that the value of the likelihood function is insensitive to permuting the labels of the states of the HMM. We interpret the states of the HMM as the qualitative states of the system and assume that there is a $1 - 1$ correspondence between states and observations but the HMM estimation gives us only the matrices and not the correspondence relationship $r : \Sigma_n \to \Sigma_n$.

Once we have completed the HMM estimation from a given dataset, we need to reconstruct that mapping from the observation probability matrix O. The problem

---

[1] It is never exactly 0, since the algorithm itself relies on the ergodicity of the Markov chain.

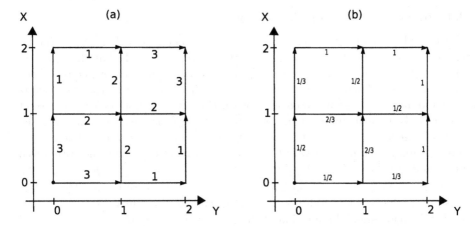

Fig. 2. Example of HMM reconstruction from non-adjacent observations. We consider a 2−gene SLN and observation set $\{\langle (0,0), (2,2)\rangle\}$. We can see the observation counts for all edges given by our procedure (a) and the resulting HMM transition probabilities (b).

of finding the most probable permutation of states given the matrix O is an instance of the well known problem of finding the maximum weight bipartite matching. We can consider the matrix O as a weight matrix in the fully connected bipartite graph between states and observations. Finding such a matching, which gives us the $1-1$ correspondence we need, can be solved in polynomial time $\mathcal{O}(n^2 \log n)$ [11]. Once we calculate the best matching between the states of the HMM and the observations, we can label the states with the observations and obtain a regular Markov Model, which can be interpreted as a SLN.

As an interesting by-product of this procedure we can calculate the matching quality

$$q(r, O) = \prod_{\sigma \in \Sigma_n} o_{\sigma, r(\sigma)}.$$

It can be interpreted as a measure (ranging from 0 to 1) of the quality of the discretization procedure used to obtain the data. The higher the score, the more confident we are that the discretization procedure matches the qualitative behaviour of the system. It is especially important in the case of real biological data, where we have no simple measures of the discretization quality. It is also possible, that different HMMs with the same likelihood have different matching quality. This leads to a modification of the algorithm: since we are interested in finding a HMM with high quality matching, we employ the multi-start procedure to find multiple locally optimal HMMs and select the one with the best matching quality.

It is important to properly discern between the matching quality q and the likelihood of the HMM model. The latter is the likelihood of the HMM model given the observed discretized data. The matching quality is the trace of the observation matrix permuted according to the matching r used only to select the best matching model among the ones with the best HMM likelihood.

3.3 Identifying Network Topology

Given a SLN model we want to uncover the dependencies between variables encoded in the discrete Markov process states. We can recall that the topology of the network was defined by the equivalence (5) using the properties of the image function. Unfortunately, without the full knowledge of the parameters of our dynamical system (such as noise ratios), we cannot recover the exact topology. In case of multidimensional SLN, the probabilities of changing the state of a gene i in a given direction while being in state σ can be influenced by different image of σ in other variables. However, when we consider the normalized probabilities $\hat{p}_i^{+}(\sigma) = p_i^{+}(\sigma)/p_i^{+}(\sigma) + p_i^{-}(\sigma)$ and $\hat{p}_i^{-}(\sigma) = p_i^{-}(\sigma)/p_i^{+}(\sigma) + p_i^{-}(\sigma)$, where $p_i^{+/-}(\sigma)$ denotes the probability of increase/decrease in variable i while in state σ, the problem reduces to an independent one-dimensional case. This leads to a reformulated definition of the network topology (5):

$$\langle i, j \rangle \in E \text{ if and only if there exist states } \sigma_1, \sigma_2$$
$$\text{such that } \sigma_1(i) \neq \sigma_2(i) \text{ and for all genes } i' \neq i \qquad (8)$$
$$\sigma_1(i') = \sigma_2(i') \text{ and } \hat{p}_j^{+}(\sigma_1) \neq \hat{p}_j^{+}(\sigma_2).$$

We can safely compare only the probabilities of increase: $\hat{p}_i^{+}$, since $\hat{p}_i^{-}(\sigma) = 1 - \hat{p}_i^{+}(\sigma)$. Using this definition would not be practical. Since we are dealing with numerical solutions we needed to weaken the sharp inequality in (8) by using a suitably chosen threshold d:

$$\langle i, j \rangle \in E \text{ if and only if there exist states } \sigma_1, \sigma_2$$
$$\text{such that } \sigma_1(i) \neq \sigma_2(i) \text{ and for all genes } i' \neq i \qquad (9)$$
$$\sigma_1(i') = \sigma_2(i') \text{ and } |\hat{p}_j^{+}(\sigma_1) - \hat{p}_j^{+}(\sigma_2)| > d.$$

Using the definition (9) we can reconstruct the topology of the whole network in a single scan of the transition matrix of a given Markov Chain.

4   In Silico Experiments

In order to test the performance of a network reconstruction method, one needs a proper dataset taken from a network with known topology. For this reasons, we decided to use artificial networks and simulate the expression profiles. This kind of evaluation can tell us exactly where are the errors, both in terms of false positives and false negatives. We have chosen three negative feedback loops (see Fig. 3 of size 2,3 and 4, presented in the book by Thomas and d'Ari [25] and used our stochastic dynamical model to simulate the expression time-series for these systems. Next we have discretized the data and sampled them in order to obtain a dataset large enough to reconstruct the topology. We tried to reconstruct the topology of regulatory networks using our method and compared it to the results obtained by the more established framework of dynamic Bayesian networks. In the following sections we describe the methodology used for artificial data generation and discuss the results of the reconstruction.

4.1   Simulating Expression Data

We have numerically simulated (using methods described by Higham [7] implemented in GNU Octave system) stochastic differential models for the three models presented in Figure 3. Our experiments with different noise-to-signal ratios (data not shown), verified that these circuits are very noise-resistant so we used the noise-to-signal ratio equal to 3/2 in our experiments. In order to make our simulations closer to reality of DNA-array experiments we follow the common practice of averaging over a number of independent trajectories of the system started from the same state. The trajectories were then discretized into binary discrete values (above and below mean observed value) and sampled (one slide after each state change) in order to obtain reasonably sized datasets.

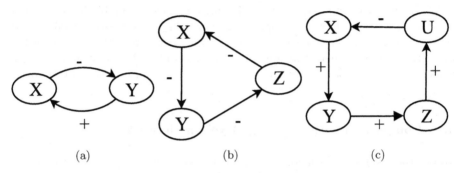

Fig. 3. Topology of the simulated networks. 2-gene (a), 3-gene (b) and 4-gene (c) negative feedback loops. Using our method, we have reconstructed exactly the same topology in all three cases.

An example of the simulated averaged trajectories is shown in Figure 4. It is interesting, that the strong effect of random noise visible in single trajectories is diminishing as the number of averaged trajectories is increased (with no noticeable change above 100). Also, another phenomenon can be observed: in the averaged trajectories the amplitude of the changes decreases in time (which is not observed in single trajectories). This is due to the lack of synchronization in time among the trajectories which corresponds to the behavior of cell lines used for production of expression time-series data.

4.2   Reconstructing Feedback Loops

Since we want to test the ability of our algorithm to reconstruct network topology from simulated time-series data, we need a network displaying oscillatory behavior. Thomas [25] observed, that negative feedback loop is a necessary feature of networks showing such behavior. We have chosen the three negative feedback loops analyzed by Thomas [25] depicted in Figure 3 as the simplest test set for our evaluation.

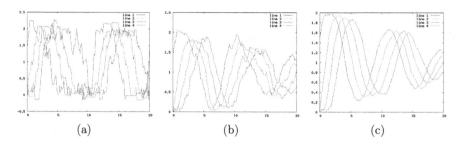

Fig. 4. Simulated trajectories of the 4-gene feedback loop with the noise to regulation ratio is 3/2 and different number of averaged cells: 1:(a), 10:(b) and 100(c)

We have constructed stochastic differential models (available in the supplemental material at `http://bioputer.mimuw.edu.pl/papers/cmsb06`) for all those systems and obtained the synthetic time-series from them as described in Section 4.1. Our algorithm was able to reconstruct the topology of all networks and assign labels to the edges correctly for the noise-to-signal ratio up to 3/2.

4.3   Comparison with Dynamic Bayesian Networks

To put the results of our algorithm into perspective we compare it with another approach. Since we are dealing with feedback loops, we cannot use the most often used formalism of Bayesian Networks because they cannot represent cyclic dependencies. For this reason we compare our results with Dynamic Bayesian Networks, a modification of Bayesian Networks approach designed specifically for this task by Murphy [13]. The problem of inferring Bayesian Networks from data is NP-hard [4] however, since the feedback loops considered are very small, we can employ the exact algorithm proposed by Ott [14]. in Figure 5 we present the models obtained using this procedure (using BDe [6] scoring function) on the same data as described in the previous section.

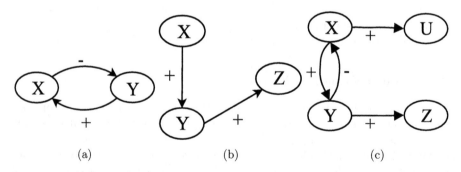

Fig. 5. Reconstruction of the networks from Figure 3 using dynamic Bayesian networks with the BDe scoring function

5   Conclusions

In this work we propose a new approach to the important problem of regulatory network reconstruction. It is based on the well established formalism of qualitative analysis introduced by Thomas extended by introducing a stochastic component. It provides us with a continuous space of possible models and a natural likelihood function allowing us to choose the one that best fits the data. The use of Hidden Markov Models to account for the uncertainty of the discretization quality gives us an external measure of the quality of discretization based on the estimated model. It may help us to identify wrong discretization of the data as well as choose the best model from many locally optimal solutions. The experiments show that this indeed leads to a better estimation of small feedback loops than it is possible with Dynamic Bayesian Networks (with the most commonly used scoring functions). It shows that this method has a potential for eliminating the need for parsimony criterion essential for Bayesian networks and, due to the close relation with dynamical systems, for analyzing the dynamics of the reconstructed models.

The main drawback of the method is currently the need to estimate the model with the number of parameters which is exponential in the number of genes. Our current work focuses on the possibilities of limiting the number of required parameters by excluding from our considerations the ones that cannot be estimated from the given data.

Acknowledgments

We would like to thank Norbert Dojer, Anna Gambin and Jacek Miekisz for fruitful discussions that contributed to this work. This work was supported by the Polish Ministry of Science and Education (grants No 3 T11F 021 28 and 3 T11F 022 29).

References

1. L. E. Baum, T. Peterie, G. Souled, and N. Weiss. A maximization technique occurring in the statistical analysis of probabilistic functions of Markov chains. *Ann. Math. Statist.*, 41(1):164–171, 1970.
2. Matthew J Beal, Francesco Falciani, Zoubin Ghahramani, Claudia Rangel, and David L Wild. A Bayesian approach to reconstructing genetic regulatory networks with hidden factors. *Bioinformatics*, 21(3):349–356, Feb 2005. Evaluation Studies.
3. KC Chen, TY Wang, HH Tseng, CY Huang, and CY Kao. A stochastic differential equation model for quantifying transcriptional regulatory network in Saccharomyces cerevisiae. *Bioinformatics*, 21(12):2883–90, Jun 2005.
4. D.M. Chickering. Learning Bayesian networks is NP-complete. *Proceedings of AI and Statistics*, 1995, 1995.
5. Norbert Dojer, Anna Gambin, Bartek Wilczynski, and Jerzy Tiuryn. Applying dynamic Bayesian networks to perturbed gene expression data. *BMC Bioinformatics*, 7, 2006.

6. Nir Friedman, Michal Linial, Iftach Nachman, and Dana Pe'er. Using Bayesian networks to analyze expression data. *Journal of Computational Biology*, 7:601–620, 2000.
7. Desmond J. Higham. An algorithmic introduction to numerical simulation of stochastic differential equations. *SIAM Reviam*, 43(3):525–546, 2001.
8. Dirk Husmeier. Sensitivity and specificity of inferring genetic regulatory interactions from microarray experiments with dynamic Bayesian networks. *Bioinformatics*, 19(17):2271–82, 2003.
9. S.A. Kauffman. Homeostasis and differentiation in random genetic control networks. *Nature*, 1969.
10. A Larrinaga, A Naldi, L Sanchez, D Thieffry, and C Chaouiya. GINsim: A software suite for the qualitative modelling, simulation and analysis of regulatory networks. *Biosystems*, Jan 2006.
11. K. Mehlhorn and St. Naeher. *The LEDA Platform of Combinatorial and Geometric Computing*. Cambridge University Press, 1999.
12. L. Mendoza, D. Thieffry, and E.R. Alvarez-Buylla. Genetic control of flower morphogenesis in Arabidopsis thaliana: a logical analyssis. *Journal of Theoretical Biology*, 1999.
13. K. Murphy and S. Mian. Modelling gene expression data using dynamic Bayesian networks. *University of California, Berkeley*, 1999.
14. S. Ott, S. Imoto, and S. Miyano. Finding optimal models for gene networks. In *Proc. of Pacific Symposium in Biocomputing*, 2004, in press.
15. Lawrence R. Rabiner. A tutorial on Hidden Markov Models and selected applications in speech recognition. *Proceedings of the IEEE*, 77(2):257–286, 1989.
16. L. Sanchez and D. Thieffry. A logical analysis of the Drosophila gap-gene system. *J Theor Biol*, 211(2):115–141, Jul 2001.
17. L Sanchez, J van Helden, and D Thieffry. Establishement of the dorso-ventral pattern during embryonic development of drosophila melanogasater: a logical analysis. *J Theor Biol*, 189(4):377–389, Dec 1997.
18. Lucas Sanchez and Denis Thieffry. Segmenting the fly embryo: a logical analysis of the pair-rule cross-regulatory module. *J Theor Biol*, 224(4):517–537, Oct 2003.
19. Ilya Shmulevich, Edward R Dougherty, Seungchan Kim, and Wei Zhang. Probabilistic Boolean Networks: a rule-based uncertainty model for gene regulatory networks. *Bioinformatics*, 18(2):261–274, Feb 2002.
20. El Houssine Snoussi. Qualitative dynamics of piecewise-linear differential equations: a discrete mapping approach. *Dynamics and stability of systems*, 4(3-4):189–207, 1989.
21. Paul T. Spellman, Gavin Sherlock, Michael Q. Zhang, Vishwanath R. Iyer, Kirk Anders, Michael B. Eisen, Patrick O. Brown, David Botstein, and Bruce Futcher. Comprehensive identification of cell cycle regulated genes of the yeast Saccharomyces cerevisiae by microarray hybridization. *Molecular Biology of the Cell*, 9(12):3273–3297, Dec 1998.
22. Denis Thieffry and Lucas Sanchez. Alternative epigenetic states understood in terms of specific regulatory structures. *Ann N Y Acad Sci*, 981:135–153, Dec 2002.
23. Denis Thieffry and Lucas Sanchez. Dynamical modelling of pattern formation during embryonic development. *Curr Opin Genet Dev*, 13(4):326–330, Aug 2003.
24. Rene Thomas. Boolean formalization of genetic control circuits. *Journal of Theoretical Biology*, 42:563, 1973.
25. Rene Thomas and Richard D'Ari. *Biological Feedback*. CRC Press, 1990.

Identifying Submodules of Cellular Regulatory Networks

Guido Sanguinetti[1], Magnus Rattray[2], and Neil D. Lawrence[1]

[1] Department of Computer Science, University of Sheffield
211 Portobello Street, Sheffield S1 4DP, UK
{guido, neil}@dcs.shef.ac.uk
[2] School of Computer Science, University of Manchester
Oxford Road, Manchester M13 9PM, UK
magnus@cs.man.ac.uk

Abstract. Recent high throughput techniques in molecular biology have brought about the possibility of directly identifying the architecture of regulatory networks on a genome-wide scale. However, the computational task of estimating fine-grained models on a genome-wide scale is daunting. Therefore, it is of great importance to be able to reliably identify submodules of the network that can be effectively modelled as independent subunits. In this paper we present a procedure to obtain submodules of a cellular network by using information from gene-expression measurements. We integrate network architecture data with genome-wide gene expression measurements in order to determine which regulatory relations are actually confirmed by the expression data. We then use this information to obtain non-trivial submodules of the regulatory network using two distinct algorithms, a naive exhaustive algorithm and a spectral algorithm based on the eigendecomposition of an affinity matrix. We test our method on two yeast biological data sets, using regulatory information obtained from chromatin immunoprecipitation.

1   Introduction

The modelling of cellular networks has undergone a revolution in recent years. The advent of high throughput techniques such as microarrays and chromatin immunoprecipitation (ChIP [1,2]) has resulted in a rapid increase in the amount of data available, so that it is possible to measure on a genome-wide scale both the expression levels of thousands of genes and the architecture (connectivity) of the regulatory network which links genes to their regulators (transcription factors). However, this data is often very noisy, and the sheer amount of data makes the development of quantitative fine grained models impossible.

Gene networks are frequently modelled in very different ways at different scales [3]. Network modelling at the genome-wide scale is often limited to the topology of networks. For example, Luscombe *et al.* used a large database constructed by integrating all available data on transcriptional regulation from a variety of sources (ChIP-on-chip, protein interaction data, *etc.*) to model the

C. Priami (Ed.): CMSB 2006, LNBI 4210, pp. 155–168, 2006.
© Springer-Verlag Berlin Heidelberg 2006

changes in the topology of the yeast regulatory network in different experimental conditions [4]. While this result was *per se* of great importance in furthering our understanding of transcriptional regulation, it is not clear how this approach could be used to model the *dynamics* of the system. At the other end of the spectrum [5], small networks consisting of a few transcription factors and their established target genes are often modelled in a realistic fine grained way, allowing for a quantitative explanation of qualitative behaviours in the cellular processes such as cycles, spatial gradients, *etc.*

While these fine grained models are often very successful in describing specific processes, they rely on rather strong assumptions. First of all, they need the regulatory links they exploit to be *true* regulations. While there is a growing number of experimentally validated regulatory relations in a number of organisms, the main techniques to study regulatory networks on a genome-wide scale are ChIP-on-chip [1] and motif conservation [6]. However, it is well known that ChIP-on-chip only measures the binding of a transcription factor to the promoter region of the gene. While binding is obviously a necessary condition for transcription to be initiated, there is abundant biological evidence [7] that shows that it is not a sufficient condition. Therefore, we may expect that interpreting ChIP-on-chip data as evidence for regulation may lead to many false positives, which would obviously be a big problem for any fine grained model. As for motif conservation, it is often difficult to assign a motif to a unique transcription factor and large numbers of false positives can be expected. Secondly, the system modelled should be reasonably isolated from the rest of the cell. Often collateral processes are simply modelled as noise in fine grained models, and this approximation would clearly break down in the presence of strong interactions with variables not included in the model.

We recently presented a probabilistic dynamical model which allowed us to infer both the active transcription factor protein concentrations and the intensity of the regulatory links between transcription factors and their target genes [8,9]. The model was computationally efficient so that the network could be modelled at the genome level, and its probabilistic nature meant that we could estimate the whole *probability distribution* of the concentrations and regulatory intensities, rather than just providing point estimates. This means that the significance level of the regulatory interactions could be assessed. This information can be used in many ways: for example, one may use it to obtain a refinement of the ChIP data, so that regulatory relations below a certain significance threshold are effectively treated as false positives. However, the information about the absolute value of the regulatory intensity is also of interest, since low intensity regulations (however significant) could be ignored when trying to obtain submodules of manageable size.

The main novelty of this paper is to present two algorithms to obtain submodules of regulatory networks. The first algorithm is a simple exhaustive search algorithm. While in principle this is applicable to any network with binary connectivity, it obtains biologically relevant submodules when applied to a network comprising significant regulations only. The second algorithm is a spectral method based on an eigenvalue decomposition of an affinity matrix and on a generalisation of the spectral clustering algorithm described in [10]. This takes into account the absolute value of the regulatory intensity and has the advantage of providing a natural way of ranking the submodules according to their importance in the global cellular network.

The paper is organised as follows: we first briefly review the probabilistic model used to infer the regulatory intensities. We then present the two algorithms to identify submodules of the regulatory network. In the results section we demonstrate our approach on two yeast data sets, the benchmark cell cycle data set of [11] and the more recent metabolic cycle data set of [12]. Finally, we discuss the relative merits of the two algorithms we proposed and their validity as an alternative approach to existing graph clustering algorithms.

2   Quantitative Inference of Regulatory Networks

Here we briefly review the probabilistic dynamical model for inference of regulatory networks proposed in [9]. This builds on the model presented in [8], which in turn extends the linear regression approach, first introduced in [13], to take into account gene-specific effects. We have (log transformed) expression level measurements y_{nt} for N genes at T time points. We assume that the binding of q transcription factors to the N genes is known (for example via ChIP-on-chip experiments), so that we have a binary matrix X whose nm entry X_{nm} is one if gene n is bound by transcription factor m and zero otherwise. We can then write down our model as

$$y_{nt} = \sum_{m=1}^{q} X_{nm} b_{nm} c_{mt} + \mu_n + \epsilon_{nt}. \tag{1}$$

Here b_{nm} represents the regulatory intensity with which transcription factor m enhances gene n (negative intensity models repression), c_{mt} models the (log) active protein concentration of transcription factor m at time t, μ_n is the baseline expression level of gene n and $\epsilon_{nt} \sim \mathcal{N}\left(0, \sigma^2\right)$ is an error term.

The model is then specified by a choice of prior distributions on the random variables b_{nm}, c_{mt} and μ_n. We assign spherical Gaussian priors to the regulatory intensities and the baseline expression level

$$b_{nm} \sim \mathcal{N}\left(0, \alpha^2\right)$$

$$\mu_n \sim \mathcal{N}\left(\tau, \beta\right).$$

The choice of prior distribution on the concentrations c_{mt} depends on the specific biological situation we wish to model. For example, for independent samples we may assume that the prior distribution on c_{mt} factorises along time t. As we are going to model time series data, an appropriate choice for the prior distribution on c_{mt} is a time-stationary Markov chain

$$c_{mt} = \gamma_m c_{m(t-1)} + \eta_{mt}$$

$$\eta_{mt} \sim \mathcal{N}\left(0, 1 - \gamma_m^2\right) \tag{2}$$

$$c_{m1} \sim \mathcal{N}\left(0, 1\right).$$

The variance in (2) is chosen so that the process is stationary, *i.e.* the expected changes over a period of time Δt depend only on the length of the time interval, not on its starting or finishing point. The parameters $\gamma_m \in [0, 1]$ model the temporal continuity of the sequence c_{mt}. Values of γ_m close to 1 lead to smoothly varying samples, with contiguous time points having very similar values of concentration. On the other hand, low values of γ_m lead to samples with little correlation among time points, so that in the limit of $\gamma_m = 0$ the modelling situation of independent time points is recovered.

Having selected prior distributions for the latent variables b_{nm}, c_{mt} and μ_n we can use equation (1) to compute a joint likelihood for all the latent and observed variables

$$\begin{aligned} p\left(y_{nt}, b_{nm}, c_{mt}, \mu_n | X\right) &= \\ &= p\left(y_{nt} | b_{nm}, c_{mt}, \mu_n, X\right) p\left(b_{nm} | \alpha\right) p\left(c_{mt} | \gamma_m\right) p\left(\mu_n | \tau, \beta\right). \end{aligned} \tag{3}$$

We can then estimate the hyperparameters α, γ_m, σ, τ and β by type II maximum likelihood. Unfortunately, exact marginalisation of equation (3) is not possible and we have to resort to approximate numerical methods. This can be done *e.g.* using a variational EM algorithm as proposed in [9], where details of the implementation are given.

Once the hyperparameters have been estimated, we can obtain the posterior distribution for the latent variables given the data using Bayes' theorem

$$p\left(b, c, \mu | y\right) = \frac{p\left(y | b, c, \mu\right) p\left(b, c, \mu\right)}{\int p\left(y, b, c, \mu\right) db dc d\mu}. \tag{4}$$

3   Identifying Submodules

3.1   Naive Approach

Given the posterior probability on the regulatory intensities b_{nm}, one can associate a significance level to each regulatory interaction by considering the ratio between the posterior means and the associated standard deviations. One can then obtain a refined network structure comprising only of significant regulatory relations by considering only relations above a certain significance threshold (which can be viewed as the only parameter in this algorithm). It is then straightforward to find submodules in a regulatory network with binary connectivity. One can start with any transcription factor and subsequently include other transcription factors which have common targets with the first one. This can be iterated and it will obviously converge to a unique set of submodules. This procedure is schematically described in Algorithm 1.

Algorithm 1. Identify submodules of a network with binary connectivity

---

Input data: set R of regulators, set G of genes, regulatory intensities b_{nm};
Construct a binary connectivity matrix X by thresholding the intensities
repeat
    Choose a regulator $r_1 \in R$. Include the set of all its target genes $G_{r1} \subset G$;
    repeat
        Include the set of regulators other than r_1 regulating genes in G_{r1}, $R_{G_{r1}} \subset R$;
        Include all genes regulated by $R_{G_{r1}}$ not included in G_{r1};
    until No new genes are found;
    Output reduced sets R_m, G_m for the submodule and $\bar{R}, \bar{G}$ for the elements not
    included in the submodule;
until $\bar{R}, \bar{G}$ are the empty set.

3.2 Introducing the Regulatory Intensities

The main drawback of the procedure outlined in Algorithm 1 is that it does not take into account the information about the regulatory intensities, apart from using it as a guideline to introduce thresholds of significance. Specifically, it only exploits the outputs of the probabilistic model in order to obtain a refinement of the network architecture, which is only a minimal part of the information contained in the posterior distribution over b_{nm}.

However, when trying to identify submodules considering all the available information on the regulatory intensities, we may find that there are few truly independent submodules, and it might be hard to manually determine which submodules are approximately independent. In practice, we would like to be able to have an automated way to obtain submodules.

Since our probabilistic model reconstructs transcription factors concentrations and regulatory intensities from time-course microarray data, we can interpret the regulatory strengths as a measure of the involvement of a transcription factor in the cellular processes in which its target genes participate. A standard technique for retrieving genes associated with (approximately independent) cellular processes is PCA (also known as SVD, [14]). However, the *eigengenes* retrieved by PCA are not necessarily disjoint in terms of gene participation, in particular the same genes can be represented in different eigengenes, mirroring the biological fact that the same genes can participate in more than one cellular process. While this constitutes an important piece of information in its own right, it could be a drawback from the point of view of identifying independent submodules. We therefore propose a modified algorithm which extends the spectral clustering algorithm developed in [10].

Given the posterior distribution over the regulatory intensities

$$p\left(b_{nm}|y\right) \sim \mathcal{N}\left(b_{nm}|\bar{b}_{nm}, \sigma^2_{b_{nm}}\right)$$

we construct an affinity matrix C between transcription factors using the formula

$$C_{ij} = \left|\langle \mathbf{b}_i^T \rangle\right| \left|\langle \mathbf{b}_j \rangle\right|. \tag{5}$$

Algorithm 2. Identifying transcription factors associated with submodules of a network using the regulatory intensities.

Input data: affinity matrix A;
repeat
    Compute the eigendecomposition of A, giving eigenvalues λ_i and eigenvectors $E = \{\mathbf{e}_i\}$, $i = 1, \ldots, q$;
    Define $B = \{\mathbf{e}_1\}$, $\bar{B} = E - B$
    If $\mathbf{e}_i \in \bar{B}$ is such that $|\mathbf{e_j}|^T |\mathbf{e}_i| = 0 \quad \forall \mathbf{e}_j \in B$, include $\mathbf{e_i}$ in B;
until No such $\mathbf{e}_i$ can be found

Here, $\langle \mathbf{b}_i \rangle$ denotes the posterior expectation of the vector containing the regulatory intensities with which transcription factor i influences all the genes in the genome (set to zero for genes that are not bound by that transcription factor). We use the absolute value of the intensity since for the purpose of identifying submodules we are not interested in the sign of the regulation. According to this formula, then, two transcription factors will have high similarity if they coregulate with high intensity a large number of target genes.

If we assume that there are p independent submodules, with strong internal links, the affinity matrix (5) will be have p blocks on the diagonal (up to a reordering of the rows and columns) showing a very high internal covariance, while the remaining off-diagonal entries will be much smaller. By identifying these blocks, one can then obtain the transcription factors involved in the submodules. The blocks can be obtained by noticing that, for a non-degenerate spectrum (which holds with probability 1), the eigenvectors of C will present a block structure too, so that eigenvectors pertaining to different blocks will have non-zero entries in different positions. By selecting exactly one eigenvector per each block we obtain a set of *clustering eigenvectors*[1], and we can obtain the transcription factors belonging to different modules by considering the nonzero entries of the clustering eigenvectors. Furthermore, the eigenvalues associated with the clustering eigenvectors are monotonically related to the total regulatory intensity associated with the submodule (the sum of all the regulatory intensities of all the links in the network). Therefore, we can use the eigenvalues to rank the various submodules in terms of their importance in the overall network. A strategy to identify the submodules can therefore be obtained as outlined in Algorithm 2.

If the modules are not exactly independent, but links between modules are characterised by low regulatory intensity, we can introduce a sensitivity parameter θ and replace step 3 in algorithm 2 by $|\mathbf{e_j}|^T |\mathbf{e}_i| < \theta \quad \forall \mathbf{e}_j \in B$. As the eigenvectors of a matrix with non-degenerate spectrum are stable under perturbations, we are guaranteed that, for suitably small choices of θ, approximately independent submodules will be found.

In practice, it is often the case in biological networks that there are few submodules of the regulatory network active in a given experimental condition, so

---

[1] The name is chosen for their analogy with spectral clustering [10].

that we may expect the submodules identified by the clustering eigenvectors with highest associated eigenvalue to be biologically relevant, while submodules associated with small eigenvalues will be less relevant.

The simplicity of the algorithm leads to several advantages. For example, by considering the eigenvectors of the dual matrix

$$K_{lp} = \sum_{i=1}^{q} |\langle b_{li}\rangle|\,|\langle b_{pi}\rangle|\,, \tag{6}$$

one can retrieve the genes involved in the submodules.

4 Results

4.1 Data Sets

We tested our method on two yeast data sets, the benchmark cell cycle data set of [11] and the recent metabolic cycle data set of [12]. These data sets were analysed in our recent studies [8,9]. The connectivity data we used in both cases was obtained using ChIP: for the metabolic cycle data, we used the recent ChIP data of [1], while for the cell cycle data we chose to use the older ChIP data of [2] since this combination has been extensively studied in the literature [15, and references therein]. The ChIP data is continuous, but, following the suggestion of [2], we binarised it by giving a one value when the associated p-value was smaller than 10^{-3}. This was shown in [8] to be a reasonable choice of cut-off, as it retained many regulatory relations while keeping the number of false positives reasonable.

4.2 Cell Cycle Data

Spellman *et al.* [11] used cDNA microarrays to monitor the gene expression levels of 6181 genes during the yeast cell cycle, discovering that over 800 genes are cell cycle-regulated. Cells were synchronised using different experimental techniques. We selected the cdc15 data set, consisting of 24 experimental points in a time sequence.

The connectivity data we used for this data set was that obtained by [2]. In this study, ChIP was performed on 113 transcription factors, monitoring their binding to 6270 genes.

We removed from the data set genes which were not bound by any transcription factor and transcription factors not binding any gene. We also removed the expression data of genes with five or more missing values in the microarray data, leaving a network of 1975 genes and 104 transcription factors.

For the purposes of identifying submodules, we are primarily interested in the regulatory intensities with which transcription factors regulate target genes. Therefore, we will use the model described in Section 2 to obtain posterior estimates for the regulatory intensities b_{nm}. Also, we will be interested primarily in nontrivial submodules, *i.e.* submodules involving more than one regulator.

Identifying submodules using the ChIP data. As ChIP monitors only the binding of transcription factors to promoter regions of genes, and not the actual regulation, we may expect that many true positives at the binding level are actually false positives at the regulatory level. For example, the ChIP data of [2], using a p-value of 10^{-3}, gives 3656 bindings involving 104 transcription factors and 1975 genes. However, if we consider the posterior statistics for the regulatory intensities, we see that most of these bindings are not associated with a regulatory intensity significantly different from zero. Specifically, only 1238 bindings are associated with a regulatory intensity greater than twice its posterior error (significant with 95% confidence), and only 749 are significant at 99% confidence level.

This large number of false positives is a serious problem when trying to identify submodules. For example, if we use the naive Algorithm 1 directly on the ChIP data, we obtain only one nontrivial[2] submodule involving 100 transcription factors and 1957 genes. Obviously, the usefulness of such information is very limited.

Identifying submodules using significant regulations. Things change dramatically if we construct a binary connectivity matrix by considering only significant regulatory relations. In order to avoid obtaining too large components, we fixed the thresholding parameter to be equal to four. At this stringent significance threshold the network size reduces significantly, as there are now 81 transcription factors regulating a total of 438 genes. More importantly, there are now nine distinct nontrivial submodules of the regulatory network, each involving between two and thirteen transcription factors.

The submodules identified are highly coherent functionally. To appreciate this, we follow [2] and group transcription factors into five broad functional categories according to the function of their target genes. These categories are cell cycle, developmental processes, DNA/RNA biosynthesis, environmental response and metabolism [2, see Figure 5 inset]. We then see that the largest submodule, consisting of 13 transcription factors regulating 117 genes, is largely made up of transcription factors functionally related to the cell cycle. In fact, all of the active transcription factors functionally related to the cell cycle (with the exception of SKN7 and SWI6 which are not involved in any nontrivial module) belong to this submodule. These are ACE2, FKH1, FKH2, MBP1, MCM1, NDD1, SWI4 and SWI5. Among the other transcription factors in the module, three (STE12, DIG1 and PHD1) are associated with developmental processes and the remaining two (RLM1 and RFX1) are associated with environmental response. The presence of these transcription factors in the same module could indicate a coupling between different cellular processes (for example, it is reasonable that cell cycle and cell development could be coupled), but it could also be due to the fact that certain transcription factors may be involved in more than one cellular process,

---

[2] There are four trivial submodules made up of a single transcription factor regulating genes with only one regulator.

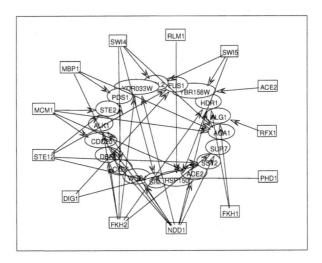

Fig. 1. Graphical representation of the nontrivial part of the cell cycle submodule of the regulatory network obtained by considering only significant regulatory relations. The boxes represent the transcription factors, the inner vertices represent the 19 genes regulated by more than one transcription factor.

hence rendering the boundaries between functional categories somewhat fuzzy. A graphical representation[3]of this submodule is given in Figure 1.

The smaller submodules exhibit similar functional coherence. For example, there are four independent submodules involving transcription factors related to cell metabolism, consisting respectively of: ARG80, ARG81 and GCN4; ARO80 and CBF1; LEU3 and RTG3 and DAL82 and MTH1. Other two submodules consist mainly of genes related to environmental response, one including CIN5, MAC1 and YAP6 together with AZF1 (related to metabolism) and the other one including CAD1 and YAP1. The remaining two submodules consist of two transcription factors belonging to different functional categories. The nontrivial part of one of these submodules is shown graphically in Figure 2. As it can be seen, this is a reasonably sized system which could be amenable to a more detailed description.

In passing, we not that widely applied heuristic methods such as k-means clustering perform very badly in identifying submodules of the network. For example, k-means applied to the columns of the effective connectivity matrix with a random initialisation returns only one cluster.

Identifying submodules using regulatory intensities. While considering only significant regulations clearly leads to a significant advantage when trying to identify submodules, a simple thresholding technique as discussed in the previous

---

[3] The graphs in this paper were obtained using the MATLAB interface for GraphViz, available at http://www.cs.ubc.ca/ murphyk/Software/GraphViz/ graphviz.html.

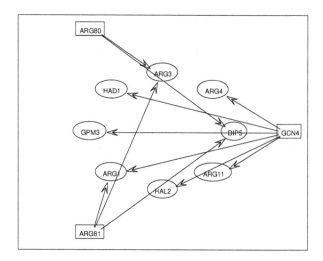

Fig. 2. Graphical representation of one of the submodules of the regulatory network obtained by considering only significant regulatory relations. This submodules is functionally related to the cell metabolism.

section clearly does not make use of the wealth of information contained in the regulatory strengths. We therefore studied the cell cycle data using the spectral algorithm described in section 3.2.

We constructed the affinity matrix as in (5) by using all regulatory intensities with a signal to noise ratio greater than 2 (95% significance level) and selecting

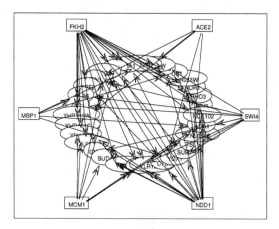

Fig. 3. Graphical representation of the principal submodule obtained by considering the regulatory intensities. All the transcription factors involved in this submodule (indicated in the outer boxes) are key regulators of the cell cycle. The inner vertices represent the genes with more than one significant regulator involved in the submodule.

only genes significantly regulated by two or more transcription factors (these are the only ones that will contribute to the off-diagonal part of the covariance). We then applied the submodule finding Algorithm 2 with a sensitivity parameter 0.01. This gave four clustering eigenvectors, yielding submodules involving between seven and two transcription factors each. Ranking these using the eigenvalues associated, we find that the submodules exhibit a remarkable functional coherence. For example, 98.7% of the mass of the first clustering eigenvector is accounted for by six transcription factors. These are ACE2, FKH2, MBP1, MCM1, NDD1 and SWI4 and are all functionally associated with the cell cycle. By considering the genes involved in this submodule, obtained by considering the eigendecomposition of the dual matrix (6), we also recognise some key genes involved in the cell cycle, such as AGA1, CLB2, CTS1, YOX1 and the transcription factor genes ACE2 and SWI5. The nontrivial part of this submodule of the regulatory network is shown in Figure 3. Similarly, the second eigenvector has 99.9% of its mass concentrated on two transcription factors, DAL82 and MTH1, which are related to carbohydrate/nitrogen metabolism, 99.3% of the third eigenvector's mass is accounted for by AZF1, CUP9 and DAL81, which are related to cell metabolism (CUP9 is also associated with response to oxidative stress), 99% of the mass of the fourth clustering eigenvector is accounted for by LEU3 and STP1, both related to cell metabolism.

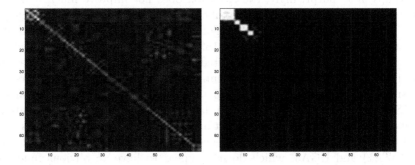

Fig. 4. Graphical representation of the affinity matrix obtained using the regulatory intensities for the cell cycle data set(*left*) and block structure obtained from the submodules found using the spectral Algorithm 2. One strongly interconnected submodule is evident in the top left corner of the affinity matrix; the other submodules are associated with much weaker interactions and are hard to appreciate at a glance.

A major difference with the naive submodule finding Algorithm 1 is the nonexhaustive nature of the spectral algorithm. Specifically, while the naive algorithm will assign each transcription factor represented in the network to exactly one (possibly trivial) submodule, most transcription factors are not included into any submodule by the spectral algorithm. This can be understood by considering the structure of the affinity matrix, which is shown graphically in Figure 4, *left*. While there is one evident block with very high internal covariance in the top left corner (representing the dominant clustering eigenvector associated with the cell cycle), the other submodules are not easily appreciated, since they are associated with much weaker regulatory intensities. The block structure given by the submodules is shown graphically in Figure 4 *right*. Notice however that most transcription factors are not associated with any submodule, indicating that they do not appear to be key in any cellular process going on during the cell cycle.

4.3   Metabolic Cycle Data

Tu *et al.* used oligonucleotide microarrays to measure gene expression levels during the yeast metabolic cycle, *i.e.* glycolitic and respiratory oscillations following a brief period of starvation. The samples were prepared approximately every 25 minutes and covered three full cycles, giving a total of 36 time points [12].

The connectivity we used to analyse this data set was obtained integrating the two ChIP experiments of Lee *et al.* [2] and Harbison *et al.* [1], resulting in a very large network of 3178 genes and 177 transcription factors. By integrating the two datasets, we capture the largest number of potential regulatory relations, which also implies we are introducing a large number of false positives. It is not surprising then that trying to identify submodules directly from the ChIP data leads to a single huge module including all transcription factors and all genes.

Perhaps more surprisingly, the situation does not improve much if we consider only regulations with a high significance level (signal to noise ratio greater than four). Although the number of significant regulations is much smaller than the number of potential regulations (1826 versus 7082), the resulting network still appears to be highly interconnected, so that the application of the naive algorithm again yields one very large submodule (134 transcription factors) and two small submodules containing two transcription factors each. These ones are CST6 and SFP1, two transcription factors which may be loosely related to metabolism (CST6 regulates genes that utilise non optimal carbon sources, while SFP1 activates ribosome biogenesis genes in response to various nutrients) and A1(MATA1) and UGA3, which do not appear to have an obvious functional relationship.

We get a completely different picture if we use the information contained in the regulatory strengths. If we again construct an affinity matrix by retaining the regulatory strengths of all regulations which are significant at 95% for genes regulated by at least two transcription factors, the spectral submodule finding Algorithm 2 (again with sensitivity parameter set to 0.01) returns seven nontrivial submodules.

Somewhat surprisingly, the first clustering eigenvector is again related to the cell cycle: 96.6% of its mass is concentrated on the ten transcription factors ACE2, FKH2, MBP1, MCM1, NDD1, SKN7, STB1, SWI4, SWI5 and SWI6, which are all well known key players of the yeast cell cycle. This seems to add support to the hypothesis, advanced by Tu *et al.*, that the metabolic cycle and the cell cycle might be coupled [12]. The functional coherence of the other submodules is less clear: while GTS1 and RIM101, which account for 99.8% of

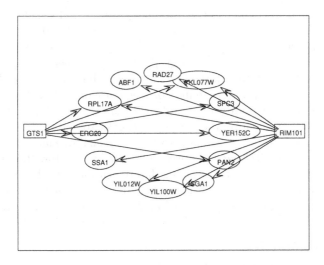

Fig. 5. Graphical representation of the submodule of the metabolic cycle given by GTS1 and RIM101, two transcription factors involved in regulating sporulation

the mass of the second clustering eigenvector, are both involved in sporulation, the functional annotations of the transcription factors involved in other submodules are less coherent. For example, the coupling between MSS11 (which regulates starch degradation) and WAR1 (which promotes acid and ammonia transporters) is plausible but may need further experimental validation before being accepted. A graphical representation of the submodule formed by GTS1 and RIM101 is given in Figure 5.

5   Discussion

In this paper we proposed two algorithms to identify approximately independent submodules of the cellular regulatory network. Both methods rely on having genome-wide information on the intensity with which transcription factors regulate their target genes, obtained for example by using the recent model proposed in [9]. While the first algorithm is a simple exhaustive search, the second is more subtle, being based on the spectral decomposition of an affinity matrix between transcription factors, and is somewhat related to the algorithm proposed in [10] for the automatic detection of non-convex clusters.

Experimental results obtained using the algorithms on two yeast data sets reveals that both methods can find biologically plausible submodules of the regulatory network, and in many cases these submodules are of small enough size to be amenable to be modelled in a more detailed fashion. The two algorithms have complementary strengths: while the naive search algorithm has the advantage of assigning each transcription factor to a unique submodule, many transcription factors are not assigned to any module by the spectral algorithm. On the other hand, the functional coherence of the submodules identified by the spectral algorithm seems to be higher in the examples studied, and sensible submodules are found even when the network is too interconnected for the naive search to yield any submodules.

Another popular method to cluster graphs which has been extensively applied to biological problems is the Markov Cluster Algorithm (MCL), which was used successfully to find families of proteins from sequence data [16]. However, this algorithm is designed for undirected graphs with an associated similarity matrix, while the graphs obtained from regulatory networks are naturally directed (with arrows going from transcription factors to genes). Even if we marginalise the genes by considering an affinity matrix between transcription factors, this is generally not a consistent similarity matrix, making the application of MCL very hard. Bearing in mind the largely exploratory nature of finding submodules of the regulatory network, we preferred to use simpler and more interpretable methods.

Acknowledgements

The authors gratefully acknowledge support from a BBSRC award "Improved processing of microarray data with probabilistic models".

References

1. C. T. Harbison et al., Nature **431**, 99 (2004).
2. T. I. Lee et al., Science **298**, 799 (2002).
3. T. Schlitt and A. Brazma, FEBS letts **579**, 1859 (2005).
4. N. M. Luscombe et al., Nature **431**, 308 (2004).
5. N. A. Monk, Biochemical Society Transactions **31**, 1457 (2003).
6. X. Xie et al., Nature **434**, 338 (2005).
7. R. Martone et al., Proceedings of the National Academy of Sciences USA **100**, 12247 (2003).
8. G. Sanguinetti, M. Rattray, and N. D. Lawrence, A probabilistic dynamical model for quantitative inference of the regulatory mechanism of transcription, To appear in *Bioinformatics*, 2006.
9. G. Sanguinetti, N. D. Lawrence, and M. Rattray, Probabilistic inference of transcription factors concentrations and gene-specific regulatory activities, Technical Report CS-06-06, University of Sheffield, 2006.
10. G. Sanguinetti, J. Laidler, and N. D. Lawrence, Automatic determination of the number of clusters using spectral algorithms, in *Proceedings of MLSP 2005*, pages 55–60, 2005.
11. P. T. Spellman et al., Molecular Biology of the Cell **9**, 3273 (1998).
12. B. P. Tu, A. Kudlicki, M. Rowicka, and S. L.McKnight, Science **310**, 1152 (2005).
13. J. C. Liao et al., Proceedings of the National Academy of Sciences USA **100**, 15522 (2003).
14. O. Alter, P. O. Brown, and D. Botstein, Proc. Natl. Acad. Sci. USA **97**, 10101 (2000).
15. A.-L. Boulesteix and K. Strimmer, Theor. Biol. Med.Model. **2**, 1471 (2005).
16. A.J.Enright, S. van Dongen, and C. Ouzounis, Nucleic Acids Research **30**, 1575 (2002).

Incorporating Time Delays into the Logical Analysis of Gene Regulatory Networks

Heike Siebert and Alexander Bockmayr

DFG Research Center MATHEON,
Freie Universität Berlin, Arnimallee 3, D-14195 Berlin, Germany
siebert@mi.fu-berlin.de, bockmayr@mi.fu-berlin.de

Abstract. Based on the logical description of gene regulatory networks developed by R. Thomas, we introduce an enhanced modelling approach that uses timed automata. It yields a refined qualitative description of the dynamics of the system incorporating information not only on ratios of kinetic constants related to synthesis and decay, but also on the time delays occurring in the operations of the system. We demonstrate the potential of our approach by analysing an illustrative gene regulatory network of bacteriophage λ.

1 Introduction

When modelling a gene regulatory network one has basically two options. Traditionally, such a system is modelled with differential equations. The equations used, however, are mostly non-linear and thus cannot be solved analytically. Furthermore, the available experimental data is often of qualitative character and does not allow a precise determination of quantitative parameters for the differential model. This eventually led to the development of qualitative modelling approaches. R. Thomas introduced a logical formalism in the 1970s, which, over the years, has been further developed and successfully applied to different biological problems (see [7], [8] and references therein). The only information on a concentration of gene products required in this formalism is whether or not it is above a threshold relevant for some interaction in the network. Furthermore, parameters holding information about the ratio of production and spontaneous decay rates of the gene products are used. The values of these parameters determine the dynamical behaviour of the system, which is represented as a state transition graph. Moreover, Thomas realized that a realistic model should not be based on the assumption that the time delay from the start of the synthesis of a given product until the point where the concentration reaches a threshold is the same for all the genes in the network. Neither will the time delays associated with synthesis and those associated with decay be the same. Therefore, he uses an asynchronous description of the dynamics of the system, i. e., a state in the state transition graph differs from its predecessor in one component only.

In order to refine the model, we would like to incorporate information about the values of the time delays. Since precise data about the time delays is not

C. Priami (Ed.): CMSB 2006, LNBI 4210, pp. 169–183, 2006.
© Springer-Verlag Berlin Heidelberg 2006

available (in biological systems the delays will not even have an exactly determined value), the information is given in the form of inequalities that pose
constraints on the time delays. So we need to keep track of time while the system evolves. A theoretical framework providing us with the necessary premises
is the theory of timed automata. Each gene is equipped with a clock which is
used to evaluate the conditions posed on the time delays of that particular gene
during the evolution of the system. The resulting transition system is in general
nondeterministic, but the additional information inserted allows for a refined
view of the dynamics. Conclusions about stability of dynamical behaviour and
restriction to certain behaviour in comparison to the predictions of the Thomas
model become possible. Moreover, the possibility of synchronous update is not
excluded under certain conditions.

In the first part of the paper we give a thorough mathematical description
of the Thomas formalism in Sect. 2 and of the modelling approach using timed
automata in Sect. 3 and 4. In Sect. 5, we show that, by using our approach, it
is possible to obtain the state transition graph of the Thomas model. Also, we
outline further possibilities our model offers. To illustrate the theoretical considerations, we analyse a simple regulatory network of bacteriophage λ in Sect. 6.
In addition to the mere formal analysis, we have implemented the network using
the verification tool UPPAAL. In the last section, we discuss the mathematical
and biological perspectives of our approach.

2   Generalised Logical Formalism of Thomas

In this section we give a formal definition of a gene regulatory network in the
sense of the modelling approach of R. Thomas (see for example [7] and [8]). We
use mainly the formalism introduced in [4].

Definition 1. *Let $n \in \mathbb{N}$. An* interaction graph *(or biological regulatory graph)
$\mathcal{I}$ is a labelled directed graph with vertex set $V := \{\alpha_1, \ldots, \alpha_n\}$ and edge set
E. Each edge $\alpha_j \to \alpha_i$ is labelled with a sign $\varepsilon_{ij} \in \{+, -\}$ and an integer
$b_{ij} \in \{1, \ldots, d_j\}$, where d_j denotes the out-degree of α_j. Furthermore, we assume
that $\{b_{ij} ; \exists\, \alpha_j \to \alpha_i\} = \{1, \ldots, p_j\}$ for all $j \in \{1, \ldots, n\}$ and $p_j \leq d_j$. We call
$\{0, \ldots, p_j\}$ the* range *of α_j. For each $i \in \{1, \ldots, n\}$ we denote by $Pred(\alpha_i)$ the
set of vertices α_j such that $\alpha_j \to \alpha_i$ is an edge in E.*

The vertices of this graph represent the genes of the gene regulatory network,
the range of a vertex the different expression levels of the corresponding gene
affecting the behaviour of the network. An edge $\alpha_j \to \alpha_i$ signifies that the gene
product of α_j influences the gene α_i in a positive or negative way depending on
ε_{ij} and provided that the expression level of α_j is equal or above b_{ij}. Note that
the values b_{ij} do not have to be pairwise distinct.

In order to describe the behaviour of a gene regulatory network we need a
formal framework to capture its dynamics.

Definition 2. *Let $\mathcal{I}$ be an interaction graph. A* state *of the system described by
$\mathcal{I}$ is a tuple $s \in S^n := \{0, \ldots, p_1\} \times \cdots \times \{0, \ldots, p_n\}$. The set of* resources *$R_i(s)$*

of α_i in state s is the set

$$\{\alpha_j \in Pred(\alpha_i)\,;\, (\varepsilon_{ij} = +\, \wedge\, s_j \geq b_{ij})\, \vee\, (\varepsilon_{ij} = -\, \wedge\, s_j < b_{ij})\}.$$

Finally, we define the set of (logical) parameters

$$K(\mathcal{I}) := \{K_{\alpha_i,\omega} \in \{0,\ldots,p_i\}\,;\, i \in \{1,\ldots,n\},\, \omega \subset Pred(\alpha_i)\}.$$

We call the pair $(\mathcal{I}, K(\mathcal{I}))$ a gene regulatory network.

The set of resources $R_i(s)$ provides information about the presence of activators and the absence of inhibitors for some gene α_i in state s. The value of the parameter $K_{\alpha_i,R_i(s)}$ indicates how the expression level of gene α_i will evolve. The product concentration will increase (decrease) if the parameter value is greater (smaller) than s_i. The expression level stays the same if both values are equal.

Thomas and Snoussi used this formalism to discretize a certain class of differential equation systems (see [5]). To reflect this, the following constraint has to be posed on the parameter values:

$$\omega \subset \omega' \Rightarrow K_{\alpha_i,\omega} \leq K_{\alpha_i,\omega'} \tag{1}$$

for all $i \in \{1,\ldots,n\}$. The condition signifies that an effective activator or a non-effective inhibitor cannot induce the decrease of the expression level of α_i. In the following we will always assume that this constraint is valid in order to compare our modelling approach with the one used by Thomas. However, in the last section of this paper we will discuss possible generalisations of the model not requiring the constraint (1).

To conclude this section, we describe the dynamics of the gene regulatory network by means of a state transition graph.

Definition 3. *The state transition graph $\mathcal{S}_N = (\mathcal{I}, K(\mathcal{I}))$ corresponding to a gene regulatory network N is a directed graph with vertex set S^n. There is an edge $s \to s'$ if there is $i \in \{1,\ldots,n\}$ such that $|s'_i - K_{\alpha_i,R_i(s)}| = |s_i - K_{\alpha_i,R_i(s)}| - 1$ and $s_j = s'_j$ for all $j \in \{1,\ldots,n\} \setminus \{i\}$.*

The above definition reflects the use of the asynchronous update rule, since a state differs from a successor state in one component only. If s is a state such that an evolution in more than one component is indicated, then there will be more than one successor of s. Note that s is a steady state if s has no outgoing edge.

A gene regulatory network comprising two genes connected with a positive and a negative edge and the resulting state transition graph are given in Fig. 1. We use this simple example to illustrate the construction of the timed automaton representing the network in Sect. 4.

3   Timed Automata

In this section we formally introduce timed automata. We mainly use the definitions and notations given in [1].

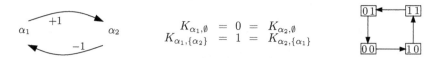

Fig. 1. Interaction graph, parameters and state transition graph of a simple gene regulatory system

To introduce the concept of time in our system, we consider a set $C :=$ $\{c_1, \ldots, c_n\}$ of real variables which behave according to the differential equations $\dot{c}_i = 1$. These variables are called *clocks*. They progress synchronously and can be reset to zero under certain conditions. We define the set $\Phi(C)$ of *clock constraints* φ by the grammar

$$\varphi ::= c \leq q \,|\, c \geq q \,|\, c < q \,|\, c > q \,|\, \varphi_1 \wedge \varphi_2 \,,$$

where $c \in C$ and q is a rational constant.

A *clock interpretation* is a function $u : C \to \mathbb{R}_{\geq 0}$ from the set of clocks to the non-negative reals. For $\delta \in \mathbb{R}_{\geq 0}$, we denote by $u + \delta$ the clock interpretation that maps each $c \in C$ to $u(c) + \delta$. For $Y \subset C$, we indicate by $u[Y := 0]$ the clock interpretation that maps $c \in Y$ to zero and agrees with u over $C \setminus Y$. A clock interpretation u satisfies a clock constraint φ if $\varphi(u) = true$. The set of all clock interpretations is denoted by $\mathbb{R}_{\geq 0}^C$.

Definition 4. *A* timed automaton *is a tuple* $(L, L^0, \Sigma, C, I, E)$, *where* L *is a finite set of* locations, $L^0 \subset L$ *is the set of* initial locations, Σ *is a finite set of* events *(or labels),* C *is a finite set of* clocks, $I : L \to \Phi(C)$ *is a mapping that labels each location with some clock constraint which is called the* invariant *of the location, and* $E \subset L \times \Sigma \times \Phi(C) \times 2^C \times L$ *is a set of* switches.

A timed automaton can be represented as a directed graph with vertex set L. The vertices are labelled with the corresponding invariants and are marked as initial locations if they belong to L^0. The edges of the graph correspond to the switches and are labelled with an event, a clock constraint called *guard* specifying when the switch is enabled, and a subset of C comprising the clocks that are reset to zero with this switch. While switches are instantaneous, time may elapse in a location. To describe the dynamics of such an automaton formally, we use the notion of a transition system.

Definition 5. *Let* A *be a timed automaton. The (labelled)* transition system T_A *associated with* A *is a tuple* $(Q, Q^0, \Gamma, \to)$, *where* Q *is the set of states* $(l, u) \in$ $L \times \mathbb{R}_{\geq 0}^C$ *such that* u *satisfies the invariant* $I(l)$, Q^0 *comprises the states* $(l, u) \in Q$ *where* $l \in L^0$ *and* u *ascribes the value zero to each clock, and* $\Gamma := \Sigma \cup \mathbb{R}_{\geq 0}$. *Moreover,* $\to \subset Q \times \Gamma \times Q$ *is defined as the set comprising*

- $(l, u) \xrightarrow{\delta} (l, u + \delta)$ *for* $\delta \in \mathbb{R}_{\geq 0}$ *such that for all* $0 \leq \delta' \leq \delta$ *the clock interpretation* $u + \delta'$ *satisfies the invariant* $I(l)$, *and*

- $(l, u) \xrightarrow{a} (l', u[R := 0])$ for $a \in \Sigma$ such that there is a switch (l, a, φ, R, l') in E, u satisfies φ, and $u[R := 0]$ satisfies $I(l')$.

The elements of $\rightarrow$ are called transitions.

The first kind of transition is a state change due to elapse of time, while the second one is due to a location-switch and is called *discrete*. Again we can visualise the object T_A as a directed graph with vertex set Q and edges corresponding to the transitions given by $\rightarrow$. We will use terminology from graph theory with respect to T_A. Note, that by definition the set of states may be infinite and that the transition system is in general nondeterministic, i.e., a state may have more than one successor. Moreover, it is possible that a state is the source for edges labelled with a real value as well as for edges labelled with events. However, although every discrete transition corresponds to a switch in A, there may be switches in A that do not lead to a transition in T_A. That is due to the additional conditions placed on the clock interpretations. Furthermore, we obtain a modified transition system by considering only the location vectors as states, dropping all transitions labelled with real values, but keeping every discrete transition of T_A. We call it the *discrete (or symbolic) transition system of A*.

4   Modelling with Timed Automata

In order to model a gene regulatory network as a timed automaton, we first introduce components that correspond to the genes of the network. They constitute the building blocks that compose the automaton, representing the network much in the same way n timed automata are integrated to represent a product automaton (see [1]).

In the following, let $N = (\mathcal{I}, K(\mathcal{I}))$ be a gene regulatory network comprising the genes $\alpha_1, \ldots, \alpha_n$.

Constructing the components. Let $i \in \{1, \ldots, n\}$. We define the component $A_i := (L_i, L_i^0, \Sigma_i, C_i, I_i, E_i)$ corresponding to α_i according to the syntax of timed automata. In addition we will label the locations with a set of switch conditions.

Locations: We define L_i as the set comprising the elements α_i^k for $k \in \{0, \ldots, p_i\}$, α_i^{k+} for $k \in \{0, \ldots, p_i - 1\}$, and α_i^{k-} for $k \in \{1, \ldots, p_i\}$. Location α_i^k represents a situation where gene α_i maintains expression level k. We call such a location *regular*. If the superscript is $k+$ resp. $k-$, the expression level is k but the concentration of the gene product tends to increase resp. decrease. Those locations are called *intermediate*. We define $L_i^0 := \{\alpha_i^k ; k \in \{0, \ldots, p_i\}\}$.

Events: The events in Σ_i correspond to the intermediate locations. We set $\Sigma_i := \{a_i^{k+}, a_i^{m-} ; k \in \{0, \ldots, p_i - 1\}, m \in \{1, \ldots, p_i\}\}$. These events will be used later on to identify certain discrete transitions starting in the intermediate locations.

Clocks: For each gene we use a single clock, so $C_i := \{c_i\}$.

Invariants: We define the mapping $I_i : L_i \rightarrow \Phi(C_i)$ as follows. Every regular location α_i^k is mapped to $c_i \geq 0$ (evaluating to *true*). For each intermediate location $\alpha_i^{k\varepsilon}$, $\varepsilon \in \{+, -\}$, we choose $t_i^{k\varepsilon} \in \mathbb{Q}_{\geq 0}$ and set $I_i(\alpha_i^{k\varepsilon}) = (c_i \leq t_i^{k\varepsilon})$. The value $t_i^{k\varepsilon}$ signifies the maximal time delay occurring before the expression level of α_i changes to $k + 1$, if $\varepsilon = +$, or to $k - 1$, if $\varepsilon = -$. During that time a change in the expression level of α_i may yet be averted if the expression levels of the genes influencing α_i change.

Switches: To specify the guard conditions on the switches, we choose constants concerning time $t_i^{(k,k+1)}, t_i^{(k+1,k)} \in \mathbb{Q}_{\geq 0}$ for all $k \in \{0, \ldots, p_i - 1\}$. There are two kinds of switches in the set E_i. For all $k \in \{0, \ldots, p_i - 1\}$, we have $(\alpha_i^{k+}, a_i^{k+}, \varphi_i^{k+}, \{c_i\}, \alpha_i^{k+1}) \in E_i$, where $\varphi_i^{k+} = (c_i \geq t_i^{(k,k+1)})$. Furthermore, for $k \in \{1, \ldots, p_i\}$, the switch $(\alpha_i^{k-}, a_i^{k-}, \varphi_i^{k-}, \{c_i\}, \alpha_i^{k-1})$ with $\varphi_i^{k-} = (c_i \geq t_i^{(k,k-1)})$ is in E_i. The given time constraints determine the minimal time delay before a change in the expression level can occur. Choosing the time constants associated with the guards smaller than those associated with the invariants of the corresponding intermediate location leads to indeterministic behaviour of the system in that location.

Switch conditions: To each location in L_i we assign certain conditions which later will be used to define the switches of the timed automaton of the gene regulatory network. These conditions concern the locations of all components $A_j, j \in \{1, \ldots, n\}$. We interpret locations as integer values by using the function $\iota : \bigcup_{j \in \{1, \ldots, n\}} L_j \to \mathbb{N}_0$ that maps the locations α_j^k, α_j^{k+} and α_j^{k-} to k.

Let $k \in \{1, \ldots, p_i - 1\}$. We define logical conditions Λ_i^k and $\overline{\Lambda}_i^k$ as follows. For every $\alpha_j \in Pred(\alpha_i)$ and l_j a location of A_j let

$$\lambda_i^{\alpha_j}(l_j) := \begin{cases} \iota(l_j) \geq b_{ij}, & \varepsilon_{ij} = + \\ \iota(l_j) < b_{ij}, & \varepsilon_{ij} = - \end{cases} \qquad \overline{\lambda}_i^{\alpha_j}(l_j) := \begin{cases} \iota(l_j) < b_{ij}, & \varepsilon_{ij} = + \\ \iota(l_j) \geq b_{ij}, & \varepsilon_{ij} = - \end{cases} .$$

Let $\omega_1, \ldots, \omega_{m_i^1}, \upsilon_1, \ldots, \upsilon_{m_i^2}$ be the subsets of $Pred(\alpha_i)$ such that the parameter inequalities $K_{\alpha_i, \omega_h} > k$ for all $h \in \{1, \ldots, m_i^1\}$ as well as $K_{\alpha_i, \upsilon_h} < k$ for all $h \in \{1, \ldots, m_i^2\}$ hold. Let $l \in L_1 \times \cdots \times L_n$. Then we define $\lambda_i^{\omega_h}(l) := \bigwedge_{\alpha_j \in \omega_h} \lambda_i^{\alpha_j}(l_j)$ and $\overline{\lambda}_i^{\upsilon_h}(l) := \bigwedge_{\alpha_j \in Pred(\alpha_i) \setminus \upsilon_h} \overline{\lambda}_i^{\alpha_j}(l_j)$. Finally, we set

$$\Lambda_i^k(l) := \bigvee_{h \in \{1, \ldots, m_i^1\}} \lambda_i^{\omega_h} \qquad \text{and} \qquad \overline{\Lambda}_i^k(l) := \bigvee_{h \in \{1, \ldots, m_i^2\}} \overline{\lambda}_i^{\upsilon_h} .$$

We define Λ_i^0 and $\overline{\Lambda}_i^{p_i}$ accordingly.

Now, we assign all locations α_i^k, $k \in \{1, \ldots, p_i - 1\}$ the conditions Λ_i^k and $\overline{\Lambda}_i^k$. The location α_i^0 resp. $\alpha_i^{p_i}$ is labelled with Λ_i^0 resp. $\overline{\Lambda}_i^{p_i}$ only. Furthermore, we associate with location α_i^{k+} the condition $\Psi_i^{k+} := \neg \Lambda_i^k$ for all $k \in \{0, \ldots, p_i - 1\}$, and allot to location α_i^{k-} the condition $\Psi_i^{k-} := \neg \overline{\Lambda}_i^k$ for all $k \in \{1, \ldots, p_i\}$.

The conditions defined above correspond to the set of resources used in the formalism of Thomas and thus play a key role in the dynamics of the system. If the condition Λ_i^k is met, the gene α_i will start producing its product at a higher rate. This is represented by a transition to the location α_i^{k+} (see the definition of the switches of the timed automaton A defined below). However, it is possible that some change in the expression levels of genes influencing α_i occurs while α_i has not yet reached the location α_i^{k+1}. If those changes are such that the condition Ψ_i^{k+} is satisfied, then the premises for α_i to reach the expression level $k + 1$ are no longer given, and it will return to the location α_i^k (again see the definition of the switches of A below). The conditions $\overline{\Lambda}_i^k$ and Ψ_i^{k-} are used similarly for the decrease of the expression level.

Note that whenever $\omega_{h_1} \subset \omega_{h_2}$ for sets ω_h defined as above, condition $\lambda_i^{\omega_{h_2}}$ can be deleted from the expression Λ_i^k due to the constraints (1) on the parameter values. A corresponding statement holds for the sets υ_h.

Formally, the components defined above are timed automata. However, it does not make sense to evaluate their behaviour in isolation from each other. This becomes apparent when looking at the graph representation. Most locations in the automaton A_i are not connected by edges. Every path in the graph contains at most one edge. Figure 2 illustrates this observation. It shows the components A_1 and A_2 corresponding to the genes α_1 and α_2 in Fig. 1. Each component comprises the regular locations α_i^0 and α_i^1 and the intermediate locations α_i^{0+} and α_i^{1-}, represented as circles in the graph. The first line below the location identifier in a circle is the corresponding invariant, the second line shows the corresponding switch condition. Since both genes have only two expression levels, each location is only labelled with one switch condition. For example, we have $2^{Pred(\alpha_1)} = \{\emptyset, \{\alpha_2\}\}$, $\lambda_1^{\alpha_2}(l_2) = (\iota(l_2) < 1)$, and $\overline{\lambda}_1^{\alpha_2}(l_2) = (\iota(l_2) \geq 1)$. Since $K_{\alpha_1,\{\alpha_2\}} = 1$, we have $\lambda_1^{\{\alpha_2\}} = (\iota(l_2) < 1)$ and $\Lambda_1^0(l) = (\iota(l_2) < 1)$. Furthermore, $K_{\alpha_1,\emptyset} = 0$ and thus $\overline{\lambda}_1^{\emptyset}(l) = (\iota(l_2) \geq 1) = \overline{\Lambda}_1^1$. The switches are represented as directed edges from the first to the last component of the switch. They are labelled with the guard, the event, and the set of clocks that are to be reset.

Modelling the network. In this paragraph, we construct the timed automaton $A_N := (L, L^0, \Sigma, C, I, E)$ representing the network N by means of components $A_1, \ldots, A_n$ in the following way. We define $L := L_1 \times \cdots \times L_n$, $L^0 := L_1^0 \times \cdots \times L_n^0$ and $\Sigma := \{a\} \cup \bigcup_{i \in \{1, \ldots, n\}} \Sigma_i$. Here a will signify a general event, which is used to indicate that the switch is defined by means of the switch conditions of the components A_i (see below). A location in L is called *regular*, if all of its components are regular, and *intermediate* otherwise. Furthermore, we define the set of clocks $C := \bigcup_{i \in \{1, \ldots, n\}} C_i$ and $I : L \to \Phi(C)$, $(l_1, \ldots, l_n) \mapsto (I_1(l_1) \wedge \cdots \wedge I_n(l_n))$. The set of switches $E \subset L \times \Sigma \times \Phi(C) \times 2^C \times L$ is comprised of the following elements:

- For every $i \in \{1, \ldots, n\}$ and every switch $(l_i, a_i, \varphi_i, R_i, l_i') \in E_i$ the tuple $(h, a_i, \varphi_i, R_i, h')$, with $h, h' \in L$, $h_j = h_j'$ for all $j \neq i$, $h_i = l_i$ and $h_i' = l_i'$, is a switch in E.
- Let $(l, a, \varphi, R, l') \in L \times \Sigma \times \Phi(C) \times 2^C \times L$ with $\varphi := true$. Let J be the largest subset of $\{1, \ldots, n\}$ such that for each l_j, $j \in J$, one of the switch conditions associated with l_j is true. Assume R comprises the clocks c_j, $j \in J$. Let $l_i = l_i'$ for all $i \notin J$, and let, for all $j \in J$,

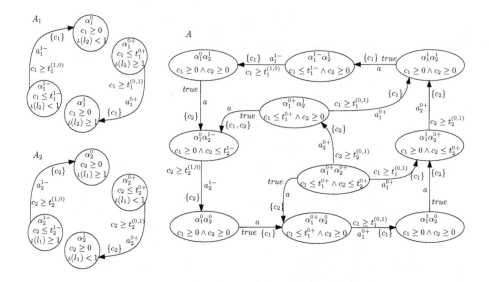

Fig. 2. On the left, components A_1 and A_2 representing the genes α_1 and α_2 in Figure 1. On the right, a section of the timed automaton A constructed from A_1 and A_2.

$$l'_j = \begin{cases} \alpha_j^{k-}, & l_j = \alpha_j^k \text{ for some } k \text{ and } \overline{\Lambda}_j^k(l) = true \\ \alpha_j^{k+}, & l_j = \alpha_j^k \text{ for some } k \text{ and } \Lambda_j^k(l) = true \\ \alpha_j^k, & l_j = \alpha_j^{k+} \text{ for some } k \text{ and } \Psi_j^{k+}(l) = true \\ \alpha_j^k, & l_j = \alpha_j^{k-} \text{ for some } k \text{ and } \Psi_j^{k-}(l) = true \end{cases} \quad (2)$$

Then (l, a, φ, R, l') is a switch in E.

Although the formal definition of the switches looks quite complicated, the actual meaning is straightforward. A location change occurs when the current state of locations allows for a change. The switch conditions Λ_j^k, $\overline{\Lambda}_j^k$, Ψ_j^{k+} and Ψ_j^{k-} carry the information which conditions, depending on the current location of A, the expression levels of the genes influencing α_j have to satisfy in order to induce a change in the expression level of α_j (see remarks on switch conditions of components A_i). Furthermore, changes in the expression level of a gene happen gradually. That is, for every two locations l, l' connected by a switch we have $|\iota(l_i) - \iota(l'_i)| \leq 1$ for all $i \in \{1, \ldots, n\}$. The event a is used to identify the switches that include the checking of the switch conditions of some location.

The timed automaton A representing the gene regulatory network in Fig. 1 is partially presented in Fig. 2. The figure includes all regular locations of A as well as all locations that are the target of an edge (representing a switch) starting in a regular location. Moreover, we chose two locations that render interesting switches. All switches of A starting in a location displayed in Fig. 2 are indicated. We show the construction of the switches exemplarily for the location $(\alpha_1^{0+}, \alpha_2^1)$. First, we note that $(\alpha_1^{0+}, a_1^{0+}, (c_1 \geq t_1^{(0,1)}), \{c_1\}, \alpha_1^1)$ is a switch in A_1.

Thus $((\alpha_1^{0+}, \alpha_2^1), a_1^{0+}, (c_1 \geq t_1^{(0,1)}), \{c_1\}, (\alpha_1^1, \alpha_2^1))$ is a switch in A. Now we check the switch conditions in $l = (\alpha_1^{0+}, \alpha_2^1)$. The condition $\Psi_1^{0+}(l) = (\iota(l_2) \geq 1)$ is true in l as is the condition $\overline{\Lambda}_2^1(l) = (\iota(l_1) < 1)$. Thus, $J = \{1, 2\}$. We obtain the target location l' of the switch according to (2). Since $\Psi_1^{0+}(l)$ is true and $l_1 = \alpha_1^{0+}$, we get $l_1' = \alpha_1^0$, and since $\overline{\Lambda}_2^1(l)$ is true and $l_2 = \alpha_2^1$, we get $l_2' = \alpha_2^{1-}$. The guard condition for the switch is $true$, it is labelled with a and both clocks c_1 and c_2 are reset.

The associated transition system. Let $T_A = (Q, Q^0, \Gamma, \rightarrow)$ be the transition system associated with A. Note that the above definition of the first kind of switch in E reflects the use of the asynchronous update of the expression levels in the transition system. More precisely, although more than one component of the discrete state may change in one step (via switches labelled with a), a change in expression level will occur in one component at most. We will refine the system in one aspect, which leads to a smaller set of possible transitions. Whenever $(l, u) \in Q$ is a state such that there is some transition $(l, u) \xrightarrow{a} (l', v)$ for some state $(l', v) \in Q$, we delete every transition of the form $(l, u) \xrightarrow{\delta} (l, u + \delta)$ regardless of the value of δ. We call a an *urgent* event. That is to say, whenever some transition is labelled with the urgent event a, it is not possible for time to elapse further in location l. However, there may be further discrete transitions starting in (l, u), which would allow for synchronous (in the temporal sense) update (see example in Sect. 6). If we want to avoid this, we delete all other transitions starting in (l, u), and call a an *overriding* event. Unless otherwise stated, we assume in the following that a is urgent.

Furthermore, note that a transition labelled with a never leads to a change in the expression levels of the genes, and that the set J in the definition of the second kind of switch is chosen maximal. Thus, if a path in T_A starts in a regular location and its first transition is labelled with a, then the second transition in the path will not be labelled with a.

Here, some discrete state $l \in L$ is called a steady state if T_A does not contain a discrete transition starting in (l, u), for all clock interpretations u.

To analyse the dynamics of the gene regulatory network we consider the paths in T_A that start in some initial state in Q^0. Questions of interest are for example if a steady state is reachable from a given initial location via some path in T_A. We will discuss the analysis of T_A in a later section.

5 Comparison of the Models

In this section, we aim to show that on the one hand the information inherent in the state transition graph as defined in Definition 3 can also be obtained from the transition system of a suitable timed automaton. On the other hand, the modelling approach via timed automata offers possibilities to incorporate information about gene regulatory networks that cannot be included in the Thomas model, and thus leads to a refined view on the dynamics of the system.

Let $\mathcal{S}_N$ be the state transition graph corresponding to N and A the timed automaton derived from N. We set $t_i^{k\varepsilon}, t_i^{(k,k+1)}, t_i^{(k+1,k)} = 0$ for all $i \in \{1, \ldots, n\}$,

$\varepsilon \in \{+, -\}$. Thus every guard condition evaluates to *true* and time does not elapse in the intermediate locations.

Now, we derive a graph G from T_A as follows. First we identify locations of A_i representing the same expression level, i.e., for $k \in \{1, \ldots, p_i - 1\}$ we define $v_k^{\alpha_i} := \{\alpha_i^k, \alpha_i^{k+}, \alpha_i^{k-}\}$, $v_0^{\alpha_i} := \{\alpha_i^0, \alpha_i^{0+}\}$ and $v_{p_i}^{\alpha_i} := \{\alpha_i^{p_i}, \alpha_i^{p_i-}\}$. Let $V^{\alpha_i} := \{v_k^{\alpha_i} ; k \in \{0, \ldots, p_i\}\}$ and $V := V^{\alpha_1} \times \cdots \times V^{\alpha_n}$ be the vertex set of G. Furthermore, there is an edge $v \to w$, if $v \neq w$ and if there is a path in T_A from some state (l, u), such that l is regular, to a state (l', u') satisfying $l'_i \in w_i$ for all i, such that every discrete state on the path other than l' is an element of $v_1 \times \cdots \times v_n$. The condition to start in a regular state l ensures that the first discrete transition occurring is labelled with a. This excludes the possibility of a change of expression level that does not correspond to the parameter values. We can drop the condition, if we declare a an overriding event.

Now, we need to show that $\mathcal{S}_N$ is contained in G. For the sake of completeness we prove the following stronger statement.

Theorem 1. *The graphs $\mathcal{S}_N$ and G are isomorphic.*

Proof. We define $f : S^n \to V$, $(s_1, \ldots, s_n) \mapsto (v_{s_1}^{\alpha_1}, \ldots, v_{s_n}^{\alpha_n})$. Then it is easy to see that f is a bijection.

Let $s \to s'$ be an edge in $\mathcal{S}_N$. We have to show that $f(s) \to f(s')$ is an edge in G. Set $v := f(s)$ and $w := f(s')$. According to the definition of edges in $\mathcal{S}_N$, there is a $j \in \{1, \ldots, n\}$ such that $|s'_j - K_{\alpha_j, R_j(s)}| = |s_j - K_{\alpha_j, R_j(s)}| - 1$ and $s_i = s'_i$ for all $i \in \{1, \ldots, n\} \setminus \{j\}$. Thus, $v_i = w_i$ for all $i \neq j$, and $v_j \neq w_j$.

First we consider the case that $s_j < K_{\alpha_j, R_j(s)}$. It follows that $s_j \neq p_j$, and thus $\alpha_j^{s_j}, \alpha_j^{s_j+} \in v_j$, and $s'_j = s_j + 1$. We choose $l \in L$ such that $l_i = \alpha_i^{s_i}$ for all $i \in \{1, \ldots, n\}$, thus $l \in v_1 \times \cdots \times v_n$ is regular. Furthermore, we choose the clock interpretation u that assigns each clock the value zero.

We have $R_j(s) \subset Pred(\alpha_j)$ and, by definition, we know that $\lambda_j^{R_j(s)}(l)$, and thus the switch condition $\Lambda_j^{s_j}(l)$, is true. It follows that there is a switch $(l, a, \varphi, R, \tilde{l}) \in E$ with $\varphi = true$, $\tilde{l}_j = \alpha_j^{s_j+}$ and $\tilde{l}_i \in v_i$ for all $i \neq j$. Thus we find a transition $(l, u) \xrightarrow{a} (\tilde{l}, u)$. Since time is not allowed to elapse in intermediate locations, and since no transition starting in $(\tilde{l}, u)$ is labelled with a according to the observations made in the preceeding section, every transition starting in $(\tilde{l}, u)$ will lead to a state that differs from $(\tilde{l}, u)$ in one component of the location vector only. Moreover, we have $(\alpha_j^{s_j+}, a_j^{s_j+}, \varphi_i^{s_j+}, \{c_j\}, \alpha_j^{s_j+1}) \in E_j$ and thus there is a transition $(\tilde{l}, u) \to (l', u)$ labelled with $a_j^{s_j+}$, with $l'_j = \alpha_j^{s_j+1} \in w_j$ and $l'_i = \tilde{l}_i \in v_i = w_i$ for $i \neq j$. It follows that $f(s) = v \to w = f(s')$ is an edge in G.

The case that $s_j > K_{\alpha_j, R_j(s)}$ and thus $s'_j = s_j - 1$ can be treated analogously.

Now let $v \to w$ be an edge in G. We set $s := f^{-1}(v)$ and $s' := f^{-1}(w)$. According to the definition there is a path $((l^1, u^1), \ldots, (l^m, u^m))$ in T_A such that l^1 is regular, $l_i^j \in v_i$ for all $i \in \{1, \ldots, n\}$, $j \in \{1, \ldots, m-1\}$ and $l_i^m \in w_i$ for all $i \in \{1, \ldots, n\}$. Since $l^1 \neq l^m$, there is some discrete transition in the path. Since every component of l^1 is regular, and thus the only discrete transition starting there is labelled by a, and since a is an urgent event, we can deduce that $(l^1, u^1) \to (l^2, u^2)$ is labelled by a. Then l^2 has at least one component which is an intermediate location. Let $J \subset \{1, \ldots, n\}$ be such that l^2_j is an intermediate location for all $j \in J$, and l^2_i is a regular location for all $i \notin J$. Then $l^2_i = l^1_i$ for all $i \notin J$. Since time is not allowed to elapse in the intermediate locations, the transition from (l^2, u^2) to (l^3, u^3) has to be discrete. Moreover, we know that the transition is not labelled by a, since the first transition of the path is already labelled that way. It follows that there is $j \in J$ such that l^3_j is regular, $l^3_j \neq l^2_j$, and $l^3_i = l^2_i$ for all $i \neq j$. Furthermore, the expression levels of gene α_j in location l^1_j and in location l^3_j differ. We can deduce that $l^3_j \notin v_j$ and thus $l^3_j \in w_j$, $m = 3$ and $w_i = v_i$ for all $i \neq j$. We have $l^1_j = \alpha^{s_j}_j$ and $l^3_j = \alpha^{s'_j}_j$ and $|s_j - s'_j| = 1$.

We first consider the case that $s'_j = s_j + 1$, i.e., $l^1_j = \alpha^{s_j}_j$, $l^2_j = \alpha^{s_j+}_j$ and $l^3_j = \alpha^{s_j+1}_j$. Since there is a transition from (l^1, u^1) to (l^2, u^2), we can deduce that the switch condition $\Lambda^{s_j}_j(l^1)$ evaluates to $true$. Thus, there exists a subset ω of $Pred(\alpha_j)$ such that $K_{\alpha_j, \omega} > s_j$ and $\lambda^\omega_j(l^1)$ is true. By definition of the resources, we have $R_j(s) \supset \omega$ and thus $K_{\alpha_j, R_j(s)} \geq K_{\alpha_j, \omega} > s_j$ It follows that $|s_j - K_{\alpha_j, R_j(s)}| - 1 = K_{\alpha_j, R_j(s)} - s_j - 1 = K_{\alpha_j, r_j(s)} - s'_j = |s'_j - K_{\alpha_j, R_j(s)}|$ and thus that $s \to s'$ is an edge in the state transition graph S_N.

The case that $s'_j = s_j - 1$ can be treated analogously. $\square$

In the above proof we used the most basic version of a timed automaton representing the network in question. Furthermore, we simplified the transition system T_A. Obviously, our modelling approach is designed to incorporate additional information about the biological system, such as information about the actual values of synthesis and decay rates. Thereby we obtain a more precise idea of the dynamics of the system. For example, we may be able to discard certain paths in the state transition graph that violate conditions involving the time delays (see the example presented in the next section). Furthermore, we can evaluate stability and feasibility of a certain behaviour, i.e., a path in the discrete transition system, in terms of clock interpretations that allow for that behaviour. The stricter the conditions the clock interpretations have to satisfy to permit a certain behaviour, the less allowance is made for fluctuations in the actual time delays of the genes involved.

The intermediate locations give supplementary information about the behaviour of the genes. For instance, it is possible to distinguish between a gene keeping the same expression level because there is no change in the expression levels of the genes influencing it, and the same behaviour due to alternating opposed influences. In the first case, the gene stays in the regular location representing the expression level, in the latter case it also traverses the corresponding intermediate locations.

Moreover, although this model uses asynchronous updates, it also allows for synchronous updates in the sense that two discrete transitions may occur at the same point in time. This may lead to paths in the transition system that are not incorporated in the state transition graph of the Thomas formalism.

To clarify the above considerations we give an illustrative example in the next section.

6   Bacteriophage λ

Temperate bacteriophages are viruses that can act in two different ways upon infection of a bacterium. If they display the lytic response, the virus multiplies, kills and lyses the cell. However, in some cases the viral DNA integrates into the bacterial chromosome, rendering the viral genome harmless for the so-called lysogenic bacterium. In [6], the formalism of Thomas is used to describe and analyse the genetic network associated with this behaviour. Figure 3 shows the simplified model they propose. We denote with X the gene cI and with Y the gene cro of the bacteriophage λ. The choice of the thresholds and parameter values is based on experimental data. They render the loop starting in X ineffective with respect to the dynamics. Thus we will omit it in the modelling of the timed automaton. The resulting state transition graph shows two possible behaviours. The steady state in $(1, 0)$ can be related to the lysogenic, the cycle comprising the states $(0, 1)$ and $(0, 2)$ to the lytic behaviour.

$$
\begin{aligned}
K_{X,\emptyset} &= 0 & K_{Y,\emptyset} &= 0 \\
K_{X,\{X\}} &= 0 & K_{Y,\{X\}} &= 1 \\
K_{X,\{Y\}} &= 1 & K_{Y,\{Y\}} &= 0 \\
K_{X,\{X,Y\}} &= 1 & K_{Y,\{X,Y\}} &= 2
\end{aligned}
$$

Fig. 3. Model of a network of bacteriophage λ in the Thomas formalism

Now let us analyse that network modelled as a timed automaton. The component corresponding to X is of the same form as A_1 in Fig. 2. Since Y influences X as well as itself, the corresponding component is slightly more complex. Both components are shown in Fig. 4. Furthermore, the figure displays graphs, which are condensed versions of the different transition systems derived from the timed automaton combining X and Y. With the exception of graph (c), the vertices of the graphs represent the expression levels of the genes, which correspond to the integer value of the location superscript. For instance, states (X^0, Y^{1-}) and (X^{0+}, Y^1) are both represented by $(0, 1)$. We analyse the dynamics of the system starting only from regular states. Thus, edges as well as paths in the graphs from some vertex $(j_1\, j_2)$ to a vertex $(i_1\, i_2)$ signify that the system can evolve from (X^{j_1}, Y^{j_2}) to a state where X and Y have expression level i_1 and i_2 respectively. Thereby it traverses states with expression levels corresponding to the vertices in the path, provided there is an actual point in time in which the genes acquire those expression levels. Again graph (c) is an exception to this representation and its analysis will clarify the distinction made.

We specify our model by choosing values for the maximal and minimal time delays. Set $t_Z^{k+} = t_Z^{k-} = 10$ and $t_Z^{(j,l)} = 5$ for all $Z \in \{X, Y\}$ and $k, j, l \in \{0, 1, 2\}$. That is to say, the time delays for synthesis and decay are all in the same range regardless of the gene and the expression level. If we declare a to be an overriding event, we avoid the possibility that there is a path from $(0, 0)$ to $(1, 1)$ in the

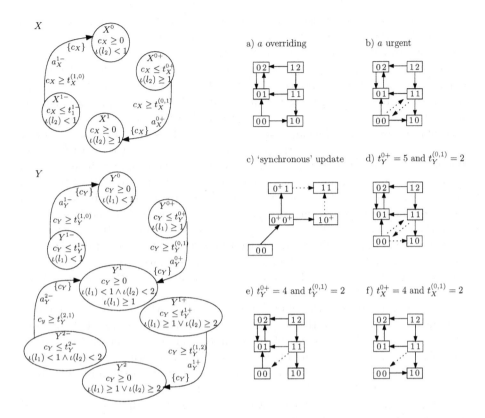

Fig. 4. The components X and Y representing the corresponding genes of the network in Figure 3. On the right, graphs representing the dynamical behaviour of the system derived from the transition systems resulting from different specifications of the model. Unless otherwise stated a is an urgent event and we set $t_Z^{k+} = t_Z^{k-} = 10$ and $t_Z^{(j,l)} = 5$ for all $Z \in \{X,Y\}$ and $k,j,l \in \{0,1,2\}$.

graph derived from the corresponding transition system. This is illustrated in Fig. 4 (a) and matches the state transition graph in Fig. 3. In (b), a is again an urgent event. We obtain two opposite edges between $(0,0)$ and $(1,1)$. However, there are very strict conditions posed on the time delays in order for the system to traverse those edges, which we drew dotted for that reason. To clarify the situation, we follow the path from $(0,0)$ to $(1,1)$ via the intermediate states shown in (c). A switch labelled with a leads to $(0^+, 0^+)$. Assuming that X reaches the next expression level faster than Y after a time delay $5 \leq r_X \leq 10$, we reach $(1, 0^+)$. In that situation two switches are enabled. One is labelled by a and leads to $(1,0)$. Since time is not allowed to pass, whenever the actual time r_Y that Y needs to reach the expression level 1 differs from r_X, that switch is taken. Only in the case that both time delays are exactly equal, the system will move via the switch labelled by a_Y^{0+} to $(1,1)$. Analogous considerations apply to the path via $(0^+, 1)$. It follows that although states $(0,0)$ and $(1,1)$ form a cycle in the graph, it is not plausible that the system will traverse that cycle. Once in the cycle, even the slightest perturbation of one of the time delays suffices for the system to leave the cycle. It is unstable.

These considerations apply not only to the edges representing synchronous update. In Fig. 4 (d) we change the values for t_Y^{0+} and $t_Y^{(0,1)}$ to express that the synthesis of Y is usually faster than that of X. The system can reach the state $(1,0)$ only if Y needs the maximal and X the minimal time to change their expression level. So, usually we would expect the system to reach the cycle comprising $(0,1)$ and $(0,2)$, corresponding to the lytic behaviour of the bacteriophage. If we know that Y is always faster than X in reaching the expression level 1, we can altogether eliminate both the edge leading from $(0,0)$ to $(1,0)$, and the one leading to $(1,1)$, as shown in (e). There is no clock interpretation satisfying the posed conditions. If we reverse the situation of X and Y, we eliminate the edges from $(0,0)$ to $(0,1)$ and $(1,1)$ as shown in (f). In this case, the system starting in $(0,0)$ will always reach the steady state $(1,0)$ representing the lysogenic response of the bacteriophage. The incorporation of data concerning the time delays can thus lead to a substantial refinement of the analysis of the dynamical behaviour.

We have implemented the above system in UPPAAL[1], a tool for analysing systems modelled as networks of timed automata (see [3]). Since UPPAAL uses product automata in the sense of the definition in [1], we had to make some modifications in the modelling of the components. Primarily, we converted the switch conditions to actual switches, which synchronise via the input of an external component that ensures the desired update mechanisms of the system. Using the UPPAAL model checking engine, we verified the above mentioned dynamical properties of the different specifications of our model.

7 Perspectives

In this paper, we introduced a discrete modelling approach that extends the established formalism of Thomas by incorporating constraints on the time delays occurring in the operations of biological systems. We addressed some of the advantages this kind of model offers, but naturally there is much room for future work. One of the most interesting possibilities the model provides is the evaluation of feasibility and stability of certain behaviours of the system by means of the constraints posed on the time delays. We may find cycles in the transition system (implying homeostatic behaviour of the real system), the persistence of which requires that equalities for time delays are satisfied. It is highly unlikely that a biological system will sustain a behaviour which does not allow for the slightest perturbance in its temporal processes. A cycle persisting for a range of values for each time delay will be a lot more stable. The merit of such considerations was already mentioned by Thomas (see [8]). It calls for a thorough analysis with mathematical methods as well as testing with substantial biological examples.

---

[1] http://www.uppaal.com

Furthermore, it seems worthwhile to relax some of the conditions posed by the Thomas formalism. Dropping constraint (1) would allow for a combination of genes to have a different influence (inhibition, activation) on the target gene than each would have on its own. It also could be advantageous to allow a gene product to influence a target gene depending on its concentration. For instance, it may be activating in low but inhibiting in high concentrations. That translates to the formalism by allowing multiple edges in the interaction graph.

We would like to close with some remarks regarding the analysis of the dynamics of our model. The theory of timed automata provides powerful results concerning analysis and verification of the model by means of model checking techniques. For example, CTL and LTL model checking problems can be decided for timed automata (see [2]). However, we face the state explosion problem and moreover the task to phrase biological questions in terms suitable for model checking. A thorough study of problems and possibilities of applying model checking techniques to answer biologically relevant questions using the modelling framework given in this paper seems necessary and profitable.

References

1. R. Alur. Timed Automata. In *Proceedings of the 11th International Conference on Computer Aided Verification*, volume 1633 of *LNCS*, pages 8–22. Springer, 1999.
2. R. Alur, T. Henzinger, G. Lafferriere, and G. Pappas. Discrete abstractions of hybrid systems. In *Proceedings of the IEEE*, 2000.
3. J. Bengtsson and W. Yi. Timed Automata: Semantics, Algorithms and Tools. In *Lecture Notes on Concurrency and Petri Nets*, volume 3098 of *LNCS*, pages 87–124. Springer, 2004.
4. G. Bernot, J.-P. Comet, A. Richard, and J. Guespin. Application of formal methods to biological regulatory networks: extending Thomas' asynchronous logical approach with temporal logic. *J. Theor. Biol.*, 229:339–347, 2004.
5. E. H. Snoussi. Logical identification of all steady states: the concept of feedback loop characteristic states. *Bull. Math. Biol.*, 55:973–991, 1993.
6. D. Thieffry and R. Thomas. Dynamical behaviour of biological regulatory networks - II. Immunity control in bacteriophage lambda. *Bull. Math. Biol.*, 57:277–297, 1995.
7. R. Thomas and R. d'Ari. *Biological Feedback*. CRC Press, 1990.
8. R. Thomas and M. Kaufman. Multistationarity, the basis of cell differentiation and memory. II. Logical analysis of regulatory networks in terms of feedback circuits. *Chaos*, 11:180–195, 2001.

A Computational Model for Eukaryotic Directional Sensing

Andrea Gamba[1], Antonio de Candia[2], Fausto Cavalli[3], Stefano Di Talia[4]
Antonio Coniglio[2], Federico Bussolino[5], and Guido Serini[5]

[1] Department of Mathematics, Politecnico di Torino, and INFN – Unit of Turin,
10129 Torino, Italia
andrea.gamba@polito.it
[2] Department of Physical Sciences, University of Naples "Federico II" and
INFM – Unit of Naples, 80126, Napoli, Italia
[3] Dipartimento di Matematica, Università degli Studi di Milano, via Saldini 50,
20133 Milano, Italia
fausto.cavalli@mat.unimi.it
[4] Laboratory of Mathematical Physics, The Rockefeller University, New York,
NY 10021, USA
stefano@orbis.rockefeller.edu
[5] Department of Oncological Sciences and Division of Molecular Angiogenesis,
IRCC, Institute for Cancer Research and Treatment, University of Torino School of
Medicine, 10060 Candiolo (TO), Italia
guido.serini@ircc.it

Abstract. Many eukaryotic cell types share the ability to migrate directionally in response to external chemoattractant gradients. This ability is central in the development of complex organisms, and is the result of billion years of evolution. Cells exposed to shallow gradients in chemoattractant concentration respond with strongly asymmetric accumulation of several signaling factors, such as phosphoinositides and enzymes. This early symmetry-breaking stage is believed to trigger effector pathways leading to cell movement. Although many factors implied in directional sensing have been recently discovered, the physical mechanism of signal amplification is not yet well understood. We have proposed that directional sensing is the consequence of a phase ordering process mediated by phosphoinositide diffusion and driven by the distribution of chemotactic signal. By studying a realistic computational model that describes enzymatic activity, recruitment to the plasmamembrane, and diffusion of phosphoinositide products we have shown that the effective enzyme-enzyme interaction induced by catalysis and diffusion introduces an instability of the system towards phase separation for realistic values of physical parameters. In this framework, large reversible amplification of shallow chemotactic gradients, selective localization of chemical factors, macroscopic response timescales, and spontaneous polarization arise.

1 Introduction

A wide variety of eukaryotic cells are able to respond and migrate directionally in response to external chemoattractant gradients. This behavior is essential for

C. Priami (Ed.): CMSB 2006, LNBI 4210, pp. 184–195, 2006.
© Springer-Verlag Berlin Heidelberg 2006

a variety of processes including angiogenesis, nerve growth, wound healing and embryogenesis. Perhaps the most distinguished chemotactic response is exemplified by neutrophils as they navigate to sites of inflammation. When exposed to an attractant gradient, these cells quickly orient themselves and move using anterior pseudopod extension together with posterior contraction and retraction. This highly regulated amoeboid motion can be achieved in the presence of very shallow attractant gradients.

The signaling factors responsible for this complex behavior are now beginning to emerge. The general picture obtained from the analysis of chemotaxis in different eukaryotic cell types indicates that, in the process of directional sensing, a shallow extracellular gradient of chemoattractant is translated into an equally shallow gradient of receptor activation [14] that in turn induces the recruitment of the cytosolic enzyme phosphatidylinositol 3-kinase PI3K to the plasmamembrane, where it phosphorylates PIP_2 into PIP_3. However, phosphoinositide distribution does not simply mirror the receptor activation gradient, but rather a strong and sharp separation in PIP_2 and PIP_3-rich phases arises, realizing a powerful and efficient amplification of the external chemotactic signal. PIP_3 acts as a docking site for effector proteins that induce cell polarization [3], and eventually cell motion [11]. Cell polarization can be decoupled from directional sensing by the use of inhibitors of actin polymerization so that cells are immobilized, but respond with the same signal amplification of untreated cells [8]. The action of PI3K is counteracted by the phosphatase PTEN that dephosphorylates PIP_3 into PIP_2 [14]. PTEN localization at the cell membrane depends upon the binding to PIP_2 of its first 16 N-terminal aminoacids [7].

2   A Phase Separation Process

In physical terms, the process of directional sensing shows the characteristic phenomenology of phase separation [12]. However, it is not clear which mechanism could be responsible for it. In known physical models, such as binary alloys, phase separation is the consequence of some kind of interaction among the constituents of a system, which can favor their segregation in separated phases [13]. However, one can show [5] that, even in the absence of direct enzyme-enzyme or phosphoinositide-phosphoinositide interactions, catalysis and phosphoinositide diffusion mediate an effective interaction among enzymes, which is sufficient to drive the system towards phase separation. To this aim, we have simulated the kinetics of the network of chemical reactions that represents the ubiquitous biochemical backbone of the directional sensing module. Since the chemical system is characterized by extremely low concentrations of chemical factors and evolution takes place out of equilibrium, we used a stochastic approach [6,2]. Indeed, rare, large fluctuations are likely to be relevant for kinetics in the presence of unstable or metastable states. Simulated reactions and diffusion processes taking place in the inner face of the cell plasmamembrane are

1. PI3K(cytosol)+Rec(i) $\rightleftharpoons$ PI3K·Rec(i)
2. PTEN(cytosol)+$PIP_2(i)$ $\rightleftharpoons$ PTEN·$PIP_2(i)$

3. PI3K·Rec(i)+PIP$_2$(i) → PI3K·Rec(i)+PIP$_3$(i)
4. PTEN·PIP$_2$(i)+PIP$_3$(i) → PTEN·PIP$_2$(i)+PIP$_2$(i)
5. PIP$_2$(i) → PIP$_2$(j)
6. PIP$_3$(i) → PIP$_3$(j)

where index i represents a generic plasmamembrane site and j one of its nearest neighbours (see also Fig. 1). The probability of performing a simulated reac-

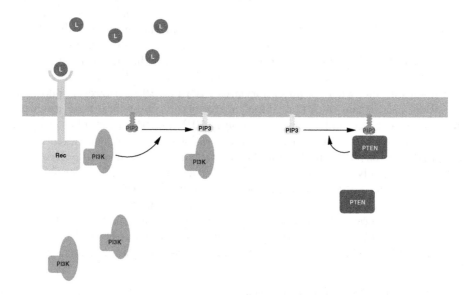

Fig. 1. Biochemical scheme of the simulated reaction network

tion on a given site is proportional to realistic kinetic reaction rates and local reactant concentrations (Tables 1, 2). In order to mimick experiments on immobilized cells treated with actin inhibitors the plasmamembrane is represented as a spherical surface of radius $R = 10\mu$m. This allows to study the phenomenon of directional sensing without the additional complexity introduced by dynamical changes in the cell morphology leading to a polarized, elongated form. The system is partitioned in $N_s = 10242$ computational sites, which are sufficiently large to host hundreds of phospholipid molecules but small enough to allow for a correct resolution of self-organized phospholipid patches. The cell cytosol is represented as an unstructured reservoir containing a variable number of PI3K and PTEN enzymes, which can bind and unbind to the cell membrane according to the rules described in Table 1. Chemical factors localized in the cytosol are indicated in Table 1 with the corresponding subscript, while factors attached to the membrane are indicated with a subscript representing the membrane site where they are localized. PIP$_2$ and PIP$_3$ molecules are assumed to freely diffuse on the cell membrane with the diffusion coefficient D specified in Table 2. Surface diffusivity of PI3K and PTEN molecules bound to phosphoinositides is

Table 1. Probabilities of chemical reactions and diffusion processes. Let X·Y denote the bound state of species X and Y, [X] the global concentration of species X in the whole cell, $[X]_{\text{cyto}}$ the cytosolic concentration, and $[X]_i$ the local concentration on plasmamembrane site i. The rate for a given reaction on site i is denoted by f_i, V is the cell volume, $\sum'$ denotes sum over nearest neighbours, and $(x)_+ = x$ for positive x and 0 otherwise. Time is advanced as a Poisson process of intensity equal to the reciprocal of the sum of the frequencies for all the processes. The simulations were performed using the values for kinetic rates and Michaelis-Menten constants given in Table 2.

Reaction	f_i
PI3K(cytosol)+Rec(i) $\rightarrow$ PI3K·Rec(i)	$\frac{V}{N_s} k_{\text{ass}}^{\text{Rec}} [\text{Rec}]_i [\text{PI3K}]_{\text{cyto}}$
PI3K(cytosol)+Rec(i) $\leftarrow$ PI3K·Rec(i)	$\frac{1}{N_s} k_{\text{diss}}^{\text{Rec}} [\text{Rec} \cdot \text{PI3K}]_i$
PTEN(cytosol)+PIP$_2$(i) $\rightarrow$ PTEN·PIP$_2$(i)	$\frac{V}{N_s} k_{\text{ass}}^{\text{PIP}_2} [\text{PIP}_2]_i [\text{PTEN}]_{\text{cyto}}$
PTEN(cytosol)+PIP$_2$(i) $\leftarrow$ PTEN·PIP$_2$(i)	$\frac{1}{N_s} k_{\text{diss}}^{\text{PIP}_2} [\text{PIP}_2 \cdot \text{PTEN}]_i$
PI3K·Rec(i)+PIP$_2$(i) $\rightarrow$ PI3K·Rec(i)+PIP$_3$(i)	$k_{\text{cat}}^{\text{PI3K}} \frac{[\text{Rec}\cdot\text{PI3K}]_i [\text{PIP}_2]_i}{K_M^{\text{PI3K}} + [\text{PIP}_2]_i}$
PTEN·PIP$_2$(i)+PIP$_3$(i) $\rightarrow$ PTEN·PIP$_2$(i)+PIP$_2$(i)	$k_{\text{cat}}^{\text{PTEN}} \frac{[\text{Rec}\cdot\text{PTEN}]_i [\text{PIP}_3]_i}{K_M^{\text{PTEN}} + [\text{PIP}_3]_i}$
PIP$_2$(i)$\rightarrow$PIP$_2$(j)	$\frac{D}{\sqrt{3}S_{\text{site}}} \sum' ([\text{PIP}_2]_i - [\text{PIP}_2]_j)_+$
PIP$_3$(i)$\rightarrow$PIP$_3$(j)	$\frac{D}{\sqrt{3}S_{\text{site}}} \sum' ([\text{PIP}_3]_i - [\text{PIP}_3]_j)_+$

Table 2. Physical and kinetic parameters used in the simulations

Parameter	Value	Parameter	Value
R	10.00 μm	$k_{\text{cat}}^{\text{PI3K}}$	1.00 s^{-1}
[Rec]	0.00-50.00 nM	$k_{\text{cat}}^{\text{PTEN}}$	0.50 s^{-1}
[PI3K]	50.00 nM	K_M^{PI3K}	200.00 nM
[PTEN]	50.00 nM	K_M^{PTEN}	200.00 nM
[PIP$_2$]	500.00 nM	$k_{\text{ass}}^{\text{Rec}}$	50.00 (s μM)$^{-1}$
D	0.10-1.00 μm^2/s	$k_{\text{ass}}^{\text{PIP}_2}$	50.00 (s μM)$^{-1}$
$k_{\text{diss}}^{\text{Rec}}$	0.10 s^{-1}	$k_{\text{diss}}^{\text{PIP}_2}$	0.10 s^{-1}

neglected, since it is expected to be much less than the diffusivity of free phosphoinositides. Reaction-diffusion kinetics is simulated according to Gillespie's method [6], generalized to the case of an anisotropic environment. For each iteration, reaction probabilities are computed for each site according to the formulae given in Table 1. Catalytic processes are described by Michaelis-Menten kinetics. The density of activated receptors is proportional to extracellular chemoattractant concentration. The probability of diffusion from a computational site to a neighboring one is assumed to be proportional to the difference in local concentrations, according to Ficks law. A site and a reaction are chosen at random, according to the computed probabilistic weights, and the reaction is performed on the chosen site, meaning that the concentration tables are adjourned according to the reaction stoichiometry. Time is then advanced as a Poisson process of intensity proportional to the reciprocal of the sum of all of the frequencies. This procedure, repeated over many different realizations, correctly approximates the stochastic process described in Table 1.

A convenient order parameter measuring the degree of phase separation of the phosphoinositide mixture is Binder's cumulant [1]

$$g = \frac{1}{2}\left(3 - \frac{\langle(\varphi - \langle\varphi\rangle)^4\rangle}{\langle(\varphi - \langle\varphi\rangle)^2\rangle^2}\right)$$

where $\varphi = \varphi_i = [\text{PIP}_3]_i - [\text{PIP}_2]_i$ is a difference of local concentrations on site i and $\langle\cdots\rangle$ denotes average over many different random realizations. The cumulant is zero when φ has a Gaussian distribution representing a uniform mixture, and becomes of order 1 when the φ distribution is given by two sharp peaks, representing separation in two well-distinct phases.

Spontaneous or signal-driven phase symmetry breaking leads to the formation of PIP_2, PIP_3 rich clusters of different sizes. Cluster sizes can be characterized by harmonic analysis. For each realization, the fluctuations $\delta\varphi = \varphi - \langle\varphi\rangle$ of the φ field can be expanded in spherical harmonics. Let us consider the two-point correlation functions

$$\langle\delta\varphi(\boldsymbol{u})\delta\varphi(\boldsymbol{u}')\rangle = \sum_{l=1}^{+\infty} C_l P_l(\boldsymbol{u} \cdot \boldsymbol{u}')$$

where P_l are Legendre polynomials. When most of the weight is concentrated on the l-th harmonic component, average phosphoinositide clusters extend over the characteristic length $\pi R/2l$. In particular, a large weight concentrated in the first harmonic component corresponds to the separation of the system in two complementary clusters, respectively rich in PIP_2 and PIP_3.

3   Dynamic Phase Diagram

We have run many random realizations of the system for different (ρ, D) pairs, where ρ is the surface concentration of activated receptors and D is phosphoinositide diffusivity. For each random realization we started from a stationary homogeneous PTEN, PIP_2 distribution. At time $t = 0$ receptor activation was switched on; either activated receptors were isotropically distributed or the isotropic distribution was perturbed with a linear term producing a 5% difference in activated receptor density between the North and the South poles. In the isotropic case, we found that in a wide region of parameter space the chemical network presents an instability with respect to phase separation, i.e. the homogeneous phosphoinositide mixture realized soon after receptor activation is unstable and tends to decay into spatially separated PIP_2 and PIP_3 rich phases.

Characteristic times for phase separation vary from the order of a minute to that of an hour, depending on receptor activation. The dynamic behavior and stationary state of the system strongly depend on the values of two key parameters: the concentration ρ of activated receptors and the diffusivity D. In the case of anisotropic stimulation, orientation of PIP_2 and PIP_3 patches clearly correlates with the signal anisotropy (see Fig. 2) In the anisotropic case, phase separation takes place in a larger region of parameter space and in times that can be shorter by one order of magnitude.

Average phase separation times as functions of receptor activation $\rho = [Rec]$ and diffusivity D are plotted for isotropic activation in Fig. 3a and for 5% anisotropic activation in Fig 3c. Light areas correspond to non phase-separating systems. In the dark areas phase separation takes place in less than 5 minutes of simulated time, while close to the boundary of the broken symmetry region phase separation can take times of the order of an hour.

Average cluster sizes at stationarity are plotted in Figs. 3b,d. In the light region, cluster sizes are of the order of the size of the system, corresponding to the formation of pairs of complementary PIP_2 and PIP_3 patches (Fig. 2).

For diffusivities smaller than 0.1 $\mu m^2/s$ the diffusion-mediated interaction is unable to establish correlations on lengths of the order of the size of the system and one observes the formation of clusters of separated phases of size much smaller than the size of the system.

For diffusivities larger than 2 $\mu m^2/s$ the tendency to phase separation is contrasted by the disordering action of phosphoinositide diffusion. Average phase separation times for the anisotropic case are plotted in Fig. 3c.

By comparing the isotropic and the anisotropic case it appears that there is a large region of parameter space where phase separation is not observed with isotropic stimulation, while a 5% anisotropic modulation of activated receptor density triggers a fast phase separation process. Cluster sizes are in the average larger in the anisotropic case than in the isotropic case.

The transition from a phase-separating to a phase-mixing regime results from a competition between the ordering effect of the interactions and the disordering effect of molecular diffusivity. The frontier between these two regimes varies continuously as a function of parameters. Importantly, we found that the overall phase separation picture is robust with respect to parameter perturbations, since it persists even for concentrations and reaction rates differing from those of Table 3 by one order of magnitude.

It is also worth noticing that both in isotropic and anisotropic conditions, signal amplification is completely reversible. Switching off receptor activation abolishes phase separation, delocalizes PI3K from the plasmamembrane to the cytosol, and brings the system back to the quiescent state.

Physically, the mechanism leading to cluster formation can be understood as follows. Receptor activation shifts the chemical potential for PI3K, which is thus recruited to the plasmamembrane. PI3K catalytic activity produces PIP_3 molecules from the initial PIP_2 sea. Initially, the two phosphoinositide species are well mixed. Fluctuations in PIP_2 and PIP_3 concentrations are however enhanced by preferential binding of PTEN to its own diffusing phosphoinositide product, PIP_2. Binding of a PTEN molecule to a cell membrane site induces a localized transformation of PIP_3 into PIP_2, resulting in higher probability of binding other PTEN molecules at neighboring sites. This positive

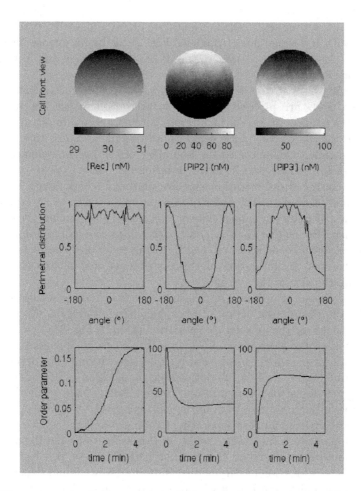

Fig. 2. Phase separation in the presence of 5% anisotropic receptor activation switched on as described in the text. The 5% activation gradient pointed in the upward vertical direction. First row: cell front view. Second row: concentrations measured along the cell perimeter and normalized with their maximum value (observe that the anisotropic component in the distribution of activated receptors is so small that it is masked by noise). Third row: time evolution of Binder's parameter g, showing its variation in time from zero (homogeneous PIP$_2$-PIP$_3$ mixture) to nonzero values (phase separation between the two species). First column: receptor activation. Second column: PIP$_2$ concentration. Third column: PIP$_3$ concentration. Observe the complementarity in the PIP$_2$ and PIP$_3$ distribution: regions rich in PIP$_2$ are poor in PIP$_3$ and vice-versa. This complementarity is biochemically necessary to trigger cell motion, and results from the fact that PIP$_2$ rich regions are also rich in the PTEN phosphatase, which dephosphorylates PIP$_3$.

feedback loop not only amplifies the inhibitory PTEN signal, but via phosphoinositide diffusion it also establishes spatio-temporal correlations that enhance the probability of observing PTEN enzymes at neighboring sites as well.

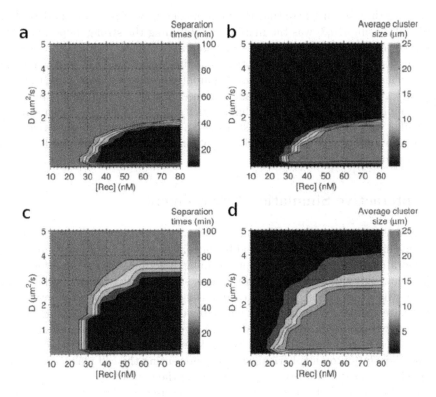

Fig. 3. Dynamic phase diagram. Average phase separation times and average cluster sizes are shown using a grayscale as functions of receptor activation [Rec] and diffusivity D, for isotropic and 5% anisotropic activation. In the isotropic case, panels show: (**a**) Average phase separation time. (**b**) Average cluster size as a function of [Rec] and D. In the anisotropic case, panels show: (**c**) Average phase separation time. (**d**) Average cluster size. For anisotropic activation phase separation is faster, takes place in a larger region of parameter space, and is correlated with the anisotropy direction.

If strong enough, this diffusion-induced interaction drives the system towards spontaneous phase separation[1]. The time needed by the system to fall into the more stable, phase-separated phase can however be a long one if the symmetric, unbroken phase is metastable. In that case, a small anisotropic perturbation in the pattern of receptor activation can be enormously amplified by the system instability.

---

[1] We speak here of "spontaneous" phase separation since in the case of isotropic stimulation no term describing the stochastic evolution of the system explicitly breaks its spherical symmetry, however, separation into asymmetric clusters occurs nevertheless as a result of random fluctuations. This differs from the situation of anisotropic stimulation, where asymmetry is introduced "by hands" from the very beginning.

It is worth observing here that the main inadequacy of previous models of directional sensing [10,8] was the inability of obtaining the strong, experimentally observed amplification without the introduction of many "ad hoc", unproved hypotheses about the structure of the signalling networks. On the other hand, in our phase separation model many apparently conflicting aspects of the phenomenology of phase separation, such as insensitivity to uniform stimulation, large amplification of the extracellular signal, and stochastic polarization, are reconciled with almost no effort, just using biochemically well-characterized factors which are already known to play a role in directional sensing and realistic diffusion and reaction rates.

4   Interactive Simulation Environment

A physical and computational modeling approach can prove useful in testing and unveiling the identity of minimal biochemical networks whose dynamics can incorporate the blueprints for complex cellular functions, such as chemotaxis.

To allow easy experimentation of the phase-separation paradigm and its comparison with other models of directional sensing we have developed a Java-based, easy to use simulation environment where physical and chemical parameters can be set at will and the surface distribution of the relevant chemical factors can be observed in real time. The underlying kinetic code simulates the stochastic chemical evolution on a spherical surface representing the cell plasmamembrane coupled to an enzymatic reservoir representing the cytosol.

The physical and chemical parameters to be used in the simulation can be assigned using input boxes (Fig. 4). By default, a constant activated receptor concentration $Crec$ with a superimposed concentration anisotropy of $Veps$ percents developing along the vertical direction, from bottom to top, is simulated. The user can choose to substitute this linear concentration gradient with an activation landscape produced by localized external sources. External sources can be added specifying their coordinates with respect to the center of the cell and the rate of chemoattractant release.

By default, during the simulation the local concentration difference between PIP_3 and PIP_2 as well as the order parameter g are visualized in real time. The user can require the visualization of other quantities of interest. The required graphs can be organized in a table containing the desired number of columns and rows. Three-dimensional graphs can be easily rotated by dragging any of them with the mouse. Simulation results are shown in a separate window (Fig. 4). When the simulation starts, a "Control" window appears showing the number of seconds of simulated time. After the end of the simulation, the buttons on the "Control" window can be used to pan the simulation movie forward or backwards in time. Most physical and kinetic parameters can be modified in real time while the simulation is running.

The simulation environment will be made publicly available under the GPL licence.

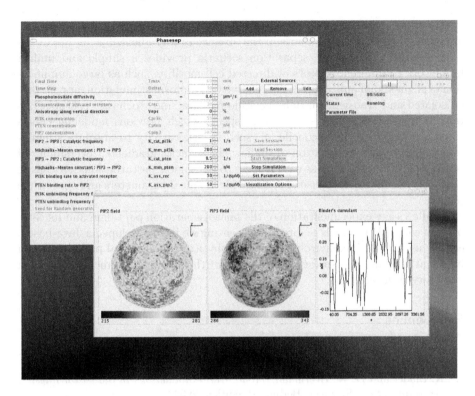

Fig. 4. Interactive simulation environment. A parameter window, a control window and a visualization window are shown. In the parameter window the user can input the values of diffusion and reaction rates, and the coordinates and intensities of one or more localized chemoattractant sources. The visualization in real time of different quantities, such as local or global concentration of chemical factors, can be required. Simulation sessions can be saved and recalled for future analysis.

5   Conclusions

Our results provide a simple physical cue to the enigmatic behavior observed in eukaryotic cells. There is a large region of parameter space where the cell can be insensitive to uniform stimulation over very large times, but responsive to slight anisotropies in receptor activation in times of the order of minutes. Accordingly, simulating shallow gradients of chemoattractant we observed PIP$_3$ patches accumulating with high probability on the side of the plasmamembrane with higher concentration of activated receptors, thus resulting into a large amplification of the chemotactic signal. Moreover, we identified an intermediate region of parameters, where phase separation under isotropic stimulation is observed on average in a long but finite time. In this case, one would predict that on long time scales cells undergo spontaneous polarization in random directions, and that the number of polarized cells grows with time. Intriguingly, this peculiar motile behavior is known as chemokinesis and is observed in cell motility experiments when cells are exposed to chemoattractants in the absence of a gradient [9].

In summary, the phase separation scenario provides a simple and unified framework to different aspects of directed cell motility, such as large amplification of slight signal anisotropies, insensitivity to uniform stimulation, appearance of isolated and transient phosphoinositide patches, and stochastic cell polarization. It unifies apparently conflicting aspects which previous modeling efforts could not satisfactorily reconcile [4], such as insensitivity to absolute stimulation values, large amplification of shallow chemotactic gradients, reversibility of phase separation, robustness with respect to parameter perturbations, stochastic character of cell response, use of realistic biochemical parameters and space-time scales.

To allow easy experimentation of the phase-separation paradigm and its comparison with other models of directional sensing we have developed a Java-based, easy to use simulation environment where physical and chemical parameters can be set at will and the surface distribution of the relevant chemical factors can be observed in real time.

References

1. K. Binder. Theory of first-order phase transitions. *Rep. Prog. Phys.*, 50:783–859, 1987.
2. K. Binder and D. W. Heermann. *Monte Carlo Simulations in Statistical Physics. An Introduction.* Springer, Berlin, 4th edition, 2002.
3. P.J. Cullen, G.E. Cozier, G. Banting, and H. Mellor. Modular phosphoinositide-binding domains–their role in signalling and membrane trafficking. *Curr Biol*, 11(21):R882–93, 2001.
4. P. Devreotes and C. Janetopoulos. Eukaryotic chemotaxis: distinctions between directional sensing and polarization. *J. Biol. Chem.*, 278:20445–20448, 2003.
5. A. Gamba, A. de Candia, S. Di Talia, A. Coniglio, F. Bussolino, and G. Serini. Diffusion limited phase separation in eukaryotic chemotaxis. *Proc. Nat. Acad. Sci. U.S.A.*, 102:16927–16932, 2005.
6. D.T. Gillespie. Exact Stochastic Simulation of Coupled Chemical Reactions. *J. Phys. Chem.*, 81:2340–2361, 1977.
7. Iijima M., Huang E. Y., Luo H. R., Vazquez F., and Devreotes P.N. Novel Mechanism of PTEN Regulation by Its Phosphatidylinositol 4,5-Bisphosphate Binding Motif Is Critical for Chemotaxis. *J. Biol. Chem.*, 16:16606–16613, 2004.
8. C. Janetopoulos, L. Ma, P.N. Devreotes, and P.A. Iglesias. Chemoattractant-induced phosphatidylinositol 3,4,5-trisphosphate accumulation is spatially amplified and adapts, independent of the actin cytoskeleton. *Proc. Natl. Acad. Sci. U. S. A.*, 101(24):8951–6, 2004.
9. D. A. Lauffenburger and A. F. Horwitz. Cell migration: a physically integrated molecular process. *Cell*, 84:359–369, 1996.
10. A. Levchenko and P.A. Iglesias. Models of eukaryotic gradient sensing: application to chemotaxis of amoebae and neutrophils. *Bioph. J.*, 82:50–63, 2002.
11. A.J. Ridley, M.A. Schwartz, K. Burridge, R.A. Firtel, M.H. Ginsberg, G. Borisy, J.T. Parsons, and A.R. Horwitz. Cell migration: integrating signals from front to back. *Science*, 302(5651):1704–9, 2003.

12. M. Seul and D. Andelman. Domain shapes and patterns: the phenomenology of modulated phases. *Science*, 267:476–483, 1995.
13. H.E. Stanley. *Introduction to Phase Transitions and Critical Phenomena*. Oxford University Press, 1987.
14. P.J.M. Van Haastert and P.N Devreotes. Chemotaxis: signalling the way forward. *Nat. Rev. Mol. Cell Biol.*, 626–624, 2004.

Modeling Evolutionary Dynamics of HIV Infection

Luca Sguanci[1,*], Pietro Liò[2], and Franco Bagnoli[1,*]

[1] Dept. Energy, Univ. of Florence, Via S. Marta 3, 50139 Firenze, Italy
`luca.sguanci@unifi.it, franco.bagnoli@unifi.it`
[2] Computer Laboratory, University of Cambridge, CB3 0FD Cambridge, UK
`pietro.lio@cl.cam.ac.uk`

Abstract. We have modelled the within-patient evolutionary process during HIV infection. We have studied viral evolution at population level (competition on the same receptor) and at species level (competitions on different receptors). During the HIV infection, several mutants of the virus arise, which are able to use different chemokine receptors, in particular the CCR5 and CXCR4 coreceptors (termed R5 and X4 phenotypes, respectively). Phylogenetic inference of chemokine receptors suggests that virus mutational pathways may generate R5 variants able to interact with a wide range of chemokine receptors different from CXCR4. Using the chemokine tree topology as conceptual framework for HIV viral speciation, we present a model of viral phenotypic mutations from R5 to X4 strains which reflect HIV late infection dynamics. Our model investigates the action of Tumor Necrosis Factor in AIDS progression and makes suggestions on better design of HAART therapy.

1 Introduction

Evolutionary biology was founded by Charles Darwin on the concept that organisms share a common origin and have subsequently diverged through time. Molecular phylogenetics has provided a statistical framework for estimating historical relationships among organisms, and it has supplied the raw data to test models of evolutionary and population genetic processes. Those have found practical uses in tracing the origins of pandemias and the routes of infectious disease transmission. Our ability to obtain molecular data has increased dramatically over the last two decades and large data sets describing a wide range of evolutionary distances are used in population genetic, phylogeny and epidemiological studies. Nevertheless, phylogenetic methods based on sequence information represent often an oversimplification when we aim at capturing the short time dynamics, i.e. the early stages of the speciation process. Population genetics focuses on this topic by investigating the behavior of mutations in populations. This discipline is related to the other important idea that Darwin expressed in The Origin of Species [1], that the exquisite match between a species and its environment is explained with natural selection, a process in which individuals with

---

* Also CSDC and INFN, sez. Firenze.

C. Priami (Ed.): CMSB 2006, LNBI 4210, pp. 196–211, 2006.
© Springer-Verlag Berlin Heidelberg 2006

beneficial mutations leave more offspring. Here we combine predictive quantitative theories of HIV evolution in the context of the selection pressure generated by the virus competition and the immune response. In particular phylogenies of the natural target of the HIV viruses, i.e. their cell receptors is combined with population genetics mathematical models. We show that combining the two leads to a better understanding of the complex molecular interaction underlying the macroscopically observable phenomena of HIV infection.

The smallest scale of molecular evolution generates genetic variability at population level. A special case is that of quasispecies which are clouds of very similar genotypes that appear in a population at mutation-selection balance [2]. Since the number of targets (the substrate) is limited, fitter clones tend to eliminate less fit mutants, which are subsequently regenerated by the mutation mechanism [3]. They are the combined result of mutations and recombination. Other sources of variability result from co-infection (simultaneous viral infection), superinfection (delayed secondary infection). On the contrary, selection and random drift decrease variability. The fact that deleterious or less fitted variants are not instantaneously counter selected allows for the coexistence and co-evolution of different strains of a virus within the same host. Although the conditions for the formation and survival of new strains have not always been understood, small scale evolution such as variability at population level may experience different mutation/selection balance than the genetic variability estimated from sequence analysis which represent fixed genotypes. Indeed, recent studies show that the rate of molecular evolution appears to accelerate when measured over evolutionary short timescales [4], which strongly contrast with substitution rates inferred in phylogenetic studies. Molecular virology studies appear the natural benchmark, given that viruses have usually very high mutation rates and large populations. We aim at modelling viral multi strain short and long term evolutionary dynamics during the immune response. The multi strains can be thought as viral populations. Since there is a tremendous lack of studies attempting at integrating population and phylogenetic studies, our work represents the efforts to link speciation at small and large evolutionary scale. This may result in a better understanding how to use the topology and branch lengths of existing species to predict future evolution.

In the next section we describe the relevant feature of the immune response which represents the selection pressure playing a key role in the speciation process. Then we use data from chemokine receptor sequences to estimate the rate of phenotype change in the virus and use this data to derive a selection-mutation model based on a set of differential equations. In the results we show that the models introduced are suited to model both short and long term evolutions. In particular we first show an example of speciation dynamics of viral population mediated by the immune sytem response. Then we model the phenotypic switch in co-receptor usage in HIV-1 infection and we also make some observations on the better design for HAART therapy. Finally we draw our conclusions.

1.1   Major Features of Within-Patient HIV Evolution

Although the process of adaptive change is difficult to study directly, natural selection has been repeatedly detected in the evolution of morphological traits (such as the beak of Darwin's finches). During HIV infection, the process of adaptation requires interaction with CD4 T cell and a chemokine receptor, either CXCR4 or CCR5. During early stages of HIV-1 infection, viral isolates most often use CCR5 to enter cells and are known as R5 HIV-1. Later in the course of HIV-1 infection, viruses that use CXCR4 in addition to CCR5 (R5X4) or CXCR4 alone (X4 variants) emerge in about 50% patients (switch virus patients) [5, 6]. These strains are syncytium-inducing and are capable of infecting not only memory T lymphocytes but also naïve CD4+ T cells and thymocytes through the CXCR4 coreceptor. The switch to use of CXCR4 has been linked to an increased virulence and with progression to AIDS, probably through the formation of cell syncytia and killing of T cell precursors. X4 HIV strains are rarely, if ever, transmitted, even when the donor predominantly carries X4 virus. CXCR4 is expressed on a majority of CD4+ T cells and thymocytes, whereas only about 5 to 25% of mature T cells and 1 to 5% of thymocytes express detectable levels of CCR5 on the cell surface [7]. It is noteworthy that X4 HIV strains stimulate the production of cellular factor called Tumor Necrosis Factor (TNF), which is associated with immune hyperstimulation, a state often implicated in T-cell depletion [8]. TNF seems able to both inhibit the replication of R5 HIV strains while having no effect on X4 HIV and to down regulate the number of CCR5 co-receptors that appear on the surface of T-cells [9].

2   Bioinformatics Analysis and Mathematical Models

We assume that the phylogenetic tree describes all sorts of genetic variants, i.e. quasispecies and species. Quasispecies appear at the leaves and are seen as single specie by the distant leaves. We make the assumptions that leaves that are very close experience the same environment, i.e. they compete for the same receptor targets. Therefore, the fitness landscape within short branch length distance is shaped by competition which decrease for longer distances.

2.1   Mutational Pathway from R5 to X4

A meaningful way to estimate the mutational pathways and phenotype difference between R5 and X4 is to use phylogenetic inference on chemokine receptors families. The statistical relationships among the species can be described using a tree. Let the phylogeny to be inferred be denoted Π. A node of Π is either a currently extant leaf node, with no descendants in Π, or an it internal node, with two or more child nodes in Π. A point of Π is defined to be any point at a node or on an edge of Π. Let t_i denote the time before present that point i was extant in Π. Let π_{ij} denote the path in Π between points i and j, and $|\pi_{ij}|$ its length. Thus, where j is an ancestor of i, $|\pi_{ij}| = t_j - t_i$. More generally, for any i and j, $|\pi_{ij}| = |\pi_{ik}| + |\pi_{kj}|$, where k is the last common ancestor (LCA) of

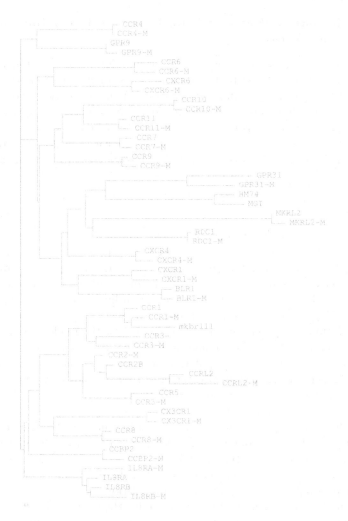

Fig. 1. The maximum likelihood phylogeny under the JTT+F+Γ model of evolution for the set of human and mouse (mouse sequences are labelled with "-M") chemokine receptors. We have considered only the external loop regions. The scale bar refers to the branch lengths, measured in expected numbers of amino acid replacements per site.

i and j. The tree parameters are topology and branch lengths. The assessment of phylogenies using distance and likelihood frameworks depend on the choice of an evolutionary model. We have computed the maximum likelihood (ML) analysis of the CRs data set using different models of evolution: Dayhoff [10], JTT [11], WAG [12], the amino acid frequencies of the data set, ('+F'), and the heterogeneity of the rates of evolution, implemented using a gamma distribution ('+Γ') [13, 14]. Bootstrap and permutation tests have been used to assess the robustness of the tree topology [15]. The tree may be used to estimate the pathways of substitutions which are supposed to have phenotype changes.

2.2   Mathematical Models of Viral Dynamics Under Immune Response Pressure

A meaningful model to study the genetics of population is that of *quasispecies*. This model, first introduced by Eigen [3] in the context of molecular evolution, describes the evolution of an infinite population of haploid individuals reproducing asexually. Each individuals has a given genotype $\sigma = (\sigma_1, \ldots, \sigma_N)$, constituted by a sequence of N symbols taken from an alphabet of size k, and is subject to the selection pressure. Mutations arise as copying errors during the reproduction process. The evolution of the concentration of a sequence, $x(\sigma, t)$, is given by:

$$\frac{dx(\sigma, t)}{dt} = \sum_{\sigma'} p(\sigma' \to \sigma) W(\sigma') x(\sigma', t) - \phi(\sigma, t) x(\sigma, t) \tag{1}$$

where $W(\sigma)$ represents the strength of the selection, $p(\sigma' \to \sigma)$ the mutation mechanism, and ϕ is a flux keeping constant the total concentration, $\sum_\sigma x(\sigma, t)$. The model can be used to model the evolution of a single quasispecies as well as extended to study the dynamics of interaction among n different populations. In the latter case we obtain a system of n first order, non-linear, differential equations.

Models using the notion of quasispecies have been adopted to study the biological evolution of populations and recently also for the modelling of the interaction between HIV-1 and the immune system [16]. Moreover, due to its intrinsic multiscale nature - indeed the population of sequences considered can be either that of genotypes or, more generally, the one of phenotypes - the model is suited to analyze both short and long range interactions.

In a phylogenetic framework a given leaf represents the common ancestor of the individuals coevolving. If we are interested in studying the short range evolution of the viral strains competing for the same co-receptor, we concentrate on a particular leaf of the phylogenetic tree. As a paradigmatic model, we may consider that introduced by Bagnoli et al. [17]. This model describes the speciation of a quasispecies population induced by competition.

In the model different individuals compete for the shared resources of a common environment, and this effect is reflected in the term corresponding to the selection strength. In particular, the growth rate W is expressed as

$$W(\sigma, t) = \exp[H_0(\sigma) - q(\sigma, t)] \tag{2}$$

where H_0 represents the static fitness (e.g. the environment) and the term $q(\sigma, t)$ accounts for the competition. We can think $q(\sigma, t)$ to be a function of the phenotypic distance between two different sequences, mimicking the fact that the competition is stronger for individuals sharing common habits. This competitive dynamics may lead to the speciation of the population. This event results in the appearance of new branches in the phylogentic tree and, as the selection pressure is continuously acting, the branches corresponding to the fittest individuals are eventually selected (see Fig. 2.2).

Now, considering the dynamics of interaction between viruses and immune system, the competition among different viral strains is induced by the immune

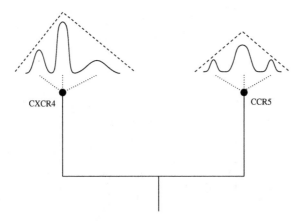

Fig. 2. Qualitative description of the short range evolution occuring on a phylogenetic tree, according to the competition model introduced by Bagnoli et al. [17]. Only two leaves (e.g. corresponding to CCR5 and CXCR4 co-receptors) are shown. In the figure we assume that only a single phenotypic character is continuosly varying and thus we assume a one-dimensional linear phenotypic space. Given the static fitness corresponding to different binding specificities (dashed line), we represent the effect of the speciation resulting from the induced competitive dynamics (solid line). The emerging new variants are then represented as dotted segments.

response. In this case a virus may escape the response by a T cell with high binding affinity, by differentiating enough. It's worth noting that this short-range dynamics alone may justify the stable multi strain infection reported in several patients (see. Sec. 3.2).

2.3   Long Range Competition and R5 to X4 Switching

Here we introduce a mathematical model to study the long range competition, mediated by the immune system response, occurring between different HIV-1 phenotypes around different leaves of the phylogenetic tree. Indeed, the viral quasispecies not only compete for using the same co-receptor (short range competition), but also for establishing a preferential chemokine signalling pathway (long-range competition). In someone who is newly infected by HIV, several variants of the virus, called R5, are often the only kind of virus that can be found. In about half of the people who develop advanced HIV disease, the virus begins to use another co-receptor called CXCR4 (X4 viral phenotype). This model supports the hypothesis that it may not be exhaustion of homeostatic responses, but rather thymic homeostatic inability along with gradual wasting of T cell supplies through hyper activation of the immune system that lead to CD4 depletion in HIV-1 infection.

We are interested in the switching in coreceptor usage and thus, by considering CD8+ cells to be at their equilibrium concentration and disregarding the effects of B cell, we concentrate on CD4 dynamics. We map the different leaves of the phylogenetic tree on a linear phenotypic space, composed by the viral

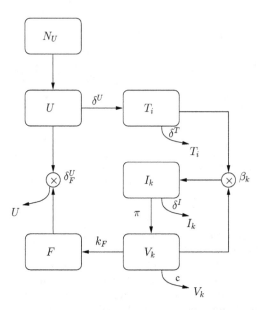

Fig. 3. Schematic description of the model for the switching from R5 to X4 viral phenotype. Naive T-cells, U, are generated at constant rate N_U and removed at rate δ^U. They give birth to differentiated, uninfected T-cells, T. These in turn are removed at constant rate δ^T and become infected as they interact with the virus. Infected T-cells, I, die at rate δ^I and contribute to the budding of viral particles, V, that are cleared out at rate c. As soon as the X4 phenotype arise, the production of the TNF starts, proportional to the X4 concentration and contribute to the clearance of naïve T-cells, via the δ_F^U parameter.

phenotypes competing for different co-receptor usage. At the beginning of the infection the only viral population present is that of R5 strains. Later on, as the infection evolves, we focus on the appearance of X4 viruses and on their subsequent interaction with R5 strains.

The model is the following:

$$\frac{dU}{dt} = N_U - \delta^U U - \delta_F^U UF \tag{3}$$

$$\frac{dT_i}{dt} = \delta^U U - \left(\sum_k \beta_k V_k\right) T_i - \delta^T T_i \tag{4}$$

$$\frac{dI_k}{dt} = \left(\sum_{k'} \mu_{kk'} \beta_{k'} V_{k'}\right)\left(\sum_i T_i\right) - \delta^I I \tag{5}$$

$$\frac{dV_k}{dt} = \pi I_k - c V_k \tag{6}$$

$$\frac{dF}{dt} = k_F \sum_{k \in X4} V_k \tag{7}$$

In the equations above, the variables modelled are the pool of immature CD4+ T cells, U, the different strains of uninfected and infected T cells (T and I, respectively), HIV virus, V, and the concentration of TNF, F. A schematic view of the model is depicted in Fig.3. The value of the parameters introduced are summarized in Table 2.3.

In particular, Equation (3) describes the constant production of immature T cells by the thymus N_U and their turning into mature T cells at rate δ^U. If X4 viruses are present, upon the interaction with TNF, immature T-cells are cleared at fixed rate δ_F^U.

Equation (4) describes how uninfected mature T cells of strain i are produced at fixed rate δ^U by the pool of immature T cells. Those cells, upon the interaction with any strain of the virus, V_k, become infected at rate $\beta_k = \beta \quad \forall k$. The infectiousness parameter, β, is not constant over time, but depends on the interplay between R5 and X4 viruses. In particular, due to the presence of TNF, the infectivity of R5 strains is reduced ($\beta_{R5}(t) = \beta - k_{R5}F(t)$), while the one of X4 viruses increases, with constant of proportionality k_{X4} ($\beta_{X4}(t) = \beta + k_{X4}\,F(t)$), mimicking the cell syncytium effect induced by the TNF molecule.

Table 1. Model for the R5 to X4 phenotypic switch: a summary of the additional parameters introduced. The value of the other parameters are medical literature referred, see also [18].

Parameter	Symbol	Value	Units of Meas.
Production of immature T cells	N_U	100	cell/μl t^{-1}
Death rate of immature T cells	δ^U	0.1	t^{-1}
Death rate of immature T cells upon the interaction with TNF	δ_F^U	10^{-5}	μl/cell t^{-1}
Decreasing infectivity of R5 phenotype due to TNF	k_{R5}	10^{-7}	(μl/cell)^2 t^{-1}
Increasing infectivity of R5 phenotype due to TNF	k_{X4}	10^{-7}	(μl/cell)^2 t^{-1}
Increasing death rate of immature T cells due to TNF	δ_{X4}^I	0.0005	μl/cell t^{-1}
Rate of production of TNF	k_F	0.0001	t^{-1}

Equation (5) describes the infection of mature T-cells. Infected T-cells of strain k arise upon the interaction of a virus of strain k with any of the mature T-cell strains. The infected cells, in turn, are cleared out at a rate δ^I. When TNF is released, this value increases linearly with constant δ_{X4}^I, $\delta^I(t) = \delta^I + \delta_{X4}^I\,F(t)$.

Equation (6) describes the budding capacity i.e. the mean number of virions produced in the unit of time by each infected T cell. We have used a value close to that reported in medical literature by [19].

Finally, in Equation (7), we model the dynamics of accumulation of TNF by assuming the increase in TNF level to be proportional, via the constant k_F, to the total concentration of X4 viruses present.

3   Results

3.1   Phenotype Change Patterns of R5 and X4 Strains

RNA viruses have been reported to have substitution rates of the order of $1 \cdot 10^{-3}$ substitution per site per replication [20]. Since a large fraction of amino acid substitutions are neutral or quasi-neutral to structural changes, they do not change dramatically the fitness of the virus [21]. Nevertheless, sometimes even a single mutation can change the fitness in a substantial way. In our model we take into consideration only non-synonymous mutations and, therefore, we explored values slight higher value than that, (i.e. 10^{-4} and 10^{-5}). These values can be compared with the phenotype changes required to bind to CCR5 or CXCR4 receptors. In other words, research into HIV dynamics has much to gain from investigating the evolution of chemokine co-receptor usage. Although CCR5 and CXCR4 are the major coreceptors used by HIV-1 a number of chemokine receptors display coreceptor activities in vitro. Also several other chemokine receptors, possibly not present on the T cell membrane, may act as targets. To date, a number of human receptors, specific for these chemokine subfamilies, have been described, though many receptors are still unassigned. Several viruses, for example Epstein-Barr, Cytomegalovirus, and Herpes Samiri, contain functional homologous to human CRs, an indication that such viruses may use these receptors to subvert the effects of host chemokines [22]. Cells different from CD4+ and CD8+ T cells, such as macrophages, express lower levels of CCR5 and CXCR4 on the cell surface [23–25], and low levels of these receptors expressed on macaque macrophages can restrict infection of some non-M-tropic R5 HIV-1 and X4 simian immunodeficiency virus (SIV) strains [26, 27].

Fundamental to the evolutionary approach is the representation of the evolution of sequences along lineages of evolutionary trees, as these trees describe the complex patterns of dependence amongst sequences that are caused by their common ancestry [12, 28, 29]. The ML tree, obtained using the JTT+F+Γ model of evolution, is shown in Figure 1. The topology clearly shows that the CCR family is not homogeneous: CCR6, CCR7, CCR9 and CCR10 are separated from the other CCRs; in particular, CCR10 clusters with CXCRs; CXCR4 and CXCR6 do not cluster with the CXCRs. The tree shows that there are many mutational steps between CCR5 and CXCR4. The phylogeny suggests that the mutations that allow the virus env to cover a wide phenotypic distance from R5 to X4, may also lead to visit other receptors. Since the external loops of CRs contain the binding specificities and have higher rates of evolution than internal loops and transmembrane segments [30], the tree Fig. 1 shows a relative longer mutational pathway between CCR5 and CXCR4 with respect to pathway linking CCR5 to other receptors.

3.2   Modeling Co-evolutive Dynamics and Speciation

Focusing on short term evolution we investigate how the co-evolutionary and competitive dynamics of viral strains, mediated by the immune response, may lead to the formation of new viral strains. In particular, if the recognition ability of viral antigens by T cells is non-uniform over different viral variants and the immune system does not discriminate among highly similar phenotypes, a competition is induced. In Fig. 4 we consider a phenotypic space composed by 25 different variants of the virus, and make a first inoculum at phenotype 15 (Fig. 4a), followed by a second delayed inoculum at phenotype 5 at time $t = 1$. The

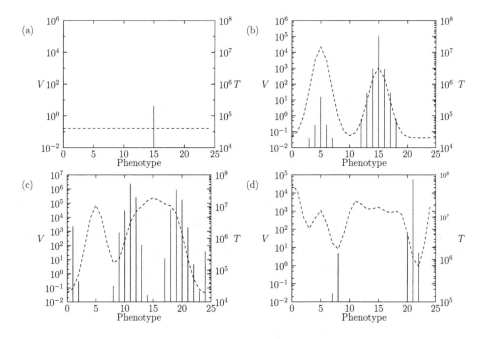

Fig. 4. Snapshots of competitive dynamics between different viral strains (vertical stems) and T lymphocites (dashed line) at four different times: $t = 0$ (a), $t = 4.5$ (b), $t = 5.25$ (c) and $t = 5.75$ (d). Virus strain 15 is present at time $t = 0$, while strain 5 is inoculated at time $t = 1$. Mutation rate $\mu = 10^{-4}$.

different interaction strength between T cells and viral phenotypes favors those viral phenotypes targeted by the weakest response. The result of the induced competition is the separation of the quasispecies centered around phenotype 15 into two clusters (quasi-speciation), Fig. 4c. It's worth noting that, due to the adaptive response by the immune system, a complex, time evolving co-evolution is established between viral populations and immune response (Figs. 4b-d).

3.3   R5 to X4 Switch and HAART Therapy

From the results derived in Sec. 3.1, it is now possible to get a better insight in the observed phenotypic switch in co-receptor usage by HIV-1 virus, by studying the coevolutive dynamics leading to X4 strain appearance by successive mutations of the ancestor R5 strain. In particular we may calculate the modelled time of switching in co-receptor usage. This time depends both on the mutation rate μ and on the phenotypic distance between R5 and X4 strains, d_P. By comparing the modelled value with the mean time inferred by the phylogenetic tree, we may tune the those model parameters to give the correct time of appearance of the X4 phenotype.

In Fig. 5 we observe the results of the stimulated production of TNF. Indeed, this regulate the interactions between immune response and the virus and

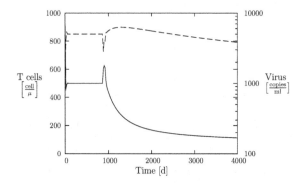

Fig. 5. Time evolution of the concentrations of uninfected T-cells (straight line) and viruses (dashed line), during R5 to X4 switch, occurring at time $t \approx 900$. The time of appearance of the X4 strains depends on the mutation rate and on the phenotypic distance between R5 and X4 viruses. After the appearance of the X4 phenotype a continuous slow decline in CD4+ T-cells level leads to AIDS phase (CD4 counts below 200cells/ml).

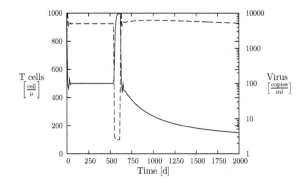

Fig. 6. The efficacy of HAART therapy may be disrupted by a sudden interruption in drugs treatment. If time has passed for mutations to populate the R5 strains closer to the X4 phenotypes, an earlier appearance of X4 strains may occur. Uninfected T-cells (straight line) and viruses (dashed line). Parameters as in Fig. 5.

between the different strains of HIV virus. The results of these interactions are a decline in T-cells level, leading to the AIDS phase of the disease, and the decline in levels of viruses using the R5 coreceptor. In the figure the temporal evolution of the infection is shown, with the appearance of the X4 strain, and the successive decline in T-cells abundances.

By using this model it is also possible to predict some scenarios in HAART treatment (see Fig. 6). This therapy is usually able to decrease the concentration of the virus in the blood and delay the X4 appearance. We have found time dynamics similar to those reported in [31]; see also [32,33]. Note that our

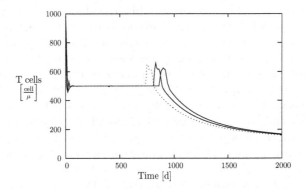

Fig. 7. CD4+ T-cells concentration during HIV-1 super-infection by a R5 viral strain. Evolution without superinfection, straight line; superinfection occurring at time $t=100$ and 400, dotted and dashed line, respectively. For a superinfection event occurring after the R5 to X4 switching the dynamics is qualitatively the same as for a single infection, (straight line). If the second delayed infection occurs before the R5 to X4 switching, the time of appearance of X4 viruses may be shorter, when the super-infecting strain is closer to the X4 phenotypes, (dotted and dashed line). Parameters as in Fig. 5.

model considers only the HIV virions which are in the blood. The clearance of virions hidden in cells or other tissues are known to be very slow [34, 35]. Now we investigate what may happen in the case of a sudden interruption in the use of the drugs. In Fig. 6 we observe how the X4 strain may appear sooner, if the different R5 strains experience the same selection pressure. In fact during the treatment the concentration of the different strains of R5 viruses is kept to a very low level while T-cell abundances increase. As the therapy is interrupted, all the strains give rise to a renewed infection. Now also the strains closer to the X4 co-receptor using viruses are populated, and a mutation leading to an X4 strain occurs sooner.

We have finally studied the case of a superinfection dynamics. In Fig. 7 we show T-cells evolution for different times of the superinfection event.

We may observe that if the superinfection occurs after the appearance of the X4, the new R5 strain does not have any effect on T-cells behavior. On the other hand is worth noting that if the new R5 inoculum take place before the X4 appearance, this may speed up the switching to the X4 phenotype if the new strain is mutationally close to the X4.

4   Discussion

Phylogenetic inference of chemokine receptors shows that there are several mutational patterns linking CCR5 to several receptors that have the same branch length of that from CCR5 to CXCR4. There is a massive abundance of signalling disruptions in the immune systems during AIDS progression, particularly after the transition R5 to X4. These disruptions may be due to variants of the virus which bind other chemokine receptors. This hypothesis also suggests that R5-late strains in not-X4 AIDS, which are known to be different from R5 early strains, may have accumulated mutations enabling them to interact with other chemokine receptors. Therefore, our model suggests the sooner the HAART the better, because the presence of a large number of R5 will increase the mutational spectra in R5 strains (late R5) and the probability of getting closer to the binding specificities of other chemokine receptors. Contrary to our phylogenetic statistical analysis, our mathematical model describes short term evolutionary dynamics through competitions among viruses at each tips of the tree. Following Kimura, we can subdivide mutations into advantageous, neutral or deleterious where the deleterious can be further subdivided into the proportions that are very slightly deleterious, and deleterious. Deleterious mutations are not expected to become fixed in large populations, but nevertheless can persist in the population for long periods of time. The average time before loss correlates with deleteriousness. Thus, as observation times diminish, we should observe a greater proportion of slightly deleterious mutations that have yet to be lost, with the most deleterious observed only in the short-term pedigree studies. For some reasons, the evolutionary continuum between variation at population genetics level and the long-term evolution has not been adequately studied. Although it is a continuum, the techniques required may change as the timescale decreases. For example, some concepts from long-term evolution (binary evolutionary trees with sequences studied only at the tips) have been extended into populations where trees are no longer binary, and ancestral sequences (at internal nodes) are still present in the population. There are hints that a formal multi-scale study is necessary.

The interest in HIV strain is motivated by concern about developing strain specific drugs. Quasispecies are likely the key for understanding the emerging infectious diseases and has implications for transmission, public health counselling, treatment and vaccine development. Moreover, the observed co-evolutionary dynamics of virus and immune response opens the way to the challenging possibility of the introduction or modulation of a viral strain to be used in therapy against an already present aggressive strain, as described by Schnell and colleagues [36]. The authors showed that the introduction of an engineered virus can achieve HIV load reduction of 92% and recovery of host cells to 17% of their normal levels (see also the mathematical model in Ref. [37]).

Different drug treatments can alter the spectrum of strains. Will R5 blocking drugs cause HIV to start using X4? And will that be worse than letting the R5-using virus stay around along at its own, slower, but no less dangerous activity?

Recent works show that TNF is a prognostic marker for the progression of HIV disease [8, 38]. We focused on both the inability of the thymus to efficiently compensate for even a relatively small loss of T cells precursors and on the role of TNF in regulating the interactions between the different strains of HIV virus. The second model we have introduced shows that keeping low the concentration of TNF, both the depletion of T-cells precursors repertoire and the R5 overcome by X4 strains slow down.

The model makes possible to investigate intermittency or switching dominance of strains and the arising of new dominant strains during different phases of therapy; how superinfection will evolve in case of replacement of drug-resistant virus with a drug-sensitive virus and acquisition of highly divergent viruses of different strains; to investigate whether antiviral treatment may increase susceptibility to superinfection by decreasing antigen load.

Let us extend the viral framework for a general understanding of the molecular evolutionary process along a tree under natural selection. If we focus on a quasispecies fitness landscape, the fitness' main component is probably related to the entrance of the virus in the cell, i.e. the interaction with the receptor. Other components are the budding characteristics and numerosity and the spectrum of mutants (hopeful monsters) generated. Therefore the height of the fitness curve mainly reflects the binding energy, while the windows of strain existence in the x axis reflects how many changes may still result in a sufficient binding. Our work may reveal relevant to phylogenetic studies on divergence date estimation which suffer from the difficulties of estimating the correct rate of molecular evolution for different branches. It is relatively straightforward to test if the data conform to a molecular clock . If the assumption of rate constancy does not hold across a tree and therefore the clock is rejected, however, the current methodology is lacking robustness in assessing the amount of relaxation from a clock hypothesis. Our approach in modeling the evolution of virus species is to investigate the different degree of competition among strains. Strains which are in the same fitness landscape have correlated rates of evolution. This agrees very well with the current use of local clock models which allow the molecular rate to vary throughout the tree, but with closely related species sharing similar rates. This approach is justified with the assumption that molecular rates are heritable because they are related to physiological, biochemical, and life-history characteristics of the species in question. That's precisely the idea of our local fitness landscape. Although the inference of rates is confounded by uncertainties in calibration points, by tree topology, and by asymmetric tree shapes, our present studies should be considered a theoretical framework for understanding how different smooth fitness landscapes which can be imagined at the leaves and nodes of a phylogenetic tree are linked by the topology and branch lengths reflecting a multiscale stepwise process of adaptation under natural selection.

References

1. Darwin, C. On the origin of species, 1859. New York University Press, 1988.
2. Biebricher C. K. and M. Eigen.2005. The error threshold. Virus Res. 107:117-27.
3. Eigen, M., and P. Schuster. 1977. The hypercycle. Naturwissenschaften 64:541-565.
4. Ho, D., A. U. Neumann, A. S. Perelson, W. Chen, J. M. Leonard M. and Markowitz. 1995. Rapid turnover of plasma virions and CD4 lymphocytes in HIV-1 infection. Nature 373:123-126.
5. Karlsson, I., L. Antonsson, Y. Shi, M. Oberg, A. Karlsson, J. Albert, B. Olde, C. Owman, M. Jansson and E. M. Fenyo. 2004. Coevolution of RANTES sensitivity and mode of CCR5 receptor use by human immunodeficiency virus type 1 of the R5 phenotype. J. Virol. 78:11807-11815.

6. Gorry, P.R., M. Churchill, S. M. Crowe, A. L. Cunningham and D. Gabuzda. 2005. Pathogenesis of macrophage tropic HIV. Curr. HIV Res. 3:53-60.

7. Gray, L., J. Sterjovski, M. Churchill, P. Ellery, N. Nasr, S. R. Lewin, S. M. Crowe, S. L. Wesselingh, A. L. Cunningham and P. R. Gorry. 2005. Uncoupling coreceptor usage of human immunodeficiency virus type 1 (HIV-1) from macrophage tropism reveals biological properties of CCR5-restricted HIV-1 isolates from patients with acquired immunodeficiency syndrome. Virology 337:384-98

8. Herbeuval, J. P., A. W. Hardy, A. Boasso, S. A. Anderson, M. J. Dolan, M. Dy, and G. M. Shearer. 2005. Regulation of TNF-related apoptosis-inducing ligand on primary CD4+ T cells by HIV-1: Role of type I IFN-producing plasmacytoid dendritic cells. Proc. Nat. Acad. Sci. USA 102:13974-13979.

9. Pavlakis, G. N., A. Valentin, M. Morrow and R. Yarchoan. Differential effects of TNF on HIV-1 expression: R5 inhibition and implications for viral evolution. Int Conf AIDS 2004 Jul 11-16; 15:(abstract no. MoOrA1048).

10. Dayhoff, M. O., R. M. Schwartz and B. C. Orcutt. 1978. A model of evolutionary change in proteins. In Atlas of Protein Sequence and Structure, (Vol. 5, Suppl. 3) (Dayhoff, M.O., ed.), pp. 345.352, National Biomedical Research Foundation.

11. Jones D. T., W. R. Taylor and J. M. Thornton. 1994. A mutation data matrix for transmembrane proteins. FEBS Lett. 339:269-75.

12. Whelan, S. and N. Goldman. 2001. A general empirical model of protein evolution derived from multiple protein families using a maximum likelihood approach. Mol Biol Evol. 18:691-699.

13. Yang, Z. (1994) Maximum likelihood phylogenetic estimation from DNA sequences with variable rates over sites: approximate methods. J. Mol. Evol. 39:306-314

14. Yang, Z. (1997) PAML: a program package for phylogenetic analysis by maximum likelihood. CABIOS 13:555-556

15. Whelan, S., P. Liò, N. Goldman. 2001. P. Molecularhylogenetics: state-of-the-art methods for looking into the past. G. Trendsenet. 17:262-272.

16. Bagnoli, F., P. Liò and L. Sguanci. 2005. Modeling viral coevolution: HIV multi-clonal persistence and competition dynamics. Phys. A, In Press, Corrected Proof, Available online 28 November 2005.

17. Bagnoli, F., and M. Bezzi. 1997. Speciation as Pattern Formation by Competition in a Smooth Fitness Landscape. Phys. Rev. Lett. 79:3302-3305.

18. Burch, C. L., and L. Chao. 2000. Evolvability of an RNA virus determined by its mutational neighborhood. Nature 406:625-628.

19. Chao, L., M.P. Davenport, S. Forrest, A.S. Perelson. 2004. A stochastic model of cytotoxic T cell responses, J. Theor. Biol. 228: 227-240.

20. Jenkins, M. G. , A. Rambaut, O. G. Pybus, and E.C. Holmes. 2002. Rates of molecular evolution in RNA viruses: a quantitative phylogenetic analysis. J Mol Evol 54:152-161.

21. Liò, P., and N. Goldman. 2004. Phylogenomics and bioinformatics of SARS-CoV. Trends in Microbiology. 12:106-111.

22. Murphy, P. M. 2001. Viral exploitation and subversion of the immune system through chemokine mimicry. Nature Immunology 2:116-122.

23. Lewin, S.R., S. Sonza, L. B. Irving, C. F. McDonald, J. Mills and S. M. Crowe. 1996. Surface CD4 is critical to in vitro HIV infection of human alveolar macrophages. AIDS Res. Hum. Retroviruses 12:877-883.

24. Ometto, L., M. Zanchetta, A. Cabrelle, G. Esposito, M. Mainardi, L. Chieco-Bianchi and A. De Rossi. 1999. Restriction of HIV type 1 infection in macrophages heterozygous for a deletion in the CCchemokine receptor 5 gene. AIDS Res. Hum. Retroviruses 15:1441-1452.

25. Wang, J., K. Crawford, M. Yuan, H. Wang, P. R. Gorry and D. Gabuzda. 2002. Regulation of CC chemokine receptor 5 and CD4 expression and human immunodeficiency virus type 1 replication in human macrophages and microglia by T helper type 2 cytokines. J. Infect. Dis. 185:885-897.

26. Bannert, N., D. S. Schenten, J. Craig and J. Sodroski. 2000. The level of CD4 expression limits infection of primary rhesus monkey macrophages by a T-tropic simian immunodeficiency virus and macrophagetropic human immunodeficiency viruses. J. Virol. 74:10984-10993.

27. Mori, K., M. Rosenzweig and R.C. Desrosiers. 2000. Mechanisms for adaptation of simian immunodeficiency virus to replication in alveolar macrophages. J. Virol. 74:10852-10859.

28. Shields, D.C. 2000. Gene conversions among chemokine genes. Gene 246:239-245.

29. Goh, C. S., A. A. Bogan, M. Joachimiak, D. Walther, F. E. Cohen. 2000. Co-evolution of proteins with their interaction partners. J. Mol. Bio. 299:283-293.

30. Samson, M., G. LaRosa, F. Libert, P. Paindavoine, G. Detheux, R. Vassay, M. Pandarmentier. 1997. The second extracellular loop of CCR5 is the major determinant of ligand specificity. J. Biol. Chem 272:24934-24941.

31. V. Muller, A. F. M. Maree, and R. J. DE Boer 2001 Release of Virus from Lymphoid Tissue Affects Human Immunodeficiency Virus Type 1 and Hepatitis C Virus Kinetics in the Blood J. Virol. 75:2597-2603.

32. A.S. Perelson, Essunger P, Cao Y, Vesanen M, Hurley A, Saksela K, Markowitz M, Ho DD. 1997 Decay characteristics of HIV-1-infected compartments during combination therapy. Nature 387:188-191;

33. Perelson A.S., A.U. Neumann A.U., Markowitz M, Leonard J.M., Ho D.D. 1996 HIV-1 dynamics in vivo: virion clearance rate, infected cell life-span, and viral generation time. Science 271:1582-1586.

34. Di Mascio M., G. Dornadula, H. Zhang, J. Sullivan, Y. Xu, J. Kulkosky, R. J. Pomerantz, and A. S. Perelson 2003 In a Subset of Subjects on Highly Active Antiretroviral Therapy, Human Immunodeficiency Virus Type 1 RNA in Plasma Decays from 50 to ¡5 Copies per Milliliter, with a Half-Life of 6 Months. J. Virol., 77: 2271 - 2275.

35. Hlavacek W.S., N. I. Stilianakis, D. W. Notermans, S. A. Danner, and A. S. Perelson. 2000 Influence of follicular dendritic cells on decay of HIV during antiretroviral therapy PNAS, 97: 10966 - 10971.

36. Schnell, J., E. Johnson, L. Buonocore and J. K. Rose. 1997. Construction of a novel virus that targets HIV-1-infected cells and controls HIV-1 infection. Cell 90:849-857.

37. Revilla, T. and G. Garcýa-Ramos. 2003. Math. Biosc. 185:191.

38. Herbeuval, J. P., Grivel, J.-C., Boasso A., Hardy, A. W. , Chougnet, C., Dolan M. J., Yagita, H., Lifson J. D., and Shearer, G. M. 2005. CD4+ T-cell death induced by infectious and noninfectious HIV-1: role of type 1 interferon-dependent, TRAIL/DR5-mediated apoptosis. Blood 106: 3524-3531.

Compositional Reachability Analysis of Genetic Networks

Gregor Gössler

POP ART project, INRIA Rhône-Alpes, France

Abstract. Genetic regulatory networks have been modeled as discrete transition systems by many approaches, benefiting from a large number of formal verification algorithms available for the analysis of discrete transition systems. However, most of these approaches do not scale up well. In this article, we explore the use of compositionality for the analysis of genetic regulatory networks. We present a framework for modeling genetic regulatory networks in a modular yet faithful manner based on the mathematically well-founded formalism of differential inclusions. We then propose a compositional algorithm to efficiently analyze reachability properties of the model. A case study shows the potential of this approach.

1 Introduction

A genetic regulatory network usually encompasses a multitude of complex, interacting feedback loops. Being able to model and analyze its behavior is crucial for understanding the interactions between the proteins, and their functions. Genetic regulatory networks have been modeled as discrete transition systems by many approaches, benefiting from a large number of formal verification algorithms available for the analysis of discrete transition systems. However, most of these approaches face the problem of state space explosion, as even models of modest size (from a biological point of view) usually lead to large transition systems, due to a combinatorial blow-up of the number of states. Even if the modeling formalism allows for a compact representation of the state space, such as Petri nets, subsequent analysis algorithms have to cope with the full state space. In practice, non-compositional approaches for the analysis of genetic regulatory networks do not scale up well.

In order to deal with the problem of state space explosion, different techniques have been developed in the formal verification community, such as partial order reduction, abstraction, and compositional approaches. In this article, we explore the use of compositionality for the analysis of genetic regulatory networks. Compositional analysis means that the behavior of a system consisting of different components is analyzed by separately examining the behavior of the components and how they interact, rather than by monolithically analyzing the behavior of the overall system. It therefore can be more efficient than non-compositional analysis, and scale better.

C. Priami (Ed.): CMSB 2006, LNBI 4210, pp. 212–226, 2006.

A precondition for compositional algorithms to be applicable, is that the model be structured. Therefore, this paper makes two contributions: first, we present a modeling framework for genetic regulatory networks in which the different components of the system (in our case, proteins or sets of proteins) and the way they constrain each other, are modeled separately and modularly. Second, we propose a compositional algorithm allowing to efficiently analyze reachability properties of the model.

Cellular functions are often distributed over groups of components that interact within large networks. The components are organized in functional modules, forming a hierarchical architecture [25,27]. Therefore, the approach of compositional analysis agrees with the modular structure of genetic regulatory networks, and may take advantage of it by using compositionality on different levels of modularity, for instance, between individual genes, sub-networks, or individual cells.

However, compositionality is not everything. The model should also faithfully represent the actual behavior of the modeled network. The approach we present is based on the mathematically well-founded formalism of qualitative simulation [14].

Related work. By now there is a large number of approaches to model and analyze genetic networks. An overview is given in the survey of [11]. The modeling approaches adopt different mathematical frameworks, which vary in expressiveness and the availability and efficiency of verification algorithms. Most of the algorithms "flatten" the model and work on the global state space, without computationally taking advantage of the modularity of the problem. The approach of [6] compositionally models gene networks in a stochastic framework.

There has been a wide variety of modeling approaches based on differential equations since the work of [19]. However, simulation and verification of the continuous model can be expensive, and many properties are not even decidable in this framework. Therefore, several ways have been investigated to discretize the continuous model defined by differential equations while preserving properties like soundness [14] and reachability [3]. [18] and [1] use predicate abstraction to automatically compute backward reachable sets of piecewise affine hybrid automata, and find a conservative approximation of reachability for linear hybrid systems, respectively. [26] addresses the *bounded reachability* problem of hybrid automata.

In order to deal with complex networks, it may be a good choice to change precision against efficiency, and directly model genetic networks in a discrete framework, such as systems of logical equations [29,5], Petri nets [22,9,28,10], or rule-based formalisms like term rewriting systems [15,16]. Formal verification can then be carried out enumeratively (for instance, [13,2,23]) or symbolically, see for example [8].

Organization of the paper. In Section 2, we introduce the modeling framework. We show how a genetic network can be modeled in a modular way in this framework, and compare the model with the qualitative model of [7]. Section 3 presents a reachability algorithm taking advantage of the modularity of the model. Section 4 illustrates our results with a case study, and Section 5 concludes.

2   Component-Based Modeling of Genetic Networks

This section briefly introduces the notions of piecewise linear system, and its qualitative simulation as defined in [7]. We then define a modularized approximation of qualitative simulation, and compare both models.

2.1   Piecewise Linear Systems

The production of a protein in a cell is regulated by the current protein concentrations, which can activate or inhibit the production, for instance by binding to the gene and disabling transcription. At the same time, proteins are degraded. This behavior of a genetic network can be modeled by a system of differential equations of the form

$$\dot{\mathbf{x}} = \mathbf{f}(\mathbf{x}, \mathbf{u}) - \mathbf{g}(\mathbf{x}, \mathbf{u})\mathbf{x} \qquad (1)$$

where $\mathbf{x}$ is a vector of protein concentrations representing the current state, $\mathbf{u}$ is a vector of *input* concentrations, and the vector-valued function $\mathbf{f}$ and matrix-valued function $\mathbf{g}$ model the production rates, and degradation rates, respectively.

The approach of [14,7] considers an abstraction where the state space of each variable x_i is partitioned into a set of intervals $\mathcal{D}_i^r$ and a set of *threshold values* $\mathcal{D}_i^s$. This induces a partition of the continuous state space into a discrete set of *domains*, in each of which Equation (1) is approximated with a system of linear differential equations.

Definition 1 (Domain). *Consider a Cartesian product* $\theta = \theta_1 \times ... \times \theta_n$ *with* $\theta_i = \{\theta_i^1, ..., \theta_i^{p_i}\}$ *an ordered set of* thresholds, *such that* $0 < \theta_i^1 < ... < \theta_i^{p_i} < max_i$. *Let*

$$\mathcal{D}_i^r(\theta) = \{[0, \theta_i^1)\} \cup \{(\theta_i^j, \theta_i^{j+1}) \mid 1 \leqslant j < p_i\} \cup \{(\theta_i^{p_i}, max_i]\}$$

and $\mathcal{D}_i^s(\theta) = \{\{\theta_i^j\} \mid 1 \leqslant j \leqslant p_i\}$. *We omit the argument* θ *when it is clear from the context. Let* $\mathcal{D}_i = \mathcal{D}_i^r \cup \mathcal{D}_i^s$, *and* $\mathcal{D} = \mathcal{D}_1 \times \mathcal{D}_2 \times ... \times \mathcal{D}_n$ *be the set of* domains. *The domains in* $\mathcal{D}^r = \mathcal{D}_1^r \times \mathcal{D}_2^r \times ... \times \mathcal{D}_n^r$ *are called regulatory domains, the domains* $\mathcal{D}^s = \mathcal{D} \setminus \mathcal{D}^r$ *are called switching domains.*

The state space $[0, max_1] \times \cdots \times [0, max_n]$ is thus partitioned into the set of domains $\mathcal{D}$.

Definition 2 (Piecewise linear system). *A piecewise linear system is a tuple* $M = (X, \theta, \boldsymbol{\mu}, \boldsymbol{\nu})$ *where*

- $X = \{x_1, ..., x_n\}$ *a set of real-valued state variables;*
- $\theta = \theta_1 \times ... \times \theta_n$, *with* $\theta_i = \{\theta_i^1, ..., \theta_i^{p_i}\}$ *such that* $0 < \theta_i^1 < ... < \theta_i^{p_i} < max_i$, *associates with each dimension an ordered set of thresholds;*

- $\mu : \mathcal{D}^r(\theta) \rightarrow \mathbb{R}^n_{\geqslant 0}$ associates with each regulatory domain a vector of production rates;
- $\nu : \mathcal{D}^r(\theta) \rightarrow diag(\mathbb{R}^n_{>0})$ associates with each regulatory domain a diagonal matrix of degradation rates.

Within a regulatory domain $D \in \mathcal{D}^r$, the protein concentrations $\mathbf{x}$ evolve according to the ratio of production rate and degradation rate:

$$\dot{\mathbf{x}} = \mu(D) - \nu(D)\mathbf{x} \tag{2}$$

and thus converge monotonically towards the *target equilibrium* ϕ, solution of $0 = \mu(D) - \nu(D)\mathbf{x}$.

Definition 3 (ϕ). *For any $D \in \mathcal{D}^r$, let $\phi(D)$ denote the* target equilibrium *of D such that*

$$\phi_i(D) = \mu_i(D)/\nu_i(D)$$

for any variable $x_i \in X$.

Hypothesis: Throughout this paper we make the assumption that for any regulatory domain D, $\exists D' \in \mathcal{D}^r . \phi(D) \in D'$, that is, all target equilibria lie within regulatory domains, as in [14].

If $\phi(D) \in D$ then the systems stays in D, otherwise it eventually leaves D and enters an adjacent switching domain. In switching domains, where μ and ν are not defined, the behavior of M is defined using differential inclusions as proposed by [17,21].

Notations. Let reg be the predicate characterizing the set of regulatory domains. For any $i \in \{1, ..., n\}$, let reg_i and $switch_i$ be predicates on $\mathcal{D}$ characterizing the regulatory intervals and thresholds of $\mathcal{D}_i$, respectively: $reg_i(D) \iff D_i \in \mathcal{D}^r_i$, and $switch_i(D) \iff D_i \in \mathcal{D}^s_i$ for any $D \in \mathcal{D}$. The *order* of a domain D is the number of variables taking a threshold value in D. Let $succ_i$ and $prec_i$ be the successor and predecessor function on the ordered set of intervals $\mathcal{D}_i$ (in the sense that for any $D_1, D_2 \in \mathcal{D}_i$, $D_1 < D_2$ if $\forall x_1 \in D_1 \ \forall x_2 \in D_2 . x_1 < x_2$). We define $succ_i\big((\theta_i^{p_i}, max_i]\big) = prec_i\big([0, \theta_i^1)\big) = \bot$.

Definition 4 ($R(D)$). *For any domain $D = (D_1, ..., D_n) \in \mathcal{D}$, let $R(D)$ be the set of regulatory domains that have D in their boundary, such that $R(D) = \{D\}$ for $D \in \mathcal{D}^r$:*

$$R(D) = \big\{(D'_1, ..., D'_n) \mid reg_i(D_i) \wedge D'_i = D_i \ \vee$$
$$switch_i(D_i) \wedge \big(D'_i = prec(D_i) \vee D'_i = succ(D_i)\big)\big\}$$

Gouzé and Sari [21] define the possible behaviors by the differential inclusion $\dot{\mathbf{x}} \in H(\mathbf{x})$ with

$$H(\mathbf{x}) = \bar{co}\big(\{\mu(D') - \nu(D')\mathbf{x} \mid D' \in R(D)\}\big)$$

where $\bar{co}(E)$ is the smallest closed convex set containing the set E. For any regulatory domain $D \in \mathcal{D}^r$ and $\mathbf{x} \in D$, $H(\mathbf{x}) = \{\mu(D) - \nu(D)\mathbf{x}\}$, that is, the behavior is consistent with Equation (2).

Definition 5 (Trajectory). *A* trajectory *of M is a solution of $\dot{\mathbf{x}} \in H(\mathbf{x})$.*

Qualitative model. The continuous behavior according to Definition 5 can be approximated by a discrete transition graph on the set of domains $\mathcal{D}$ [14,7] (where the qualitative model of [7] is more precise than [14]). This graph simulates the behavior of the underlying genetic network.

Example 1. Consider the example of two proteins a and b inhibiting each other's production [14], as shown in Figure 1. The respective production rates of proteins a and b are defined by

$$\mu_a = \begin{cases} 20 & \text{if } 0 \leqslant x_a < \theta_a^2 \wedge 0 \leqslant x_b < \theta_b^1 \\ 0 & \text{otherwise} \end{cases}$$

$$\mu_b = \begin{cases} 20 & \text{if } 0 \leqslant x_a < \theta_a^1 \wedge 0 \leqslant x_b < \theta_b^2 \\ 0 & \text{otherwise} \end{cases}$$

with $\theta_a^1 = \theta_b^1 = 4$ and $\theta_a^2 = \theta_b^2 = 8$. The degradation rate ν of both proteins is always 2. The example is thus modeled by the piecewise linear system $M = \left(\{x_a, x_b\}, \{\theta_a^1, \theta_a^2\} \times \{\theta_b^1, \theta_b^2\}, (\mu_a, \mu_b)^t, diag(\nu, \nu) \right)$.

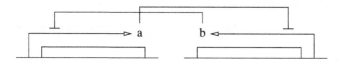

Fig. 1. Two proteins inhibiting each other

2.2   Transition Systems and Constraints

In the following, we present a simplified version of the component model adopted in [20]. For a set of variables X, let $V(X)$ denote the set of valuations of X, and let $\mathcal{P}(X) = 2^{V(X)}$ be the set of predicates on $V(X)$.

Definition 6 (Transition system). *A transition system B is a tuple (X, A, G, F) where*

- *X is a finite set of variables;*
- *A is a finite set of actions;*
- *$G : A \rightarrow \mathcal{P}(X)$ associates with every action its guard specifying when the action can occur;*
- *$F : A \rightarrow \left(V(X) \rightarrow V(X) \right)$ associates with every action its transition function.*

For convenience, we write G^a and F^a for $G(a)$ and $F(a)$, respectively.

Definition 7 (Semantics of a transition system). *A transition system $B = (X, A, G, F)$ defines a transition relation $\rightarrow : V(X) \times A \times V(X)$ such that: $\forall \mathbf{x}, \mathbf{x}' \in V(X) \; \forall a \in A . \mathbf{x} \xrightarrow{a} \mathbf{x}' \iff G^a(\mathbf{x}) \wedge \mathbf{x}' = F^a(\mathbf{x})$.*

We write $\mathbf{x} \rightarrow \mathbf{x}'$ for $\exists a \in A . \mathbf{x} \xrightarrow{a} \mathbf{x}'$, and $\rightarrow^*$ for the transitive and reflexive closure of $\rightarrow$. Given states $\mathbf{x}$ and $\mathbf{x}'$, $\mathbf{x}'$ is *reachable* from $\mathbf{x}$ if $\mathbf{x} \rightarrow^* \mathbf{x}'$.

Definition 8 (Predecessors). *Given a transition system $B = (X, A, G, F)$ and a predicate $P \in \mathcal{P}(X)$, let the predicate $pre_a(P)$ characterize the predecessors of P by action a:* $pre_a(P)(\mathbf{x}) \iff G^a(\mathbf{x}) \wedge P(F^a(\mathbf{x}))$. *Let* $pre(P) = \bigvee_{a \in A} pre_a(P)$, $pre^0(P) = P$, *and* $pre^{i+1}(P) = pre(pre^i(P))$, $i \geqslant 0$.

The predicate $pre_a(P)$ (resp. $pre(P)$) characterizes the states from which execution of a (resp. execution of some action) leads to a state satisfying P.

We define two operations on transition systems: composition and restriction. The composition of transition systems is a transition system again, and so is the restriction of a transition system.

Definition 9 (Composition). *Let $B_i = (X_1, A_i, G_i, F_i)$, $i = 1, 2$, with $X_1 \cap X_2 = \emptyset$ and $A_1 \cap A_2 = \emptyset$. $B_1 \| B_2$ is defined as the transition system $(X_1 \cup X_2, A_1 \cup A_2, G_1 \cup G_2, F_1 \cup F_2)$.*

This is the standard asynchronous product. Restrictions allow to constrain the behavior of a transition system.

Definition 10 (Action constraint). *Given a transition system $B = (X, A, G, F)$, an action constraint is a tuple of predicates $U = (U^a)_{a \in A}$ with $U^a \in \mathcal{P}(X)$.*

Definition 11 (Restriction). *The restriction of $B = (X, A, G, F)$ with an action constraint $U = (U^a)_{a \in A}$ is the transition system $B/U = (X, A, G', F)$ where for any $a \in A$, $G'(a) = G(a) \wedge U^a$ is the (restricted) guard of a in B/U.*

Example 2. Consider two transition systems $B_i = (\{x_i\}, \{inc_i, dec_i\}, G_i, F_i)$ where x_i are variables on $\{low, high\}$, $G_i(inc_i) = (x_i = low)$, $G_i(dec_i) = (x_i = high)$, $F_i(inc_i) = (x_i := high)$, and $F_i(dec_i) = (x_i := low)$, $i = 1, 2$. The composition is $B_1 \| B_2 = (\{x_1, x_2\}, \{inc_1, inc_2, dec_1, dec_2\}, G_1 \cup G_2, F_1 \cup F_2)$.

Further suppose that we want to prevent B_1 from entering state $x_1 = high$ if $x_2 = high$, and vice versa. This can be done by restricting $B_1 \| B_2$ with action constraint $U = (U^{inc_1}, U^{inc_2}, U^{dec_1}, U^{dec_2})$ where $U^{inc_1} = (x_2 = low)$, $U^{inc_2} = (x_1 = low)$, and $U^{dec_1} = U^{dec_2} = true$. The restricted system is $(B_1 \| B_2)/U = (\{x_1, x_2\}, \{inc_1, inc_2, dec_1, dec_2\}, G', F_1 \cup F_2)$ with $G'(inc_1) = G_1(inc_1) \wedge (x_2 = low)$, $G'(inc_2) = G_2(inc_2) \wedge (x_1 = low)$, $G'(dec_1) = G_1(dec_1)$, and $G'(dec_2) = G_1(dec_2)$.

Definition 12 (incr, decr). *Given a predicate P on $\mathcal{D}$ and $i \in \{1, ..., n\}$, we define the predicates $incr_i(P)$ and $decr_i(P)$ such that for any domain $D = (D_1, ..., D_i, ..., D_n) \in \mathcal{D}$, $incr_i(P)(D) = P(D_1, ..., succ_i(D_i), ..., D_n)$ if $succ_i(D_i) \neq \perp$, and $incr_i(P)(D) = false$ otherwise. Similarly, let $decr_i(P)(D) = P(D_1, ..., prec_i(D_i), ..., D_n)$ if $prec_i(D_i) \neq \perp$, and $decr_i(P)(D) = false$ otherwise.*

Intuitively, $incr_i(P)$ and $decr_i(P)$ denote the predicate P "shifted" by one domain along the i-th dimension, towards lower and higher values, respectively. For instance, consider predicate $P = (x_a = \theta_a^2)$ on the state space of Example 1. Then, $incr_a(P) = (\theta_a^1 < x_a < \theta_a^2)$ and $decr_b(P) = (x_a = \theta_a^2 \wedge \theta_b^1 \leqslant x_b \leqslant max_b)$.

2.3   Component Model of Genetic Networks

We now propose the construction of a component-based model from a piecewise linear system.

Definition 13 (*eq*). *Given* $\theta = \theta_1 \times ... \times \theta_n$, *we define predicates* $eq_i^{\#}$ *on* $\mathcal{D}$, $i \in \{1, ..., n\}$, $\# \in \{<, \leqslant, \geqslant, >\}$ *such that for any domain* $D = (D_1, ..., D_n) \in \mathcal{D}$,

$$eq_i^{\#}(D) \iff \exists D' \in R(D) \; \forall \mathbf{x} \in D_i' \; . \; \phi_i(D') \# \mathbf{x}_i \text{ for } \# \in \{<, >\}$$
$$eq_i^{=}(D) \iff \exists D' \in R(D) \; \exists \mathbf{x} \in D_i' \; . \; \phi_i(D') = \mathbf{x}_i$$

and $eq_i^{\leqslant} = eq_i^{<} \lor eq_i^{=}$, $eq_i^{\geqslant} = eq_i^{=} \lor eq_i^{>}$.

The predicates $eq_i^{\#}$ reflect the relative position of target equilibria of the adjacent regulatory domains. The predicates $eq_i^{<}(D)$ and $eq_i^{>}(D)$ specify when some adjacent regulatory domain has its target equilibrium "left" of D_i and "right" of D_i, respectively.

Definition 14 ($\check{C}(M)$). *Given a piecewise linear system* $M = (X, \theta, \boldsymbol{\mu}, \boldsymbol{\nu})$ *with* $|X| = n$, *we define the transition system* $\check{C}(M) = (B_1 \| B_2 \| ... \| B_n)/U$ *as follows.*

- $\forall i = 1, ..., n$. $B_i = \mathtt{counter}(\mathcal{D}_i)$, *where* $\mathtt{counter}(\mathcal{D}_i)$ *is a bounded counter defined on* $\mathcal{D}_i(\theta)$ *by the transition system* $\mathtt{counter}(\mathcal{D}_i) = (\{level_i\}, \{inc_i, dec_i\}, \{G^{inc_i} = level_i \leqslant \theta_i^{p_i}, G^{dec_i} = level_i \geqslant \theta_i^{1}\}, \{F^{inc_i} = (level_i := succ_i(level_i)), F^{dec_i} = (level_i := prec_i(level_i))\})$.
- U *is an action constraint such that* $U(inc_i) = V_i^{>}$ *and* $U(dec_i) = V_i^{<}$ *with*

$$V_i^{<} = reg \land eq_i^{<} \; \lor \; decr_i(reg \land eq_i^{\leqslant}) \tag{3}$$
$$V_i^{>} = reg \land eq_i^{>} \; \lor \; incr_i(reg \land eq_i^{\geqslant}) \tag{4}$$

Actions inc_i (dec_i) correspond to an increase (decrease) by one of the discretized concentration $level_i$ of protein i. The predicates $V_i^{<}$ and $V_i^{>}$ specify when a transition decrementing $level_i$ and incrementing $level_i$, respectively, is enabled. More precisely, the first term in the disjunctions of lines (3) and (4) specifies that there is a transition from a regulatory domain to a first-order switching domain in the direction of the target equilibrium of the source domain. The second term gives the conditions for transitions decreasing the order: they must be compatible with the relative position of the target equilibrium of the destination domain. Definition 14 limits the behavior of the model to transitions between regulatory and first-order switching domains. The generalization to the set of domains $\mathcal{D}$ is not presented here due to space limitation.

Remark 1. Since $\|$ is associative, Definition 14 leaves open how the system is actually partitioned into components (in the sense of sets of transition systems). The two extreme cases are that each B_i is considered as one component, or that $B_1 \| B_2 \| ... \| B_n$ is considered as one single component. This choice will usually depend on the degree of interaction between the modeled proteins. Putting all proteins in one component amounts to a non-modular model leading to non-compositional analysis. Representing each protein with a separate component may lead to a too heavy abstraction of the behavior. A good choice may gather closely interacting proteins, for instance proteins in the same cell, in one component, while modeling neighboring cells as separate components.

Notice that the above modeling framework enforces *separation of concerns* by making a clear distinction between the behaviors of the individual components, and constraints between the components.

Example 3. Figure 2 shows the transition relations of $counter(\mathcal{D}_a)$, $counter(\mathcal{D}_b)$, and $\check{C}(M)$ for the piecewise linear system M of Example 1.

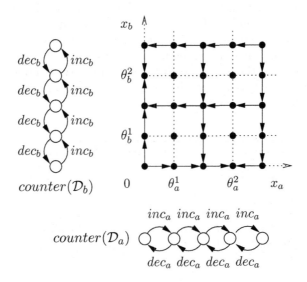

Fig. 2. The transition relations of $counter(\mathcal{D}_a)$, $counter(\mathcal{D}_b)$, and $\check{C}(M)$

Theorem 1 (Correctness). *Consider a piecewise linear system* $M = (X, \theta, \mu, \nu)$. *The behavior of* $\check{C}(M)$ *under-approximates qualitative simulation as defined in [14,7].*

3   Compositional Reachability Analysis

Based on the transition system $\check{C}(M)$, the compositional algorithm shown below can be used to check for reachability of a goal domain, or set of domains, from an initial domain. The algorithm exhibits a path, if one is found, that solves the reachability problem.

In the sequel we consider a system $B = (X, A, G, F) = (B_1 \| \ldots \| B_N)/U$ with $B_i = (A_i, X_i, G_i, F_i)$, $i \in K = \{1, \ldots, N\}$, and U an action constraint. That is, we suppose the n proteins to be modeled with N $(1 \leqslant N \leqslant n)$ components, according to Remark 1. Given a conjunction $c = c_1 \wedge ... \wedge c_N$ of predicates $c_i \in \mathcal{P}(X_i)$, $i = 1, ..., N$, let $c[i] = c_i$ denote the *projection* of c on X_i.

Let $path_k : V(X_k) \times \mathcal{P}(X_k) \rightarrow 2^{A_k}$ be a function on component k telling which action to take in order to get from some current component state towards a state satisfying some predicate. This function can be computed locally: for any predicate $P \in \mathcal{P}(X_k)$ and domain D, let

$$path_k(D[k], P) = \{a \in A_k \mid \exists i \geqslant 0 \,.\, pre_a\big(pre_k^i(P)\big)(D[k]) \,\wedge$$
$$\forall j \in \{0, ..., i\} \,.\, \neg pre_k^j(P)(D[k])\}$$

That is, $path_k(D[k], P)$ contains an action a if and only if executing a from $D[k]$ will bring component k closer to P.

For a set of actions A, let $enabling(A)$ be a list of predicates enabling some action in A: $\forall c \in enabling(A) \,.\, c \implies \bigvee_{a \in A} G(a)$. We suppose each of these predicates to be a conjunction of predicates on the components. Let $\oplus$ denote list concatenation. Given a non-empty list l, we write $l = e.l'$ where e is the first element, and l' the rest of the list. Given a list A of actions and a domain D, let $first\_enabled(A, D)$ be the first action a of A such that $G(a)(D)$, and $first\_enabled(A, D) = \bot$ if all actions are disabled.

Algorithm 1. Initial call to construct a path σ from domain D_{init} to predicate P: $(D', \sigma, success) = \mathbf{move}(D_{init}, P, \langle\rangle)$.

$\mathbf{move}\,(D, c.l, good) =$

$$\begin{cases} (D, \langle\rangle, true) & \text{if } c(D) & (1) \\ (F(a)(D), \langle a\rangle, true) & \text{if } \neg c(D) \wedge a \neq \bot & (2) \\ (D'', \sigma \oplus \sigma', true) & \text{if } \neg c(D) \wedge a = \bot \wedge goal \neq \emptyset \wedge ok \wedge ok' & (3) \\ \mathbf{move}(D, l, good) & \text{if } \neg c(D) \wedge a = \bot \wedge (goal = \emptyset \vee \neg(ok \wedge ok')) \wedge l \neq \langle\rangle & (4) \\ (D, \langle\rangle, false) & otherwise & (5) \end{cases}$$

where

$$a = first\_enabled(good, D)$$
$$goal = \bigcup_k path_k(D[k], c[k]) \smallsetminus good$$
$$(D', \sigma, ok) = \mathbf{move}\big(D, enabling(goal), good \oplus goal\big)$$
$$(D'', \sigma', ok') = \mathbf{move}(D', c, good)$$

Algorithm 1 is constructive, that is, it establishes reachability from some initial domain D_{init} to a set of domains P by constructing a path from D_{init} to P. Function **move** works as follows. It takes as arguments the current domain D, a predicate to be reached in the form of a list d of conjunctions, and a list $good$ of all actions requested to be executed, and returns a new domain, the part of the path constructed so far, and a boolean indicating whether a path was found. The five cases are (1) if the current domain satisfies the predicate to be reached, then we are done. (2) Otherwise, execute the first action in *good* that is enabled. If there is none, compute the set *goal* of actions not considered so far that bring the system closer to the first element c of d. (3) If *goal* is non-empty, recursively call **move** so as to reach some domain D' enabling some action in *goal*, then call **move** once more to continue moving towards c. (4) If reaching c fails, try the next conjunction of d. (5) If all above fails, then this call of **move** failed. It can be shown that Algorithm 1 is guaranteed to terminate. It is not guaranteed to find a path even if one exists, though. If a path is found on $\check{C}(M)$, then Theorem 1 ensures that the same path exists in the qualitative model of [7].

Algorithm 1 is compositional in the sense that it independently computes local paths through the state spaces of the components (line $goal = \bigcup_k path_k(D[k], c[k]) \setminus good$). A global path is then constructed from the local paths and the constraints between the components: when an action a to be executed is blocked by a constraint involving other components, the algorithm is called recursively to move the blocking components into a domain where a is enabled.

Example 4 (Example 3 continued.). The functioning of Algorithm 1 is illustrated by the path construction from domain $D_{init} = (\theta_a^1 < x_a < \theta_a^2 \wedge \theta_b^1 < x_b < \theta_b^2)$ to domain $D_{goal} = (x_a = \theta_a^2 \wedge 0 \leqslant x_b < \theta_b^1)$ representing a stable equilibrium. The subsequent calls of **move** are

$$
\begin{aligned}
&\textbf{move } (D_{init}, \langle D_{goal} \rangle, \langle \rangle) \\
&\quad a = \bot, goal = \{inc_a, dec_b\} \\
&\qquad \textbf{move } (D_{init}, \langle \theta_a^1 < x_a < \theta_a^2, \ldots \rangle, \langle inc_a, dec_b \rangle) \\
&\qquad = (D_1 = (\theta_a^1 < x_a < \theta_a^2 \wedge x_b = \theta_b^1), \langle dec_b \rangle, true) &&(2) \\
&\quad \textbf{move } (D_1, \langle D_{goal} \rangle, \langle \rangle) \\
&\qquad a = \bot, goal = \{inc_a, dec_b\} \\
&\qquad\quad \textbf{move } (D_1, \langle \theta_a^1 < x_a < \theta_a^2, \ldots \rangle, \langle inc_a, dec_b \rangle) \\
&\qquad\quad = (D_2 = (\theta_a^1 < x_a < \theta_a^2 \wedge 0 \leqslant x_b < \theta_b^1), \langle dec_b \rangle, true) &&(2) \\
&\qquad \textbf{move } (D_2, \langle D_{goal} \rangle, \langle \rangle) \\
&\qquad\quad a = \bot, goal = \{inc_a\} \\
&\qquad\qquad \textbf{move } (D_2, \langle x_a < \theta_a^2 \wedge 0 \leqslant x_b < \theta_b^1 \rangle, \langle inc_a \rangle) \\
&\qquad\qquad = (D_{goal}, \langle inc_a \rangle, true) &&(2) \\
&\qquad\quad \textbf{move } (D_{goal}, \langle D_{goal} \rangle, \langle \rangle) = (D_{goal}, \langle \rangle, true) \\
&\qquad\quad = (D_{goal}, \langle inc_a \rangle, true) &&(3) \\
&\qquad = (D_{goal}, \langle dec_b, inc_a \rangle, true) &&(3) \\
&= (D_{goal}, \langle dec_b, dec_b, inc_a \rangle, true) &&(3)
\end{aligned}
$$

Thus, D_{goal} is reached from D_{init} by decrementing $level_b$ twice and then incrementing $level_a$.

4   Case Study: Delta-Notch Cell Differentiation

Cell differentiation by delta-notch lateral inhibition is a well-studied genetic network [24,18]. Cell differentiation is an important step in embryonic development, as it causes initially uniform cells to assume different functions.

For each cell we consider the concentrations of two trans-membrane proteins, *Delta* and *Notch*. Following the model provided in [24], high concentrations of *Delta* and *Notch* inhibit each other's production within the same cell. High *Delta* levels activate further *Delta* production in the same cell and *Notch* production in the neighboring cells. Figure 3 illustrates these interactions.

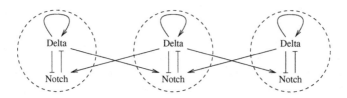

Fig. 3. Interactions within and between neighbor cells

For our case study, we consider a network consisting of 19 cells with the layout shown in Figure 4, a network of 37 cells with a similar layout, and the network of 49 cells shown in Figure 4.

For each protein we partition the continuous state space into two intervals and one threshold value: $\mathcal{D}_\Delta = \{[0, \theta_\Delta), \{\theta_\Delta\}, (\theta_\Delta, max_\Delta]\}$ and $\mathcal{D}_N = \{[0, \theta_N), \{\theta_N\}, (\theta_N, max_N]\}$. Cells with low *Delta* and high *Notch* levels ($0 \leqslant \Delta < \theta_\Delta$, $\theta_N < Notch \leqslant max_N$) are undifferentiated, whereas cells with high *Delta* and low *Notch* concentrations ($\theta_\Delta < \Delta \leqslant max_\Delta$, $0 \leqslant Notch < \theta_N$) are differentiated. We are not interested in the actual production and degradation rates of the proteins but require the target equilibria ϕ_{Δ_i} and ϕ_{Notch_i} to satisfy

$$0 \leqslant \phi_{\Delta_i} < \theta_\Delta \text{ if } Notch_i > \theta_N$$
$$\theta_\Delta < \phi_{\Delta_i} \leqslant max_\Delta \text{ if } Notch_i < \theta_N$$
$$0 \leqslant \phi_{Notch_i} < \theta_N \text{ if } \max\{\Delta_j \mid j \in neighbors(i)\} < \theta_\Delta$$
$$\theta_N < \phi_{Notch_i} \leqslant max_N \text{ if } \max\{\Delta_j \mid j \in neighbors(i)\} > \theta_\Delta$$

Considering only regulatory and first-order switching domains for a system modeling n cells, the $2n$-dimensional global state space encompasses 4^n regulatory domains and $2n \times 2^{2n-1}$ first-order switching domains, that is, 5.5×10^{12} states for 19 cells, 7.2×10^{23} states for 37 cells, and 1.6×10^{31} states for 49 cells.

We have implemented Algorithm 1 in the compositional verification tool PROMETHEUS. To start, we choose to represent each cell by one component, and check reachability of a given stable equilibrium from the initial state where all cells are non differentiated. The results reported by PROMETHEUS are consistent with the actual, experimentally observed behavior [24]. For the case of 49 cells and the state shown in Figure 4, PROMETHEUS finds a path of length 32 reaching the state.

Table 1 shows the execution times for the models of cell differentiation with 19, 37, and 49 cells, and for models of the nutritional stress response of E. coli [4] and sporulation initiation of B. subtilis taken from [12]. The subsequent columns

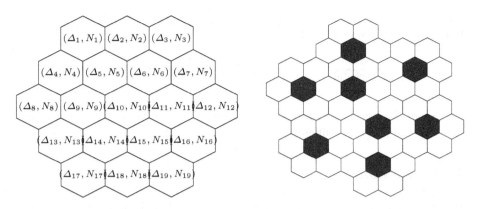

Fig. 4. Model of 19 communicating cells (left); a stable equilibrium state involving 49 cells where dark cells are differentiated (right)

show the number of domains of the model, and the times for constructing the component model and a path to the final state using Algorithm 1. All measurements have been made on the same machine, a Pentium4 at 3 GHz with 512 MB of memory.

Table 1. Performance on different models

	state space	model	reachability
E. coli	7.8×10^3	< 10 ms	0.02 s
B. subtilis	2.7×10^4	< 10 ms	0.28 s
Delta-Notch 19	5.5×10^{12}	0.01 s	1.06 s
Delta-Notch 37	7.2×10^{23}	0.05 s	10.8 s
Delta-Notch 49	1.6×10^{31}	0.13 s	7.5 s

In order to evaluate the performance increase due to compositionality, we compare the compositional approach with a non-compositional reachability analysis, using the same framework. More precisely, we use Algorithm 1 to find a path from the initial, undifferentiated state to the state of Figure 4, on different instances of the Delta-Notch model with 49 cells. The only parameter that varies is the size of the components, where extreme cases are given by the model of 98 components each modeling one protein, and the model consisting of one single component. The measured performance is shown in Table 2. For this example, the optimal degree of modularity lies around one component per cell. It should be noted that the optimal partitioning of proteins into components depend on the system, and cannot be easily generalized. For a higher degree of modularity (1 component per protein), the algorithm performs somewhat slower, probably due to an overhead in coordination between closely interacting components. As the component size increases, complexity of the (non compositional) path construction within the components exponentially blows up. Although the algorithm used for path construction within

Table 2. Benchmarks for different levels of modularity of Delta-Notch 49. (*): computation interrupted after 12 hours.

cells per component	0.5	1	3/4	7	9/10	49
reachability	10.7 s	7.5 s	8.4 s	35.5 s	(*)	(*)

a component is not designed to be optimal for large state spaces, it allows to compare the complexity for different degrees of granularity.

5 Discussion

We have presented a novel approach for component-based modeling and reachability analysis of genetic regulatory networks. The model discretizes the network dynamics defined by a system of piecewise linear differential equations. On this model, a compositional algorithm constructively analyzes reachability properties, allowing to deal with complex, high-dimensional systems. A case study and several benchmarks show the potential of this approach. In spite of the conservative approximation, our approach has yielded the expected results in the case studies carried out so far, and confirmed its efficiency.

We intend to apply the technique to genetic networks involving a hierarchy of communicating functional modules, and to models of not yet fully understood networks. We are currently investigating compositional analysis of further properties like equilibria and cyclic behavior, based on the same component model. In order to further improve precision, we intend to study the integration of the qualitative model of [3] using piecewise *affine* differential equations in our framework.

Acknowledgment. The author thanks Hidde de Jong for many fruitful discussions and comments on earlier versions of this work.

References

1. R. Alur, T. Dang, and F. Ivancić. Reachability analysis of hybrid systems via predicate abstraction. *Trans. on Embedded Computing Systems*, 2004.
2. G. Batt, D. Bergamini, H. de Jong, H. Garavel, and R. Mateescu. Model checking genetic regulatory networks using GNA and CADP. In *proc. SPIN'04*, volume 2989 of *LNCS*, pages 158–163. Springer-Verlag, 2004.
3. G. Batt, H. de Jong, J. Geiselmann, M. Page, D. Ropers, and D. Schneider. Symbolic reachability analysis of genetic regulatory networks using qualitative abstraction. Research Report 5362, INRIA, France, 2004.
4. G. Batt, D. Ropers, H. de Jong, J. Geiselmann, M. Page, and D. Schneider. Qualitative analysis and verification of hybrid models of genetic regulatory networks: Nutritional stress response in escherichia coli. In *proc. HSCC'05*, volume 3414 of *LNCS*, pages 134–150. Springer-Verlag, 2005.

5. G. Bernot, J.-P. Comet, A. Richard, and J. Guespin. Application of formal methods to biological regulatory networks: Extending Thomas' asynchronous logical approach with temporal logic. *Journal of Theoretical Biology*, 229(3):339–348, 2004.

6. R. Blossey, L. Cardelli, and A. Phillips. A compositional approach to the stochastic dynamics of gene networks. *Trans. on Comput. Syst. Biol.*, 4:99–122, 2006.

7. R. Casey, H. de Jong, and J.-L. Gouzé. Piecewise-linear models of genetic regulatory networks: Equilibria and their stability. *Mathematical Biology*, 52(1):27–56, 2006.

8. N. Chabrier and F. Fages. Symbolic model checking of biochemical networks. In *proc. CMSB'03*, 2003.

9. C. Chaouiya, E. Remy, P. Ruet, and D. Thieffry. Qualitative modelling of genetic networks: From logical regulatory graphs to standard petri nets. In J. Cortadella and W. Reisig, editors, *proc. ICATPN'04*, volume 3099 of *LNCS*, pages 137–156. Springer-Verlag, 2004.

10. J.-P. Comet, H. Klaudel, and S. Liauzu. Modeling multi-valued genetic regulatory networks using high-level petri nets. In G. Ciardo and P. Darondeau, editors, *proc. ICATPN'05*, volume 3536 of *LNCS*, pages 208–227. Springer-Verlag, 2005.

11. H. de Jong. Modeling and simulation of genetic regulatory systems: A literature review. *Journal of Computational Biology*, 9(1):69–105, 2002.

12. H. de Jong, J. Geiselmann, G. Batt, C. Hernandez, and M. Page. Qualitative simulation of the initiation of sporulation in Bacillus subtilis. *Bulletin of Mathematical Biology*, 66(2):261–300, 2004.

13. H. de Jong, J. Geiselmann, C. Hernandez, and M. Page. Genetic Network Analyzer: Qualitative simulation of genetic regulatory networks. *Bioinformatics*, 19(3):336–344, 2003.

14. H. de Jong, J.-L. Gouzé, C. Hernandez, M. Page, T. Sari, and J. Geiselmann. Qualitative simulation of genetic regulatory networks using piecewise-linear models. *Bulletin of Mathematical Biology*, 66:301–340, 2004.

15. S. Eker, M. Knapp, K. Laderoute, P. Lincoln, and C. Talcott. Pathway logic: Executable models of biological networks. *ENTCS*, 71, 2002.

16. F. Fages, S. Soliman, and N. Chabrier-Rivier. Modelling and querying interaction networks in the biochemical abstract machine BIOCHAM. *J. of Biological Physics and Chemistry*, 4(2):64–72, 2004.

17. A.F. Filippov. Differential equations with discontinuous righthand side. *Mathematics and its Applications*, 18, 1988.

18. R. Ghosh, A. Tiwari, and C. Tomlin. Automated symbolic reachability analysis; with application to delta-notch signaling automata. In O. Maler and A. Pnueli, editors, *proc. HSCC'03*, volume 2623 of *LNCS*, pages 233–248. Springer-Verlag, 2003.

19. L. Glass and S.A. Kauffman. The logical analysis of continuous, non-linear biochemical control networks. *Journal of Theoretical Biology*, 39(1):103–129, 1973.

20. G. Gössler and J. Sifakis. Component-based construction of deadlock-free systems (extended abstract). In *proc. FSTTCS'03*, volume 2914 of *LNCS*. Springer-Verlag, 2003.

21. J.-L. Gouzé and T. Sari. A class of piecewise linear differential equations arising in biological models. *Dynamical Systems*, 17(4):299–316, 2003.

22. M. Heiner and I. Koch. Petri net based model validation in systems biology. In J. Cortadella and W. Reisig, editors, *proc. ICATPN'04*, volume 3099 of *LNCS*, pages 216–237. Springer-Verlag, 2004.

23. A. Larrinaga, A. Naldi, L. Sánchez, D. Thieffry, and C. Chaouiya. GINsim: a software suite for the qualitative modelling, simulation and analysis of regulatory networks. *Biosystems*, 2005.
24. G. Marnellos, G.A. Deblandre, E. Mjolsness, and G. Kintner. Delta-Notch lateral inhibitory patterning in the emergence of ciliated cells in *Xenopus*: Experimental observations and a gene network model. In *proc. PSB'00*, volume 5, pages 326–337. World Scientific Publishing, 2000.
25. Z.N. Oltvai and A.L. Barabási. Life's complexity pyramid. *Science*, 298:763–764, 2002.
26. C. Piazza, M. Antoniotti, V. Mysore, A. Policriti, F. Winkler, and B. Mishra. Algorithmic algebraic model checking I: Challenges form systems biology. In *proc. CAV'05*, volume 3576 of *LNCS*, pages 5–19. Springer-Verlag, 2005.
27. O. Resendis-Antonio, J.A. Freyre-González, R. Menchaca-Méndez, R.M. Gutiérrez-Ríos, A. Martínez-Antonio, C. Avila-Sánchez, and J. Collado-Vides. Modular analysis of the transcriptional regulatory network of e. coli. *Trends in Genetics*, 21(1):16–20, 2005.
28. E. Simão, E. Remy, D. Thieffry, and C. Chaouiya. Qualitative modelling of regulated metabolic pathways: application to the tryptophan biosynthesis in e.coli. *Bioinformatics*, 21(Suppl. 2):ii190–ii196, 2005.
29. R. Thomas. Boolean formalisation of genetic control circuits. *J. Theor. Biol.*, 42:565–583, 1973.

Randomization and Feedback Properties of Directed Graphs Inspired by Gene Networks

M. Cosentino Lagomarsino[1,2], P. Jona[3], and B. Bassetti[2,4]

[1] UMR 168 / Institut Curie, 26 rue d'Ulm 75005 Paris, France
mcl@curie.fr
[2] Università degli Studi di Milano, Dip. Fisica, Milano, Italy
[3] Politecnico di Milano, Dip. Fisica, Pza Leonardo Da Vinci 32, 20133 Milano, Italy
[4] I.N.F.N. Milano, Italy
Tel.: +39 - (0)2 - 50317477; Fax: +39 - (0)2 - 50317480
bassetti@mi.infn.it

Abstract. Having in mind the large-scale analysis of gene regulatory networks, we review a graph decimation algorithm, called "leaf-removal", which can be used to evaluate the feedback in a random graph ensemble. In doing this, we consider the possibility of analyzing networks where the diagonal of the adjacency matrix is structured, that is, has a fixed number of nonzero entries. We test these ideas on a network model with fixed degree, using both numerical and analytical calculations. Our results are the following. First, the leaf-removal behavior for large system size enables to distinguish between different regimes of feedback. We show their relations and the connection with the onset of complexity in the graph. Second, the influence of the diagonal structure on this behavior can be relevant.

1 Introduction

Gene regulatory networks are graphs that represent interactions between genes or proteins. They are the simplest way to conceptualize the complex physico-chemical mechanisms that transform genes into proteins and modulate their activity in space and time. In the network view, all these processes are projected in a static, purely topological picture, which is sometimes simple enough to explore quantitatively [1]. Thanks to the systematic collection of many experimental results in databases, and to new large scale experimental and computational techniques that enable to sample these interactions, these graphs are now accessible to a significant extent. Some examples are the undirected graphs of protein-protein interactions, and the directed graphs of transcription and metabolic networks [1,2,3,4]. The availability of such large-scale interaction data is extremely important for post-genomic biology, and has provided for the first time a whole-system overview on the global relationships among players in a living system.

The hope is to study these graphs together with the available information on the genes and the physics/chemistry of their interactions to infer information on

C. Priami (Ed.): CMSB 2006, LNBI 4210, pp. 227–241, 2006.

the architecture and evolution of living organisms. In this program, the simplest possible approach to take is to study the topology of these networks. For instance, order parameters such as the connectivity and the clustering coefficient have been considered [5]. Other investigators have focused on the relations of gene-regulatory graphs with other observables, such as spatial distribution of genes, genome evolution, and gene expression [6,7,8,9,10]

Typically, in an investigation concerning a topological feature of a biological network, one generates so called "randomized counterparts" of the original data set as a null model. That is, random networks which conserve some topological observables of the original. The main biological question that underlies these studies asks to establish when and to what extent the observed biological topology, and thus loosely the living system, deviate from the "typical case" statistics. To answer this question, the tools from the statistical mechanics of complex systems are appropriate. For example, a topological feature that has lead to relevant findings is the occurrence of small subgraphs - or "motifs" [11].

The study presented here focuses on the topology, and in particular on the problem of evaluating and characterizing the feedback present in the network. On a generic biological standpoint, this is an important issue, as it is related to the states and the dynamics that a network can exhibit. Roughly speaking, the existence of feedback in the network topology is a necessary condition for the dynamics of the network to show multistability and cycles [12]. In presence of feedback, the relations between internal variables play an important role, as opposed to situations where the network is tree-like, and the external conditions determine completely the configurations and the dynamics. Recently, we came to similar conclusions analyzing the structure of the compatible gene expression patterns (fixed points) in a a Boolean model of a transcription network [13]. This model exhibits a transition between a regime of simple gene control, and a regime of complex control, where the internal variables become relevant and dynamically non-trivial solutions are possible. These regimes correspond to the SAT, and HARD-SAT phases of random-instance satisfiability problems. For random Boolean functions, the two regimes can be understood completely in terms of feedback in the network topology. A selection of the Boolean functions can change this outcome [14].

Rather than dealing with specific experimental networks, this is meant as a theoretical study on a model graph ensemble [1]. Our purpose here is twofold. First, to introduce some "order parameters", i.e. functions that describe the relevant feedback properties, connected to algorithms that can be used to evaluate the feedback without enumerating the cycles. Second, to study an ensemble of random graphs, or randomization technique, with structured adjacency matrices, that conserve the number of entries in their diagonal. This choice, which we will justify, leads to a distinct behavior. The two problems are introduced in section 2. We show the connections between different points of view on the problem, using simple algebraic, graph theoretical, and statistical mechanical tools. The first ap-

---

[1] By the word ensemble, we mean here a family of graphs with a, typically uniform, probability distribution.

proach is an application of a decimation algorithm called "leaf-removal" [15,16]. This algorithm links the feedback to the existence of a percolating "core" in the network, containing cycles. The numbers of core variables and edges can then be used as order parameters for the feedback. Here, we formulate three variants of the leaf-removal algorithm, and discuss the statistical meaning and the relations between them and different levels of feedback. Namely, for an oriented graph, one can use these algorithms to define and distinguish "simple" from "complex" feedback. Furthermore, we discuss how one can connect feedback to the satisfiability-like optimization problem of counting the solutions of a random linear system on the Galois field GF2 [17]. This can also be seen as a linear algebra problem concerning the kernel and rank of the connectivity matrix. The theoretical motivation for the choice of an ensemble with structured diagonal will follow naturally from this discussion. In section 3 we present our main results, as a series of "phase diagrams", which describe the typical feedback of random realizations of the graphs. In the unstructured case, the phase diagrams obtained by leaf-removal show the existence of five regimes, or "phases", characterizing the feedback in the limit of infinite graph size. Some of these regimes are connected to complexity transitions for the associated random GF2 optimization problem. Moreover, we show that the choice of a structured diagonal leads to a quantitatively different behavior, and thus to a significantly different amount of feedback in the graph. These differences are greatly enhanced if the degree distribution is scale-free.

2   Formulation of the Problem and Algorithms

The problem we want to address consists in evaluating the feedback in a random ensemble of graphs. While the range of application is more general, to avoid excess of ambiguity we choose a specific ensemble of graphs that will be treated in detail throughout the paper. We consider oriented graphs, where each node has p incoming links. The graph ensemble can be specified through a $M \times N$ Boolean matrix B (having elements 0 or 1). B represents the input-output relationships in the network. If x_i are network nodes, $B_{ji} = 1$ if $x_i \rightarrow x_j$, and zero otherwise. The matrix is rectangular because only $M < N$ nodes have an input. We allow for self links, or diagonal elements. For a simple directed graph one can say that feedback exists as soon as closed paths of directed edges emerge. Having in mind the fact that, while here we consider only topological properties, the incoming links are "inputs", that is, they encode for some conditions on the nodes (for example, on gene expression), we can also use a separate graphical representation for the nodes, or "variables", and the "functions" regulating these variables. This representation is a bipartite graph (Fig. 1). Each graph has N variables and M functions, and thus on average $\gamma = M/N$ functions per variable.

An important point concerning randomization, is that the choice of what feature to conserve and what not to conserve in the randomized counterpart is quite delicate and depends on specific considerations on the system. In the words of statistical mechanics, the typical case scenario can vary greatly with the choice of the ensemble. For instance, the network motifs shown by randomizing a

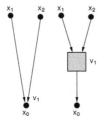

Fig. 1. Different representations of interactions in the graph G. Left: oriented graph Right: a bipartite oriented graph. V_1 is a function and x_i are variables, x_0 is the output.

network with an Erdos-Renyi random graph differ from the usual ones, for which the degree sequence is used as a topological invariant [18]. In studies of biological networks, the diagonal of B is normally disregarded, or assumed to have the same probability distribution as a row or a column. The use of considering it is a matter of the nature of the graph and the property under exam. For the case of transcription networks, an ensemble with structured diagonal might have some relevance. For example, for motifs discovery, sometimes one puts the diagonal to zero, and considers degree-conserving randomizations that do not involve the diagonal [21]. In our earlier work on transcription networks, we have considered the autoregulators as a structured diagonal [13]. We will show, for our model graph ensemble, that this leads to considerably different results for the feedback. There are other biological examples where a structured adjacency matrix emerges naturally. The simplest example are mixed interaction graphs. For instance, one can consider a composition of a transcription network with a protein-interaction network (which is a non directed graph) and pose the question of evaluating the feedback on a global scale compared to randomized counterparts.

Leaf-removal algorithms. A straightforward way to measure the amount of feedback in a graph is to count cycles. However, this is in general computationally as costly as enumerating all the paths. For this reason, it is desirable to use algorithms and order parameters that allow a quicker evaluation. To this aim, we describe three variants of a decimation algorithm, termed "leaf-removal", that is able to remove the tree-like parts of the graph, leaving the components with feedback. We define a leaf as a variable having only incoming links, and a "free" variable, or a root, a variable having only outgoing links (Fig 2). γ is a measure for the fraction of regulated variables, as opposed to external variables which only enter functions as inputs. The three variants of the leaf-removal iteratively remove links and nodes from the graph, using the following prescriptions (Fig 2).

1. LRa. Remove leaves and their incoming links.
2. LRb. As above. Additionally, remove incoming links of nodes whose incoming links are all connected to roots, which are also removed.
3. LRc. As LRa. Additionally, remove all the incoming links (together with their associated nodes) of nodes whose incoming links are connected to at least one root.

This is an iterative nonlinear procedure, where more variables may disappear in a single move. LRc works naturally on directed and undirected bipartite graphs. In fact, viewing the system as a bipartite graph, one can verify that LRc is equivalent to removing all the functions connected to a single node, ignoring directionality. Instead, LRa and LRb are thought for a directed graph, such as the ones we consider here.

There are two possible outcomes for the leaf-removal. Removing the whole graph, or stopping at a core subgraph that contains cycles. The core is composed of N_C genes and M_C functions. We want to use these as order parameters for the feedback. Equivalently, we can use $\Delta_C = \frac{N_C - M_C}{N}$ and $\gamma_C = M_C/N_C$. The difference between LRa and LRb is that LRb is able to remove tree-like parts of the graph that are upstream of a simple cycle. LRc is also able to do this. On the other hand, LRc might break some of these cycles because it disregards the orientations of the edges (Fig. 2). LRc cannot break "complex" cycles, defined as cycles where each node is connected to at least two functions.

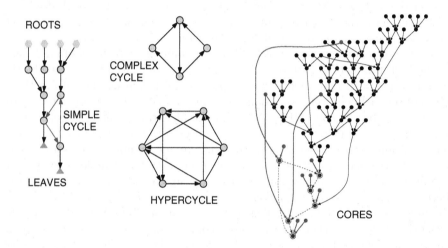

Fig. 2. Left: example of roots (free variables) and leaves for the leaf-removal algorithm. This graph contains a simple cycle (in red), which is not removed by LRa and LRb, but is removed by LRc. Middle: examples of a complex cycle and a hypercycle. A complex cycle (top) is not removed by LRc, but does not belong to the kernel of of A^t. A hypercycle (bottom) is an element of the kernel of A^t, because each variable appears in an even number of functions. Right: example of cores for the different leaf-removal variants, applied on the same initial graph. The image refers to a random graph with $p = 3$, $\gamma = 0.5$, $N = 600$. The cores are represented as a directed graph, and superimposed. The LRa core (whole figure) contains feedback loops and tree-like regions (black) upstream of the loops. The LRb core (red) does not contain the treelike parts, but all the feedback is preserved. The LRc core is empty, as this algorithm is able to break simple cycles connected to single free variables. The cycle of the original graph is indicated by circled nodes and dashed edges (blue).

Connections with Random Systems in GF2 and Adjacency Matrix Algebra. To investigate the feedback properties of the graph, one can also consider the following linear system in the Galois field GF2 (the set $\{0, 1\}$ with the conventional operations of product, and sum modulo 2).

$$Ax = v \ . \tag{1}$$

Here, v is a random vector of $GF2^M$, that represents the functions, and $A = B + I_{MN}$, where I_{MN} is the truncated $M \times N$ identity matrix, and the sums are in GF2. In other words, we imagine that each output variable is subject to a random XOR constraint, and the idea is to use this as a probe for the feedback. Each XOR constraint, or GF2 equation corresponds to a function. In the language of statistical mechanics, the random linear system (1) maps to a p-spin model on the graph [16]. The important point is that feedback translates into algebraic properties of the matrix A in GF2, and in solutions of Eq. (1). A feedback loop, or a cycle, corresponds to the pair A^o, h^o, where A^o is a $l \times l$ submatrix of A, and h^o is a l-component vector such as $h^o A^o = 0$. Indeed, the functions and variables selected by the nonzero elements of h^o are such that each variable appears in an even number of constraints.

We can also define a "hypercycle" as an M component vector h of GF2, such as the right product $hA = 0$, because the functions and variables selected by the ones in h are such that each variable appears in an even number of functions. Graphically, a hypercycle is a connected cluster made of functions that share an even number of nodes (Fig. 2). From the algebraic point of view, it is an element of the kernel of A^t, and is then connected to the solvability of Eq. (1). This consideration enables to evaluate the average number $\overline{N}$ of solutions of Eq. 1. Perhaps surprisingly, one can prove that $\overline{N} = 2^{N-M}$ under very general conditions. However, this average ceases to be significant when the hypercycles become extensive (i.e., the number of nodes they involve has order N), as the fluctuations become dominant. This is discussed in detail in Appendix A.1. The exact threshold for γ where hypercycles become extensive is a phase transition in the thermodynamic limit $N \to \infty, \ M \to \infty$ at constant γ. Precisely, it is called the SAT-UNSAT transition for Eq. (1) [19]. The UNSAT threshold depends on the graph ensemble, and has been determined in some cases [20]. In some instances, there may exist also an intermediate "HARD-SAT" or glassy phase, where 2^{N-M} solutions exists, but they belong to basins of attractions whose distance from each other [19] is order N. For a p-spin problem on a graph, this glassy phase corresponds to the presence of complex cycles [16].

Structured diagonal. As a hypercycle is a particular realization of a complex cycle, it is easy to understand how the core of a leaf-removal algorithm will in general (but not always) contain hypercycles: none of the algorithms is able to break these structures. This is shown in Appendix A.2, which discusses the relation of the leaf-removal "moves" with operations on the rows and columns of A, related to the solution of Eq. (1). As explained there, for a directed graph, the extensive hypercycle, or UNSAT region may exist only at $\gamma = 1$. In the case where the diagonal is structured, the situation is quite different, and the hypercycle phase can appear for $\gamma < 1$ [13,14]. The above consideration justifies from an abstract standpoint the intermediate situations, with a fixed fraction of ones on the diagonal of A. In considering this ensemble of matrices with structured diagonal, we can introduce an additional parameter χ, that represents the fraction of ones on the diagonal of A. It is important to note that the introduction of a structured diagonal in A changes the adjacency matrix, and thus the graph ensemble. This change can have different interpretations. Rather than focusing on a particular one, the objective here is to show on an abstract standpoint how the phase behavior of Eq. (1) is perturbed by χ.

3   Regimes of Feedback

In this section, we discuss numerical and analytical results for the leaf-removal algorithms that support the general considerations above. We considered mainly the ensemble of graphs with fixed indegree p and Poisson-distributed outdegree k, $p(k) = \frac{(p\gamma)^k}{k!} e^{-p\gamma}$. The diagonals are thrown with independent probability, to ensure that the average fraction of ones is $\chi \in [0, 1]$. The choice of a structured diagonal does not perturb the marginal probability distributions of columns or rows. One can connect χ to the notion of "orientability". If $M > N$, it is impossible to orient a graph assigning one single output per function. On the other hand, a graph with a structured diagonal can be seen as a partially oriented one, where some directed constraints coexist with some undirected ones. In this interpretation $\chi = 1$ is the simple directed graph with no self-links. The case $\chi = 0$ can be seen as a totally undirected graph, a similar ensemble to that used in [16].

We start with the totally orientable case $\chi = 1$. For each value of γ, at fixed network size N, one can generate randomized graphs and evaluate their cores numerically. This procedure is exemplified in Fig. 3 for the case of LRb. The figure shows a transition to a regime where the core is nonempty and all the graphs are sharply distributed around the average core size. Equivalently, one can evaluate the core order parameter Δ_C, which vanishes when the core is empty or $M_C = N_C$. The same order parameter is negative when $M_C > N_C$. Each LR has two critical values. The first, γ_d^x, is associated to the emergence of a nonempty (extensive) core. The second γ_s^x, to the condition $N_C < M_C$. Based on our results, γ_s is always the same for all three leaf-removals, and corresponds to the onset of the UNSAT phase of extensive hypercycles. From simulations and analytical work, $\gamma_s = 1$. γ_d^x, instead, depends on the ability to remove parts of the graph of the different algorithms.

As we have seen, LRa can remove less than LRb, because the latter is able to deal with the tree-like parts of the graph that lay upstream of the loops. Also, LRb can remove less than LRc, because LRc can break feedback loops if they are connected to a single free variable. Thus, one can expect $\gamma_d^a < \gamma_d^b < \gamma_d^c$. This is indeed our observation (Fig. 4).

Based on these results, we can distinguish the following five regimes of feedback: (1) all cores are empty, (2) only the LRa core is nonempty, (3) both the LRa

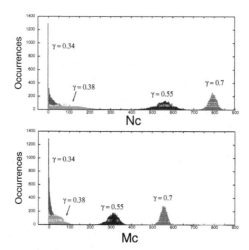

Fig. 3. Histogram of the core dimensions N_C and M_C as a function of γ for LRb. The data refer to 10^4 random networks with $p = 3$ and initial size $N = 1000$. For low γ, the cores are clustered towards the empty graph. At $\gamma \simeq 0.38$ the core distribution becomes wide. Successively, the mean values grow and the histogram acquires again a sharp single peak at increasing M_C, N_C. This is reminiscent of a second order phase transition. For LRc, this transition is much sharper (first order), and marks the onset of complexity in the core solutions γ_d^c.

and the LRb core are nonempty, (4) all the cores are nonempty with $N_C > M_C$, (5) all the cores have $N_C < M_C$. These last two regimes can be seen as thermodynamic phases connected with the SAT-UNSAT transition of the associated linear system.

1. There are no feedback loops in the typical case.
2. Feedback loops emerge, that form a core having an extensive treelike component upstream. The cycles are intensive (i.e. the core contains a number of nodes negligible with respect to N, or $o(N)$), but the tree upstream becomes extensive $(O(N))$. Analytically, one can compute that γ_d^a corresponds to the percolation-like threshold $1/p$ (see Appendix A.3 and Fig. 4). Intuitively, as soon as the graph percolates, even in the presence of a small region containing cycles, the tree upstream of the feedback loops can span an extensive part of the graph.
3. There is an extensive core of simple loops. LRb erases the tree upstream of the feedback loops, thus it can only have its threshold when the region of cycles itself becomes extensive. So far, we have not been able to compute the threshold γ_d^b analytically. However, our simulations indicate that it lies higher than γ_d^a (Fig. 4).
4. HARD-SAT phase. Intensive hypercycles, and extensive complex cycles form the core, where each variable appears in 2 or more functions. This gives a clustered structure to the space of solutions in the corresponding random

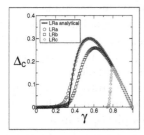
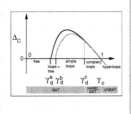
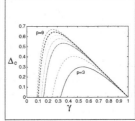

Fig. 4. Left: $\Delta_C(\gamma)$ for $\chi = 1$, $p = 3$. The solid line corresponds to the analytical calculation (Appendix A.3). The symbols are numerical results for 10^3 realizations of graphs with $N = 1000$. $\gamma_d^a < \gamma_d^b < \gamma_d^c$ mark the transition to an extensive core for the three leaf-removal algorithms. $\gamma_s = 1$ for all three algorithms to the point where Δ_C becomes negative. Middle: A scheme of the resulting phase diagram. Right: Analytical ($N \to \infty$) values of the order parameter Δ_C for the LRa algorithm, $\chi = 1$ and different values of p. The order parameter deviates from zero at the threshold $\gamma_d^a = 1/p$, and crosses again at $\gamma_c = 1$. The calculation is described in Appendix A.3.

linear system. Δ_C is proportional to the complexity Σ of the space of solutions, defined by the relation $\log \mathcal{N} \sim N(\Sigma + S)$. Here S, the entropy, measures the width of each cluster, while Σ counts the number of clusters.
5. UNSAT phase. The hypercycles become extensive. The threshold $\gamma_s = 1$ can be compute analytically (see Appendix A.3, and Fig. 4)

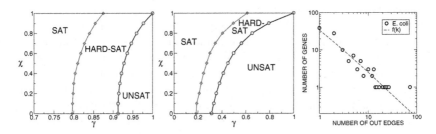

Fig. 5. Left: Phase diagram for $p = 3$ and structured diagonals (varying χ). There are quantitative changes with respect to $\chi = 1$. Middle: Phase diagram for scale-free distribution of the outdegree k. γ_d^c and γ_s move with the same trend and undergo a notable quantitative drift with increasing χ. Right: The exponent for the outdegree distribution is a fit from data on the transcription network of E. coli [21].

Considering now ensembles with a structured diagonal, one can carry the same analysis at fixed values of γ and χ. As we discussed above, LRc is not sensitive to graph orientation, and graphs with a structured diagonal can be seen as partially oriented ones. Thus, the simplest choice is to forget the other variants of the algorithms and focus on LRc. At fixed χ, there are three phases SAT, HARD-SAT, and UNSAT. On the other hand, as we argued above, because of the structure of the core matrices, these regions vary with χ, and a new phase diagram can be generated. The interesting result is that this ensemble can show quantitatively different thresholds, while leaving the marginal distributions for the row and column connectivity unchanged. We have addressed this question numerically, computing the thresholds $\gamma_d^c(\chi)$ and $\gamma_s(\chi)$. The results for the fixed p ensemble are shown in Fig. 5. The value for both thresholds increases with increasing χ. In particular, $\gamma_s(\chi)$ becomes exactly 1 in the directed case. On the other hand, the phenomenology of the transition does not vary with χ, with a discontinuous jump at the onset of a complex cycles phase, as in a first order phase transition. Thus, in the fixed p ensemble, there is a marked quantitative change in the thresholds. One may wonder whether the impact is the same for ensembles of graphs where the connectivity distributions are wider. Throughout the paper we have considered only the ensemble with fixed p and Poisson distributed k. Notably, the effect of a structure diagonal becomes larger for scale-free distributions of k. This is illustrated in Fig. 5, where we show the phase diagram for a power-law distribution for k with exponent 1.22 fitted from data from E. coli [21], and independently thrown columns for A. In this case, the influence of the diagonal can bring the hypercycle threshold γ_c down by a factor of three.

4   Discussion and Conclusions

We presented a theoretical study focused on the evaluation of feedback and the typical behavior of graphs taken from a random ensemble. The study focuses specifically on the ensemble of directed graphs with fixed indegree and Poisson outdegree. On the other hand, it is inspired by examples of biological graphs. Detecting feedback in large biological graphs and their randomized counterparts is important to understand their functioning. The use of our technique is that it allows for a quick evaluation and, more importantly, it provides some quantitative large-scale observables that can be used to measure the weight and the complexity of feedback loops. In order to do this, we introduce different variants of the leaf-removal algorithm, which naturally carry the definition of simple order parameters, depending on the properties of the core. We showed how the three algorithms relate to graph properties, algebraic operations on the adjacency matrix, and to solutions of the associated linear systems of equations in GF2. This analysis naturally leads to the abstract introduction of structured random graphs that conserve the number of entries in the diagonal of the adjacency matrix, which might be relevant in some biological situation.

Our two main results are the following. First, a phase diagram of different regimes of feedback depending on the fraction of free variables for an oriented graph. It shows a quite rich behavior of phase transitions that are interesting from the statistical physics viewpoint. These include the thresholds observed in diluted spin systems and XOR-like satisfiability problems. As already observed in [16], the onset of the complex phase is deep in the region where cycles exist and they involve a subgraph of the order of the graph size. On the other hand, the less intricate feedback regimes of intensive simple cycles connected to extensive trees, and of extensive simple cycles, might be relevant to characterize the dynamics in biological instances. The leaf-removal algorithms enable to analyze these different forms of feedback, that can be "weaker" than the complex cycles and hypercycles that are relevant for the associated GF2 problem. The second result is that the introduction of a structured diagonal, which can be interpreted as a partial orientation in the graph, has some influence on the thresholds. This is particularly true in presence of scale-free degree distribution, where we showed a phase diagram inspired by the connectivity in the E. coli transcription network [21]. The algorithms described here can be readily applied to biological data sets and their randomized counterparts. We are currently addressing this question in relation with the Darwinian evolution of some transcription and mixed transcription- and protein-interaction graphs. Finally, while this analysis is loosely inspired to graphs related to gene regulation, the need to evaluate the feedback arises in different contexts, where the tools described here could prove useful.

References

1. Uetz, P., Finley, Jr, R.: From protein networks to biological systems. FEBS Lett **579**(8) (2005) 1821–7
2. Babu, M., Luscombe, N., Aravind, L., Gerstein, M., Teichmann, S.: Structure and evolution of transcriptional regulatory networks. Curr Opin Struct Biol **14**(3) (2004) 283–91
3. Davidson, E., Rast, J., Oliveri, P., Ransick, A., Calestani, C., Yuh, C., Minokawa, T., Amore, G., Hinman, V., Arenas-Mena, C., Otim, O., Brown, C., Livi, C., Lee, P., Revilla, R., Rust, A., Pan, Z., Schilstra, M., Clarke, P., Arnone, M., Rowen, L., Cameron, R., McClay, D., Hood, L., Bolouri, H.: A genomic regulatory network for development. Science **295**(5560) (2002) 1669–78
4. Price, N., Reed, J., Palsson, B.: Genome-scale models of microbial cells: evaluating the consequences of constraints. Nat Rev Microbiol **2**(11) (2004) 886–97
5. Yook, S., Oltvai, Z., Barabasi, A.: Functional and topological characterization of protein interaction networks. Proteomics **4**(4) (2004) 928–42
6. Hurst, L., Pal, C., Lercher, M.: The evolutionary dynamics of eukaryotic gene order. Nat Rev Genet **5**(4) (2004) 299–310
7. Kepes, F.: Periodic epi-organization of the yeast genome revealed by the distribution of promoter sites. J Mol Biol **329**(5) (2003) 859–65
8. Teichmann, S., Babu, M.: Gene regulatory network growth by duplication. Nat Genet **36**(5) (2004) 492–6
9. Hahn, M., Conant, G., Wagner, A.: Molecular evolution in large genetic networks: does connectivity equal constraint? J Mol Evol **58**(2) (2004) 203–11
10. Luscombe, N., Babu, M., Yu, H., Snyder, M., Teichmann, S., Gerstein, M.: Genomic analysis of regulatory network dynamics reveals large topological changes. Nature **431**(7006) (2004) 308–12
11. Milo, R., Itzkovitz, S., Kashtan, N., Levitt, R., Shen-Orr, S., Ayzenshtat, I., Sheffer, M., Alon, U.: Superfamilies of evolved and designed networks. Science **303**(5663) (2004) 1538–42
12. Thomas, R.: Boolean formalization of genetic control circuits. J Theor Biol **42**(3) (1973) 563–85

13. Lagomarsino, M., Jona, P., Bassetti, B.: Logic backbone of a transcription network. Phys Rev Lett **95**(15) (2005) 158701
14. Correale, L., Leone, M., Pagnani, A., Weigt, M., Zecchina, R.: Core Percolation and Onset of Complexity in Boolean Networks. Phys. Rev. Lett. **96** (2006) 018101
15. Bauer, M., Golinelli, O.: Core percolation in random graphs: a critical phenomena analysis. Eur. Phys. J. B **24** (2001) 339–352
16. Mezard, M., Ricci-Tersenghi, F., Zecchina, R.: Alternative solutions to diluted p-spin models and XORSAT problems. J. Stat. Phys **505** (2003)
17. Levitskaya, A.A.: Systems of Random Equations over Finite Algebraic Structures. Cybernetics and System Analysis **41**(1) (2005) 67
18. Itzkovitz, S., Milo, R., Kashtan, N., Ziv, G., Alon, U.: Subgraphs in random networks. Phys Rev E Stat Nonlin Soft Matter Phys **68**(2 Pt 2) (2003) 026127
19. Mezard, M., Parisi, G., Zecchina, R.: Analytic and algorithmic solution of random satisfiability problems. Science **297**(5582) (2002) 812–5
20. Kolchin, V.F.: Random Graphs. Cambridge University Press, New York (1998)
21. Shen-Orr, S., Milo, R., Mangan, S., Alon, U.: Network motifs in the transcriptional regulation network of Escherichia coli. Nat Genet **31**(1) (2002) 64–8
22. Weigt, M.: Dynamics of heuristic optimization algorithms on random graphs. Eur. Phys. J. B **28** (2002) 369

A   Appendix

A.1   Solutions of the Random System in GF2

Evaluating the average number of solutions of Eq. 1 for large N at constant γ gives information in the feedback of the associated graph. We denote the kernel of a matrix by K, its range by R, and their dimensions by κ and ρ respectively. If the probability measure for v is flat, the average number of solutions for fixed A is the probability that $v \in R(A)$, i.e.

$$\text{prob}(v \in R(A)) = \frac{2^{\rho(A)}}{2^M} = 2^{-\kappa(A^t)} \ ,$$

times the number of elements in $K(A)$ (i.e. $2^{K(A)}$. The average number of solutions is thus

$$\overline{\mathcal{N}} = \langle 2^{-\kappa(A^t)} 2^{\kappa(A)} \rangle_A = 2^{N-M} \ , \tag{2}$$

where we have used the relations $\rho(A) + \kappa(A) = N$, $\rho(A^t) + \kappa(A^t) = M$, and $\rho(A) = \rho(A^t)$. Moreover, with the same reasoning, the fluctuations in the number of solutions are

$$\overline{\mathcal{N}^2} = (\overline{\mathcal{N}})^2 \langle 2^{\kappa(A^t)} \rangle_A \ ,$$

meaning that when the average $\langle 2^{\kappa(A^t)} \rangle_A$ is $O(1)$, an average number of solutions $\overline{\mathcal{N}} = 2^{N-M}$ are typically found, while this is not the case if $\langle 2^{\kappa(A^t)} \rangle_A$ is an extensive quantity. In fact, when this "selfaveraging" property breaks down, typically no solutions are found, because $\overline{\mathcal{N}}$ is supported only by the multiplicity of very rare $v \in R(A)$. This connects the solvability of the system to the topology of the hypercycles.

There are phase transitions between the two above regimes, tuned by the order parameter γ. The standard approach is to take the thermodynamic limit $N \to \infty$, $M \to \infty$ at constant γ. These transitions depend on the ensemble of graphs considered [20].

A.2 Adjacency Matrix and Leaf-Removal

Let us try to visualize the leaf-removal procedure, for instance LRc, on a generic adjacency matrix. Consider a general Boolean matrix A $M \times N$, and apply LRc. Each time we find a leaf, we assign it and its corresponding constraint a progressive number, and we use that number as a label for the rows. With these permutations, we construct a hierarchy for the leaves, as the leaves of layer a cannot appear in the clauses of layer $b \geq a$. In the tree-like case, reordering the lines of A, we obtain

$$
\begin{pmatrix}
\text{layer} & N & ... & ... & ... & ... & N-M & 1 \\
\hline
(1) & I & ... & ... & ... & ... & ... & ... \\
(2) & 0 & I & ... & ... & ... & ... & ... \\
(...) & 0 & 0 & ... & ... & ... & ... & ... \\
(m-1) & 0 & .. & 0 & I & ... & ... & ... \\
(m) & 0 & 0 & .. & 0 & I & ... & ...
\end{pmatrix}
$$

where (1) is the set of first layer leaves, (2) the second, etc. The last $N - M$ entries of each row correspond to free variables. We have thus obtained a triangulation of A, where the diagonal is made of blocks (the layers) of identity matrices.

In the presence of a core, the triangulation can be carried only until a the core is reached, and the the matrix can be rearranged to show the core in the lower right corner. If the core has hypercycles, in the UNSAT phase, the matrix structure is

$$
\begin{pmatrix}
\text{layer} & N... & ... & ... & ... & N_c & \longleftrightarrow & 1 \\
\hline
(M) & 1! & ... & .... & ... & ... & & \\
(..) & 0 & 1! & ... & ... & ... & & \\
(...) & 0 & 0 & 1! & ... & ... & & \\
\hline
(M_c) & 0 & .. & 0 & 0! & \text{core} & & \\
(..) & 0 & .. & 0 & 0 & `` & & \\
(1) & 0 & .. & 0 & 0 & `` & &
\end{pmatrix}
$$

Here, $M_C > N_C$, so typically it will not possible to find solutions to the core linear system on GF2, or the core does not contain sufficient free variables. When the ensemble for A is specified, one has to apply this procedure to all the realizations. Naturally, the outcome depends on the matrix ensemble. It also depends in general on the variant of leaf-removal that one applies.

Structured diagonal. Focusing on the diagonal of A, we note that in presence of hypercycles, one has necessarily to have some zeros in the $M \times M$ submatrix of A to realize the condition $N_c < M_c$. This can be seen in the sketch above, where the diagonal elements are followed by an exclamation mark. In particular, the diagonal contains an extensive number of ones. Thus, following the above argument, it is easy to realize that $M_c \leq N_c$, and the hypercycle phase may exists only marginally at $\gamma = 1$. For our main choice of ensemble, this is the case, as each variable can have only one input, so each constraint can always be labeled by the name of its output variable, which will appear as a one in the diagonal of A. In the case where the diagonal contains an extensive number of zeros, the situation is quite different, and the hypercycle phase can appear for $\gamma < 1$ [13,14].

A.3 Analytical Results for LRa

We present here the analytical calculation for LRa. If f_k is the probability to have k outputs, LRa defines a dynamics for it, associated by the cancellations of leaves at each time step. For every time t, one can write

$$N = N \sum f_k(t) \;;$$
$$N(t) = N \sum_{k\geq 1} f_k(t) = N\,(1 - f_0(t)) \;;$$
$$M(t)\,p = N \sum k f_k(t) \;.$$

The fraction of nonempty columns is given by the probability $1 - f_0(t)$. Writing the increments as, $\Delta N_k = N \Delta f_k = N \frac{\partial f_k}{\partial t} \Delta t$, one can choose $\Delta t = \frac{1}{M}, t \in [0:1]$, and obtain intensive equations of the kind $\frac{\partial f_k}{\partial t} = I(t)_{k,h,f_h(t)}$, where I is the matrix that represents the flux generated by a move [22].

We now separate A in the blocks S and T of constrained and free variables respectively, writing $A = [S|T]$. The variables that appear in T have an outgoing edge but no incoming ones. S has γN columns, while T has $(1 - \gamma) N$ columns. All the rows of A have p ones. The distribution for the ones appearing in the columns, i.e. for the outdegree k, is Poisson for both S and T, $f_k(0) = \frac{\lambda^k}{k!} e^{-\lambda}$, with $\lambda(0) = p\gamma$. We impose $s_{i,i} = 1$. The lines of A contain on average $p\gamma$ elements in S and $(1 - \gamma)p$ elements in T, thus after one move there are on average $p\gamma + 1$ elements in S. Defining $p' = p\,\gamma + 1$. The flux equations can be written as

$$\frac{df_k^S}{dt} = \frac{p'-1}{<k>^S(t)-1}[k f_{k+1}^S - (k-1)f_k^S]; \quad \text{for } k > 1$$
$$\frac{df_1^S}{dt} = -1 + \frac{p'-1}{<k>^S(t)-1}[f_2^S] \;,$$
$$\frac{df_0^S}{dt} = 1 \;,$$

where $< k > (t) = \sum k f_k(t)$. Summing the above equations, one obtains the evolution equation for the normalization factor $m^S := \sum p_k^c = < k >^S -1$.

$$\frac{dm^S}{dt} = -\frac{p'-1}{m^S(t)-1} \sum (k-1) f_k^S = -p\gamma$$

With initial condition $m^S(0) = \lambda(0) = p\gamma$, the solution is $m^S(t) = p\gamma\,(1-t)$. $m^S(t)$ can then be identified with $\lambda(t)$ appearing in the (Poisson) distribution

$f_k(t)$. $-\frac{\frac{d\lambda(t)}{dt}}{\lambda(t)} = \frac{p'-1}{m^S(t)} = \frac{p\gamma}{p\gamma}\frac{1}{1-t}$, from which $\frac{\lambda(t)}{\lambda(0)} = [1-t]$. Thus, for $k > 1$,

$$f_k^S = e^{\lambda(t)}\frac{\lambda(t)^{k-1}}{(k-1)!} \quad .$$

For $k = 1$, one can then write $\frac{\partial}{\partial t}f_1^S = -1 - \frac{\frac{d\lambda}{dt}}{\lambda}(\lambda e^{-\lambda})$, so that

$$f_1^S(t) = -t + e^{\lambda(t)} = -t + e^{p\gamma(t-1)} \quad .$$

The stop time t^* of the algorithm is then a solution of the equation $t^* = e^{p\gamma(t^*-1)}$. This last equation implies that if $p\gamma < 1$ the lowest solution for the stop-time is $t^* = 1$, or, in other words, all the graph is removed. On the other hand, when $p\gamma > 1$, there is a finite stop time $t^* < 1$, and thus a core. This determines the critical value $\gamma_d^a = 1/p$. The size of the portion of the core matrix contained in S is given by $M_{stop}^S = N_{stop}^S = \gamma N(1-t^*)$.

In order to evaluate the full core matrix and the order parameters, the same analysis has to be carried out for the matrix of the free variables, T. In this case, one has $p_k^T = k\frac{f_k^T}{m^T(t)}$, where $m^T(t) = \sum kf_k^T$. Again,

$$\Delta N^T = N(1-\gamma)\frac{\partial}{\partial t}f_k^T \Delta t = \frac{1-\gamma}{\gamma}\frac{\partial}{\partial t}f_k^T \quad ,$$

and the flux equations are

$$\frac{1-\gamma}{\gamma}\frac{\partial}{\partial t}f_0^T = \frac{p(1-\gamma)}{m^T(t)}f_1^T \quad ,$$
$$\frac{1-\gamma}{\gamma}\frac{\partial}{\partial t}f_1^T = \frac{p(1-\gamma)}{m^T(t)}[2f_2^T - f_1^T] \quad ,$$
$$\frac{1-\gamma}{\gamma}\frac{\partial}{\partial t}f_k^T = \frac{p(1-\gamma)}{m^T(t)}[(k+1)f_{k+1}^T - kf_k^T].$$

The last equation can be rewritten as $\frac{\partial}{\partial t}f_k^T = \frac{p\gamma}{m^T(t)}[(k+1)f_{k+1}^T - kf_k^T]$. As above, summation yields the evolution of the normalization constant $\frac{\partial}{\partial t}m^T(t) = -p\gamma$. Thus, $\frac{\frac{\partial}{\partial t}\lambda(t)^T}{\lambda^T} = \frac{p\gamma}{m^T(0)-p\gamma t}$, which gives

$$\frac{\lambda(t)^T}{\lambda^T(0)} = \frac{m^T(0) - p\gamma t}{m^T(0)} \quad ,$$

$$f_0^T(\lambda) = e^{-\lambda} \quad .$$

In conclusion, the stop time t^* is a function of $(p\gamma)$, determined by the relation $t^* = e^{p\gamma(t^*-1)}$. The transition value to an extensive core is then given by $\gamma_d^a = 1/p$. The core dimensions can be written as $M_C^S = N\gamma(1-t^*)$, and $N_C^T = (1-\gamma)(1-f_0^T) = (1-\gamma)(1-t^*)$. This last quantity gives the core order parameter $\Delta_C = (1-\gamma)(1-t^*)$. Δ_C is zero for $\gamma < \gamma_d^a$, and becomes nonzero at this critical value, in a continuous, non-differentiable way (with an infinite jump). The other threshold is easily calculated, as, for any finite $p\gamma$, $t^* > 0$, thus γ_c is given by the prefactor $1-\gamma$ in Δ_C crossing zero and becoming negative: $\gamma_c = 1$.

Computational Model of a Central Pattern Generator

Enrico Cataldo[2], John H. Byrne[1], and Douglas A. Baxter[1]

[1] Department of Neurobiology and Anatomy
The University of Texas Medical School at Houston
Houston, TX 77225
[2] Department of Biology – General Physiology Unit
Faculty of Science
University of Pisa
Via San Zeno 31 Pisa
Italy
Enrico.Cataldo@uth.tmc.edu,
John.H.Byrne@uth.tmc.edu,
Douglas.Baxter@uth.tmc.edu

Abstract. The buccal ganglia of *Aplysia* contain a central pattern generator (CPG) that mediates rhythmic movements of the foregut during feeding. This CPG is a multifunctional circuit and generates at least two types of buccal motor patterns (BMPs), one that mediates ingestion (iBMP) and another that mediates rejection (rBMP). The present study used a computational approach to examine the ways in which an ensemble of identified cells and synaptic connections function as a CPG. Hodgkin-Huxley-type models were developed that mimicked the biophysical properties of these cells and synaptic connections. The results suggest that the currently identified ensemble of cells is inadequate to produce rhythmic neural activity and that several key elements of the CPG remain to be identified.

1 Introduction

Feeding behavior of *Aplysia* is a useful model system with which to study the neural control of a relatively complex and adaptive behavior (for recent reviews see [1], [2]). The behavior has been described as a sequence of appetitive and consummatory activities. Appetitive activities (e.g., locomotion and head waving) help bring the animal in contact with food, and consummatory activities (e.g., biting, swallowing, rejection) mediate the movement of food into and out of the foregut. Consummatory activities involve rhythmic movements of structures in the foregut such as the buccal mass, the radula (the toothed grasping surfaces of the odontophore) and the jaws. All consummatory movements consist of two phases, radula protraction and retraction. Protraction and retraction movements of the radula are synchronized with movements of the lips and jaws, as well as with radula opening and closing movements, to produce a variety of functionally different consummatory movements. For example, during a bite, the jaws open as the odontophore rotates forward (i.e., protraction). Initially, the two

C. Priami (Ed.): CMSB 2006, LNBI 4210, pp. 242–256, 2006.

halves of the radula are separated (i.e., open) during protraction. Before the peak of protraction, however, the two halves of the radula begin to close and grasp the food. The radula remains closed as the odontophore retracts (i.e., backward rotation), which brings the food into the buccal cavity, and the jaws close [3]. Rejection differs from ingestion movements in that the two halves of the radula are closed as the odontophore protracts and open as it retracts, which ejects the unwanted material from the buccal cavity.

Elements of a CPG that control the rhythmic feeding movements reside primarily in the buccal ganglia, and the rhythmic patterns of neural activity underlying feeding movements have been characterized (Fig. 1) (e.g., [4], [5], [6]). For example, during a BMP, neural activity corresponding to protraction and retraction, can be monitored as bursts of spikes in identified neurons such as B63 and B64, respectively [7], [8]. iBMPs can be distinguished from rBMPs, in part, by the timing of activity in radula-closer motor neurons (e.g., B8) relative to the protraction and retractions phases of the BMP (e.g., [9], [10], [11]).

The synaptic interconnections of many of these cells have been characterized as have been their firing properties, activity during BMPs and responses to transmitters. The large body of knowledge relating to the neural circuitry that mediates feeding behavior of *Aplysia* indicates that a comprehensive and quantitative model would help explain the function of individual components of the circuit and their role in organizing behavior. Such a model would also help organize this expanding body of knowledge into a modifiable framework that would allow additional analysis and facilitate future empirical investigations.

The present study extends previous models of the feeding CPG [12], [13], [14], [15], [16], by developing a neural network that included Hodgkin-Huxley-type models and by including additional cells and their synaptic connections (Fig. 2). The properties of the models were based on both previously published empirical studies and on unpublished observations. The models were based on the premise that the functional properties of the network as a whole could be investigated if *i)* the active and passive membrane properties of each cell, *ii)* the magnitude and time course of the monosynaptic connections, and *iii)* the overall pattern of synaptic connectivity could be matched to the available physiological data. The model was robust and it provided key insights into our current level of understanding of the feeding CPG.

2 Methods

The simulations were performed with version 8 of SNNAP (Simulator for Neural Networks and Action Potentials; [16], [17], [18]). The software was run under the Microsoft Windows XP operating system on a Pentium 4 computer. The forward Euler method with a fixed time step of 45 μsec was used for numerical integration. The model will be added to ModelDB website (senselabe.med.yale.edu), and to SNNAP website (snnap.uth.tmc.edu) where it will be possible to download and run the simulations and view parameters and equations.

The initial network contained Hodgkin-Huxley-type models of nine neurons and their synaptic connections (see Fig. 2). Each cell in the network was modeled as a single, isopotential compartment. The equivalent electrical circuit for each cell consisted of a membrane capacitance (C_M) in parallel with a leakage conductance (g_L) and its associated equilibrium potential (E_L). In each neuronal model, the values for C_m, g_L, and E_L were adjusted to reflect the average empirically observed resting membrane potential, input resistance, membrane time constant and the relative size of each cell. In addition to these passive elements, one or more voltage- and time-dependent conductances and an associated equilibrium potential were added to the equivalent circuit model for each cell.

The number and type of variable conductances that were incorporated into each cell model were adjusted to reflect the unique firing characteristics of the individual cells. Several cells included conductances in addition to the fast Na^+ and K^+ conductances. The kinetics and voltage-dependencies of the various conductances were adjusted to reflect the empirically observed threshold for initiating an action potential in each cell, level of spiking activity generally observed in these cell in response to stimuli, and any unusual cellular properties, such as delayed responses to stimuli or plateau potentials. Chemical synaptic conductances (g_{cs}) were incorporated into the neuronal models by adding time-dependent conductance changes and associated reversal potentials (E_r). Five general features of synaptic connections were considered in the simulations. First, the reversal potential and the magnitude of each synaptic conductance were adjusted to reflect whether a given synaptic connection was excitatory or inhibitory and its average amplitude, respectively. Second, the time constant for each synaptic conductance was adjusted to match the general time course of the empirically observed PSPs. Third, some of the synaptic connections are multiaction and included both fast and slow components and/or both excitatory and inhibitory components. Thus, the simulated connections between some cells had multiple components that reflected the empirical observations. Fourth, empirical data indicate that the slow synaptic connection from B34 to B8 is mediated by a conductance decrease [7]. Thus, this synaptic connection was modeled as producing a decrease in the membrane conductance of B8. Fifth, some of the synaptic connections expressed homosynaptic plasticity (i.e., depression or facilitation). Thus, the simulated connections between some cells manifest homosynaptic plasticity. Electrical synaptic conductances (g_{es}) were incorporated into cell models by adding a linear conductance between any two cells. The simulated network contained four electrical connections and the parameters of these connections were adjusted to match the available empirical data. The robustness of the model was tested by running a series of simulations in which the values for some parameters were either randomly altered at the beginning of a simulation (see Results) or subjected to stochastic fluctuations throughout a simulation.

3   Results

3.1   Patterns of Fictive Feeding in Isolated Buccal Ganglia Preparations

Rhythmic patterns of neural activity that underlie feeding movements have been characterized in freely behaving animals (e.g., [6]), in reduced preparations (e.g. [9]) and in isolated ganglia (e.g. [5]). Consistent and similar patterns of neural activity have been recorded in these vastly different types of preparations, which suggests that the isolated ganglia retain sufficient circuitry to reproduce a substantial proportion of the behaviorally relevant neural activity. Thus, the BMPs that are recorded in isolated ganglia preparations can be considered fictive representations of feeding movements. Two examples of BMPs recorded from the isolated buccal ganglia are illustrated in Fig. 1. The BMPs have two phases: a protraction phase followed by a retraction phase. A number of cells have been identified whose spiking activity occurs primarily during either the protraction or retraction phases of the BMP. For example, B31/32 and B63 are active during the protraction phase [7], [8], whereas B4/5 and B64 are active during the retraction phase [19], [20]. To distinguish fictive ingestion from fictive rejection, it is necessary to monitor activity in cells that mediate the closure of the radula (e.g., B8; [9]). Fictive ingestion is characterized, in part, by activity in B8 occurring primarily during the retraction phase (Fig. 1A), whereas fictive rejection is characterized, in part, by activity in B8 occurring primarily during the protraction phase.

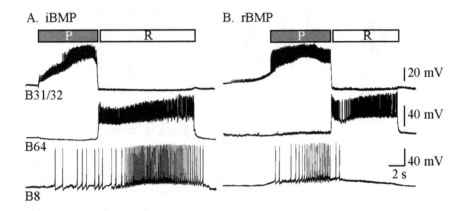

Fig. 1. Examples of BMPs. Simultaneous intracellular recordings monitored activity in several cells during spontaneously occurring BMPs. The protraction phase (indicated by the shaded bar labeled P) was monitored via activity in B31/32. The retraction phase (indicated by the open bar labeled R) was monitored via activity in B64. A: iBMPs were characterized, in part, by activity in the radula-closer motor neuron B8 occurring primarily during retraction. B: rBMPs were characterized, in part, by activity in B8 occurring primarily during protraction (Baxter and Byrne, unpublished observations).

3.2   Key Element of CPG Has Yet to Be Identified

A previous model [13], [16], which included B4/5, B31/32, B35, B51 and B52, was able to simulate rhythmic activity similar to a BMP by including a hypothetical cell (referred to cell I) that received excitation from B35. Cell I, in turn, made a mixture of excitatory and inhibitory connections with the other cells in the network. Subsequently, a cell, B64, was identified with many of the properties predicted by cell I [19]. B64 appears to terminate the protraction phase of a BMP while maintaining the retraction phase. The synaptic connections to and from B64, however, do not match all of the predictions from the previous model. Thus, the first goal of the present study was to incorporate B64 and its synaptic connections and re-examine the ability of the network to produce rhythmic activity.

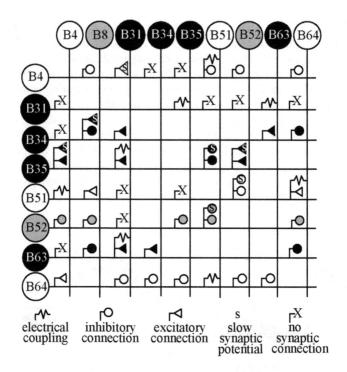

Fig. 2. Summary of synaptic connections within the CPG of the buccal ganglia. Multi-action synaptic connections (e.g, connections with both EPSPs and IPSPs) are indicated by plotting more than one type of synaptic symbol. Cells are also coded with respect to their firing pattern during BMPs. Cells that fire primarily during the protraction phase are filled with black and have white lettering. Cells that primarily fire during the retraction phase are filled with white and have black lettering. Cells that can fire during either protraction and/or retraction phases (e.g., B8, see Fig. 1) are filled with gray and have black lettering. Synaptic symbols with the letter S indicate a slow component, and synaptic symbols with the letter X indicate the confirmed absence of a synaptic connection.

In addition to incorporating B64, the network was expanded to include three additional cells B63, B34 and B8 (Fig. 2). B63 was included because computational and empirical studies indicate that a positive feedback loop between B31/32 and B63 is critical for initiating a BMP and producing the protraction phase [2], [7]. B34 was included because empirical studies suggest that it may be involved in switching between iBMPs and rBMPs [7]. B8 was included as an indicator of iBMPs versus rBMPs.

An initial attempt to simulate a BMP with this nine-cell network is illustrated in Fig. 3A. The protraction phase of the simulated BMP was monitored via activity in B31/32 and the retraction phase was monitored via activity in B64. The BMP was initiated by stimulating a brief burst of action potentials in B63 (not shown), which produced EPSPs in B31/32. The EPSPs in B31/32, in turn, elicited the sustained depolarization of B31/32 (i.e., the protraction phase). The simulated BMP failed to switch from the protraction to the retraction phase. Although it is possible to terminate the depolarization of B31/32 by including a K^+ conductance in the model of B31/32, a mechanism must still be included that initiates activity in B64. This result suggested that the circuit of Fig. 2 is insufficient to simulate the switch between protraction and retraction phases and that an additional element(s) has yet to be identified.

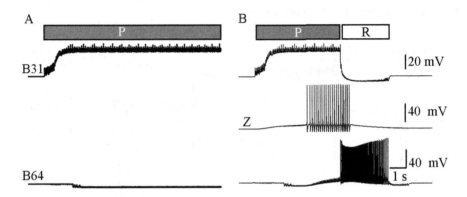

Fig. 3. Hypothetical cell Z mediates the transition from protraction to retraction phases of a BMP. A: initially, the simulated network contained only the identified cells and synaptic connections illustrated in Fig. 2. A brief depolarizing current pulse (1 s, 2 nA) was injected into B63 (not shown). The resultant activity in B63 elicited a plateau potential in B31/32 (i.e., a protraction phase, box labeled P), but no retraction phase (i.e., activity in B64). B: the transition from protraction to retraction phases (boxes labeled P and R, respectively) was accomplished by incorporating a hypothetical cell into the CPG (cell Z).

The switch between protraction and retraction phases is characterized by a hyperpolarization in cells that are active during protraction (see Fig. 1), which terminates their spike activity, and a depolarization in cells that are active during retraction, which initiates their spike activity (e.g., [4]). The hyperpolarization is mediated, in large part, by B64, which makes extensive monosynaptic inhibitory connections with cells active during the protraction phase (e.g., B31/32, B34, B35, B63; see Fig. 2). Thus, the process that mediates the switch between protraction and retraction phases is very likely to be the same process that initiates spiking in B64.

Empirical evidence supports the suggestion that an unidentified element excites B64 and thereby initiates the retraction phase. If hyperpolarizing current is injected into B64, thereby blocking activity in B64, the protraction phase is prolonged (Fig. 4A). Moreover, while B64 is hyperpolarized, an EPSP is observed in B64 at the same point in time when the switch from protraction to retraction would normally have occurred (Fig. 4A2). One interpretation of this observation is that an unidentified element (either a synaptic connection from a previously identified cell or a yet to be identified cell) excites B64, causing it to spike and thereby terminating the protraction phase and initiating the retraction phase (Fig. 4B1). Thus, these data suggest that recurrent inhibition mediates the switch from protraction to retraction phases of activity in a BMP.

To examine this hypothesis, the model was extended to include a tenth cell. This hypothetical cell, which was referred to as cell Z, was excited by B63 and, in turn, it excited B64. The amplitude and time course of this Z-mediated EPSP in B64 was adjusted to match the EPSP that was unmasked in B64 while it was hyperpolarized. As illustrated in Fig. 3B, after cell Z was incorporated into the network, stimulation of B63 elicited a protraction phase that was followed by a retraction phase. Thus, by incorporating recurrent inhibition, the ten-cell network was able to simulate the switch between protraction and retraction phases.

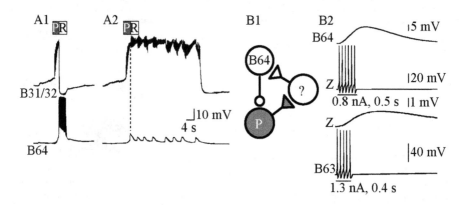

Fig. 4. Activity in B64 terminates the protraction phase of a BMP. A1: the durations of the protraction and retraction phases were indicated by the boxes labeled P and R, respectively. Note, the burst of spike activity in B64 coincides with the hyperpolarization of B31/32. A2: spike activity in B64 was blocked by a negative bias current. Blocking spike activity in B64 dramatically prolonged the duration of the protraction phase. The boxes labeled P and R indicate the protraction and retraction phases, respectively, that were recorded in Panel A. Note that a depolarization in B64 was observed at the point in time when the transition from protraction to retraction phases should have occurred (dashed line) (Baxter and Byrne, unpublished observations). B1: B64 inhibits cells that are typically active during the protraction phase of a BMP. The empirical observations illustrated in Panel A suggested that an unidentified cell (cell labeled ?) provides excitation to B64 and is responsible for initiating the retraction phase. B2: the network illustrated in Fig. 2 was extended to include a hypothetical cell labeled Z. Cell Z received a slow excitatory input from B63 and it excited B64.

3.3 Simulating Fictive Rejection

The results described above indicated that the ten-cell network could produce a pattern activity that exhibited the two essential phases of activity during a BMP (i.e., a protraction phase followed by a retraction phase). To determine whether this pattern of activity had features similar to fictive ingestion or rejection, it was necessary to monitor activity in the other cells (Fig. 5).

Brief stimulation of B63 elicited a plateau potential in B31/32 and bursts of activity in cells B35 and B34 (i.e., a protraction phase). The protraction phase was followed by bursts of activity in cells B64 and B4/5 (i.e., the retraction phase). Cell B52 produced one burst of activity that coincided with the protraction phase and second burst of activity at the end of the retraction phase. Cell B8 produced a burst of activity that coincided with the protraction phase. Thus, the general features of this simulated pattern of activity were similar to those of fictive rejection (e.g., Fig. 1B).

In addition to investigating the ways in which the protraction phase was terminated and retraction phase initiated, the model was used to investigate processes that might regulate the duration of the retraction phase (Fig. 6). For example, because of its extensive inhibitory synaptic connections, B52 may be responsible for terminating bursting. Alternatively, terminating the plateau potential in B64 through its slowly activating K$^+$ current may play a role in terminating the retraction phase. Several simulations were run to assess the ways in which various features of the model contributed to terminating the retraction phase. Although many features of the model were investigated, no single manipulation was found to substantially prolong the retraction phase. Rather, a combination of manipulations was necessary.

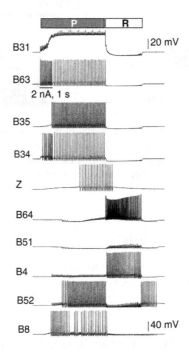

Fig. 5. Simulating a rBMP. A brief depolarization of B63 (bar) elicited a complex pattern of bursting in the CPG. Activity during the protraction phase was initiated and maintained by interactions between B31/32, B34, B35 and B63. Activity in B34 also provided suprathreshold excitation to B8 during the protraction phase. Thus, the pattern had the characteristics of a rBMP. The Z cell became active late in the protraction phase and excited B64 and thereby initiated the retraction phase. B64, in turn, inhibited cells that were active during protraction and thereby terminated the protraction phase. B64 expresses a plateau potential that maintained activity during the retraction phase. B52 expressed rebound excitation and the retraction phase was terminated when B52 escaped from its inhibitory inputs and began to fire. The protraction and retraction phases are indicated by the boxes labeled P and R, respectively.

The two processes that in combination terminated the retraction phase were the rebound excitation in B52 and the slow K⁺ conductance in B64 (Fig. 6B). Reducing the rebound excitation in in B52 to 60% of its control value blocked activity in B52 at the end of the retraction phase. Although B52 inhibits B64, blocking the second burst of spikes in B52 alone had no effect on the duration of the retraction phase (not shown). Similarly, reducing the slow K⁺ conductance in B64 to 60 % of its control value prolonged the plateau potential in an isolated model of B64 but had only a modest effect on the duration of the retraction phase in the network (not shown). If both manipulations were combined, however, the retraction phase failed to terminate (Fig. 6B). These results illustrate the ways in which the overall behavior of network emerge from the interactions of the elemental processes.

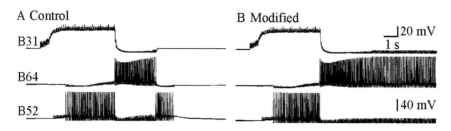

Fig. 6. Mechanisms contributing to the termination of the retraction phase. BMPs were elicited by brief stimulation (1 s, 2 nA) of B63 (not shown). A: for the control simulation, all parameter values were as in Fig. 5. B: for the modified simulation, the maximum conductances for the slow K⁺ current in B64 and for the H-type current in B52 were reduced by 40%. As a result of these two changes, B52 failed to rebound from inhibition during the retraction phase and the retraction phase failed to terminate.

3.4 Parameter Sensitivity Analysis

Each element of the model was designed to mimic the empirically measured properties of the cells and synaptic connections within the buccal ganglia. Nevertheless, there are no detailed voltage-clamp data with which to constrain the parameters. Despite this lack of detailed empirical data, a model emerged that reproduced several key features of a rBMP. It was not clear, however, to what extent the pattern generating capabilities of the neural network might be linked to a specific value or set of values for a parameter(s). To assess the quality of the model in terms of its consistency and robustness, a parameter sensitivity analysis was undertaken.

This analysis consisted of three groups of simulations. The first two groups of simulations assessed the values selected for membrane and synaptic conductances. In these two groups of simulations, all 40 synaptic conductances or all 37 membrane conductances were randomly assigned new values that were between ±15% of their control values. The ±15% value was arbitrary but was similar to values used by others to perform sensitivity analyses of computational models (e.g., [21], [22], [23]). After these randomly assigned values were incorporated, stimuli (both brief and prolonged) were applied to B63 and the ability of the modified network to generate both a single pattern of activity and continuous rhythmic activity were determined. This procedure of randomly altering all membrane or synaptic conductances and attempting to generate patterned activity was repeated ten times for each group. All twenty variants of the neural network produced both single patterns of activity and continuous rhythmic activity that were similar to that generated by the control circuit (i.e., Fig 5).

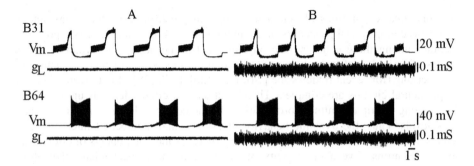

Fig. 7. Stable rhythmic activity generated by a model CPG with stochastic fluctuations in values for all membrane, synaptic and coupling conductances. Stochastic fluctuations were incorporated simultaneously into all 37 membrane conductances, all 40 synaptic conductances, and all 8 coupling conductances. The simulated neural activity was monitored by displaying the membrane potential (V_m) of two representative cells: B31 and B64. The stochastic fluctuations in the conductances were monitored by displaying the values for a representative conductance in these two cells: i.e., the leakage conductance (g_L). Each panel illustrates 30 s of simulated time. A: the magnitude of the S.D. was set to 5% of the mean. The stochastic fluctuations in the values for leakage conductances can be seen as noise in the traces labeled g_L. B: the magnitude of the S.D. was increased to 30% of the mean. Despite continual, random and relatively large fluctuations in 85 key parameters, the model CPG produced stable rhythmic activity, similar to control simulations.

A third group of simulations examined the impact of stochastic fluctuations on the ability of the model to generate rhythmic activity. Random numbers from a Gaussian distribution were simultaneously added to all 37 membrane conductances, all 40 synaptic conductances and all 8 coupling conductances. A new set of 85 random numbers was generated every 4.5 ms throughout 60 s of simulated neural activity. The mean values for these 85 stochastically fluctuating conductances were their respective control values. During successive simulations, the magnitude of the standard deviation (S.D.) of the Gaussian distribution was progressive increased in increments of 5% of the mean until the model failed to generate stable rhythmic activity. Examples of two such simulations are illustrated in Fig. 7. During the simulation in Panel A, the magnitude of the S.D. was set to 5% of the mean, whereas during the simulation in Panel B, the magnitude of the S.D. was set to 30% of the mean. Both variants of the model generated stable rhythmic activity. Moreover, the sequence of simulated activity in the other eight cells (not shown) was similar to that illustrated in Fig. 5 and resembled fictive rejection. In the models that incorporated noise, however, the durations of the protraction and retraction phases, and the inter burst intervals fluctuated. Overall, the model continued to generate stable rhythmic activity until the magnitude of the S.D. was increased to >30% of the mean and all patterns of activity resembled rBMPs.

These results indicate that the ability to generate pattern activity was a robust property that emerged from the neural network as a whole.

4 Discussion

The functional capabilities of neural networks emerge from the interactions among the intrinsic biophysical properties of the individual cells, the pattern of synaptic connections among these cells and the physiological properties of the synapses. From studies of a number of well characterized neural circuits, several general conclusions are emerging (for recent reviews see [24], [25], [26], [27], [28]). First, even relatively simple neural circuits are complex. The operation of a circuit depends upon interactions among multiple nonlinear processes at the molecular, cellular, synaptic and network levels. Thus, the dynamic behavior of even a small number of interconnected cells is not necessarily intuitive. Second, the structures and components of circuits are diverse. Neurons have a multiplicity of ionic conductance mechanisms that allow them to generate many disparate and complex patterns of activity. Similarly, synapses are not simply excitatory or inhibitory but possess a wide array of diverse properties. Thus, circuits with similar architectures can produce dramatically different responses and patterns of activity, or conversely, neural circuits that underlie similar functions can have very dissimilar components and structures. Third, the functional organization of neural circuits is dynamic. Modulation of the cellular and synaptic properties can reorganize a circuit and alter its operation. This enables a circuit to adapt and allows a single network to underlie several different functions. Thus, the nonlinearity, diversity and dynamic nature of neural circuits provide formidable challenges to arriving at a synthetic understanding of how circuits operate and adapt. Quantitative, biologically-realistic models can provide insights into the dynamics of complex nonlinear systems, such as neural circuits, that are impossible to achieve in any other way (for recent reviews see [27], [29], [30], [31], [32]).

The present study developed a computational model of a CPG that underlies aspects of feeding *Aplysia*. The model provided a quantitative summary of a large body of knowledge related to the cells, synaptic connections and their physiological properties. The models were used to explore the completeness of our knowledge of this circuit, and the functional role of specific circuit elements.

Initial simulations were unable to produce rhythmic patterns of activity similar to empirically observed BMPs. Specifically, the network was unable to switch from the protraction to retraction phases of activity. This shortcoming suggested that an element of the CPG has yet to be identified. Additional simulations and modifications to the network helped to predict some of the features of this missing element (i.e., the hypothetical cell Z). The distinguishing feature of this missing element was recurrent inhibition. To date, only two cells have been identified that make monosynaptic, excitatory connections with B64: B21 and B65. In isolated ganglia, B21 receives rhythmic depolarizations during BMPs, but it does not spike. Thus, B21 is unlikely to be the Z cell. Although B65 is active during protraction and forms an excitatory connection with B64, the B65-mediated EPSP in B64 is subthreshold anddecrements rapidly during repetitive activity. Thus, B65 is unlikely to be the Z cell. At present, none of the identified cells or synaptic connections match the predicted characteristics of the missing element, which suggest that additional empirical studies should be directed at locating cells that excite B64.

4.1 Basic Building Blocks of BMPs

The simulations are providing insights into the physiological properties that contributed to the genesis of BMPs. Excitatory feedback loops and intrinsic plateau potentials played a key role in initiating BMPs. Intrinsic plateau potentials and electrical coupling also played key roles in pattern generation. Plateau potentials represent a bistability in the membrane potentials of cells. Switch-like transitions between hyperpolarized and depolarized states can be induced by transient excitatory and inhibitory synaptic inputs. The plateau potentials in B31/32 and B64 provided the excitatory drive for the protraction and retraction phases, respectively, of the BMP. This excitatory drive was transmitted to other cells primarily via electrical coupling. In addition, electrical coupling among cells helped to coordinate burst formation during the protraction and retractions phases of a BMPs. Finally, postinhibitory rebound excitation played a role in terminating BMPs. BMPs were terminated, in part, by a burst of activity in B52 as it escaped from B64-mediated inhibition: a process that has been termed intrinsic escape.

4.2 Discrepancies Between Empirical and Simulation BMPs

Although the simulated patterns of activity matched many of the key features of empirically observed BMPs, there were some discrepancies between the two. For example, the simulated network generated only rBMPs, whereas empirically the CPG rapidly switches between generating iBMPs and rBMPs (e.g., [5], [10]). The genesis of different patterns of activity reflects, in part, the dynamic recruitment of specific subsets of cells into the CPG. For example, activity in B51 contributes to the genesis of iBMPs [33], [34], but it is silent during fictive rejection. Conversely, activity in B34 contributes to the genesis of fictive rejection. The genesis of different patterns of activity is also related to different levels of activity in some cells within the CPG [35], [36]. For example, high levels of activity in B4/5 are related to the genesis of fictive rejection, whereas low levels of activity are related to the genesis of fictive ingestion. Future simulations may be able to simulate the random switching between the genesis of fictive ingestion and rejection by incorporating additional cells, synaptic connections and/or modulatory processes. Addressing this issue will be important for understanding the adaptive responses of this circuit, in part, because the mechanisms underlying switching appear to be the target for modification by associative learning in this system.

4.3 Comparison to Previous Models of the CPG

Several previous studies have developed models of the CPG underlying aspects of feeding in *Aplysia*. Kupfermann et al. [14] developed a theoretical neural network that incorporated three layers of neurons (an input or sensory layer, a hidden or interneuron layer, and an output or motor layer) and a back-propagation algorithm. The neural network was trained to solve several behavioral selection problems that related to finding and consuming food. Although a highly abstract model, this initial simulation study suggested that the command units in the hidden layer had more complex roles in behavior than was previously appreciated and that the final behavior expressed by the system was due to the combined actions of multiple command units. Deodhar et al. [12] developed a neural network of just two neurons (and two muscles) and used a genetic algorithm to investigate synaptic parameters that allowed the system to generate efficient protraction/retraction movements of a simulated radula. This simulation was similar to a CPG network in that the solution did not depend on inputs from higher-order, sensory or modulatory cells. The simulations explored the oscillatory properties of the neural circuit and found that reciprocal inhibition between the cells appeared to function well for generating rhythmic activity. In contrast, results of the present study suggest that rhythm generation emerges from a recurrent inhibition rather than reciprocal inhibition.

4.4 Expanding the Neural Network

The present study illustrated that a ten-cell network can produce the basic pattern of activity observed during a BMP. This ten-cell network does not represent all of the identified elements of the CPG, however. For example, B20 and B65 are both believed to be part of the CPG [20], [37]. The firing properties of these cells have been characterized as have many of their postsynaptic targets. It was not possible to incorporate these cells into the present model, however, because no monosynaptic inputs have been described from other elements of the CPG to either cell. Additional cells and synaptic connections within the CPG are continually being identified and characterized. This computational model will provide a quantitative and modifiable framework with which to investigate how these newly identified elements contribute to the overall function of the CPG. Future studies will expand the simulated neural network to include additional cells and synaptic connections and will investigate functional properties of the feeding circuitry. In addition to providing a tool with which to investigate the CPG, the computational model can be expanded to include higher-order cells (e.g., the command neurons) and modulatory processes. Thus, the continual expansion and development of a computational model of the feeding circuitry should provide a useful tool for analyses of the neuronal mechanisms underlying behavior and behavioral plasticity.

Acknowledgments

This work was supported by NIH grants R01-RR11626, R01-MH58321 and P01-NS38310.

References

1. Cropper, E.C., Evans, C.G., Hurwitz, I., Jing, J., Proekt, A., Romero, A., Rosen, S.C.: Feeding Neural Networks in the Mollusc *Aplysia*. Neurosignals **13** (2004) 70-86
2. 2.Elliott, C., Susswein, A.J.: Comparative Neuroethology of Feeding Control in Mollusc. J.E- xp. Biol. **205** (2002) 877-896
3. Brembs, B., Lorenzetti, F., Baxter, D.A., Byrne, J.H.: Operant Reward Learning in *Aplysia:* Neuronal Correlates and Mechanisms. Science **296** (2002) 1706-1709

4. Church, P.J., Lloyd, P.E.: Activity of Multiple Identified Motor Neurons Recorded In-tracel-lularly During Evoked Feeding-Like Motor Programs in *Aplysia*. J. Neurophysiol.**72** (1994) 1794-1809

5. Kabotyanksi, E.A., Baxter, D.A., Cushman, S.J., Byrne, J.H.: Modulation of Fictive Fee-ing by Dopamine and Serotonin in *Aplysia*. J. Neurophysiol. **83** (2000) 374-392

6. Morton, D.W., Chiel, H.J.: *In Vivo* Buccal Nerve Activity that Distinguishes Ingestion from Rejection Can Be Used to Predict Behavioral Transitions in *Aplysia*. J. Comp. Physiol. [A] **172** (1993a)17-32

7. Hurwitz, I., Kupfermann, I., Susswein, A.J.: Different Roles of Neurons B63 and B34 that Are Active During the Protraction Phase of Buccal Motor Programs In *Aplysia Califor-nica*. J. Neurophysiol. **78** (1997) 1305-1319

8. Hurwitz, I., Neustadter, D., Morton, D.W., Chiel, H.J., Susswein, A.J.: Activity Patterns of the B31/32 Pattern Initiators Innervating the I2 Muscle of the Buccal Mass During Normal Feeding Movements in *Aplysia Californica*. J. Neurophysiol. **75** (1996) 1309-1326

9. Morton, D.W., Chiel, H.J.: The Timing of Activity in Motor Neurons that Produce Radula Movements Distinguishes Ingestion from Rejection in *Aplysia*. J. Comp. Physiol. [A] **173** (1993b) 519-536

10. Nargeot, R., Baxter, D.A., Byrne, J.H.: Contingent-Dependent Enhancement of Rhythmic Motor Programs: an *in Vitro* Analog of Operant Conditioning. J. Neurosci. **17** (1997) 8093- 8105

11. Nargeot, R., Baxter, D.A., Byrne, J.H.: Correlation between Activity in Neuron B52 and T- wo Features of Fictive Feeding in *Aplysia*. Neurosci. Lett., **328** (2002) 85-88

12. Deodhar, D., Kupfermann, I., Rosen, S.C., Weiss, K.R.: Design Constraints in Neuronal Ci- rcuits Underlying Behavior. In: Greenspan, R.J., Kyriacou, C.P., (eds.): Flexibility and Co- nstraint in Behavioral Systems. John Wiley & Sons Ltd., New York (1993) 41-52

13. Kabotyanski, E.A., Baxter, D.A., Byrne, J.H.: Experimental and Computational Analyses of a Central Pattern Generator Underlying Aspects of Feeding Behavior of *Aplysia*. Netherlan- ds J. Zool. **44** (1994) 357-373

14. Kupfermann, I., Deodhar, D., Teyke, T., Rosen, S.C., Nagahama, T., Weiss, K.R.: Behavio- ral Switching of Biting and of Directed Head Turning in *Aplysia*: Explorations Using Neur- al Network Models. Acta Biol. Hung. **43** (1992) 315-328

15. Susswein, A.J., Hurwitz, I., Thorne, R., Byrne, J.H., Baxter, D.A.: Mechanisms Underly-ing Fictive Feeding in *Aplysia:* Coupling Between a Large Neuron with Plateau Potentials and a Spiking Neuron. J. Neurophysiol. **87** (2002) 2307-2323

16. Ziv, I., Baxter, D.A., Byrne, J.H.: Simulator for Neural Networks and Action Potentials: De-scri- ption and Application. J. Neurophysiol. **71** (1994) 294-308

17. Av-Ron, E., Byrne, J.H., Baxter, D.A.: Teaching Basic Principles of Neuroscience with Co- mputer Simulations. J. Undergrad. Neurosci. Edu. in press, (2006).

18. Baxter, D.A., Byrne, J.H.: Simulator for Neural Networks and Action Potentials (SNNAP): Description and Application. In: Crasto, C. (ed.): Methods in Molecular Biology: Neuro-inf- ormatics. Humana Press, Totowa, N.J., in press (2006)

19. Hurwitz, I., Susswein, A.J.: B64, a Newly Identified Central Pattern Generator Element Pr-oducing a Phase Switch from Protraction to Retraction in Buccal Motor Programs of *Aplys- ia Californica*. J. Neurophysiol. **75** (1996) 1327-1344

20. Kabotyanski, E.A., Baxter, D.A. Byrne, J.H.: Identification and Characterization of Catech- olaminergic Neuron B65 That Initiates And Modifies Patterned Activity in the Buccal Gan- glia of *Aplysia*. J. Neurophysiol. **79** (1998) 605-621

21. Bernard, C., Axelrad, H., Giraud,B.G.: Effects Of Collateral Inhibition in a Model of the I-mmature Rat Cerebellar Cortex: Multineuron Correlations. Brain Res. Cogn. Brain Res. **1**: (1993) 100-122

22. Bhalla, U.S., Bower, J.M.: Exploring Parameter Space in Detailed Single Neuron Models: Simulations of the Mitral and Granule Cells of the Olfactory Bulb. J. Neurophysiol. **69** (1993) 1948-1965

23. Edwards, J.A., Palsson, B.O.: Robustness Analysis of the *Escherichia Coli* Metabolic Network. Biotechnol. Prog. **16** (2000) 927-939

24. Arshavsky, Y.I., Deliagina, T.G., Orlovsky, G.N.: Pattern Generation. Curr. Opin. Neurobiol. **7** (1997) 781-789,

25. Cropper, E.C., Weiss, K.R.: Synaptic Mechanisms in Invertebrate Pattern Generation. Curr. Opin. Neurobiol. **6** (1996) 833-841

26. Marder, E., Calabrese, R.L.: Principles of Rhythmic Motor Pattern Generation. Physiol. Re- v. **76** (1996) 687-717

27. Marder, E., Bucher, D., Schulz, D.J., Taylor, A.L. Invertebrate Central Pattern Generation Moves Along. Current Biology. 15: R685-R699, 2005.

28. Stein, P.S.G., Grillner, S., Selverton, A.I., Stuart, D.G. (eds.): Neurons, Networks and Mot- or Behaviors. The Mit Press, Cambridge (1998)

29. Abbott, L.F., Marder, E.: Modeling Small Networks. In: Koch, C., Segev, I. (eds.): Methods in Neuronal Modeling: from Ions to Networks, Second Edition. The Mit Press, Cambridge (1998) 361-409

30. Calabrese, R.L.: Oscillation in Motor Pattern-Generating Networks. Curr. Opin. Neurobiol. **5** (1995) 816-823

31. Marder, E., Abbott, L.F.: Theory in Motion. Curr. Opin. Neurobiol. **5** (1995) 832-840

32. Marder, E., Kopell, N., Sigvardt, K.: How Computation Aids in Understanding Biological Networks. In: Stein,P.S.G., Grillner, S., Selverston, A.I., D.G. Stuart, D.G. (eds.): Neurons, Networks, And Motor Behaviors. The Mit Press, Cambridge (1998) 139-149

33. Nargeot, R., Baxter, D.A., Byrne, J.H.: *In Vitro* Analogue of Operant Conditioning n *Aplys- ia*: I. Contingent Reinforcement Modifies the Functional Dynamics of an Identified Neuron. J. Neurosci. **15** (1999a) 2247-2260

34. Nargeot, R., Baxter, D.A., Byrne, J.H.: *In Vitro* Analogue of Operant Conditioning in *Aply- sia*: II. Modifications of the Functional Dynamics of an Identified Neuron Mediate Motor Pattern Selection. J. Neurosci. **15** (1999b) 2261-2272

35. Jing, J., Weiss, K.R.: Neural Mechanisms of Motor Program Switching in *Aplysia*. J. Neurosci. **15** (2001) 7349-7362

36. Warman, E.N., Chiel, H.J.: A New Technique for Chronic Single-Unit Extracellular Recor- ding in Freely Behaving Animals Using Pipette Electrodes. J. Neurosci. Methods **57** (1995) 161-169

37. Teyke, T., Rosen, S.C., Weiss, K.R., Kupfermann, I.: Dopaminergic Neuron B20 Gen rates Neuronal Activity in the Feeding Motor Circuitry of *Aplysia*. Brain Res. **630** (1993) 226- 237

Rewriting Game Theory as a Foundation for State-Based Models of Gene Regulation

Chafika Chettaoui[1], Franck Delaplace[1,*], Pierre Lescanne[2],
Mun'delanji Vestergaard[3], and René Vestergaard[3,*]

[1] IBISC - FRE 2873 CNRS, Evry, France
delapla@lami.univ-evry.fr
[2] LIP - UMR 5668, Ecole Normale Supérieure, Lyon
[3] JAIST, Nomi, Ishikawa, Japan
vester@jaist.ac.jp

Abstract. We present a game-theoretic foundation for gene regulatory analysis based on the recent formalism of *rewriting game theory*. Rewriting game theory is discrete and comes with a graph-based framework for understanding compromises and interactions between players and for computing Nash equilibria. The formalism explicitly represents the dynamics of its Nash equilibria and, therefore, is a suitable foundation for the study of steady states in discrete modelling. We apply the formalism to the discrete analysis of gene regulatory networks introduced by R. Thomas and S. Kauffman. Specifically, we show that their models are specific instances of a *C/P game* deduced from the *K* parameter.

1 Introduction

Gene regulation concerns the mutual inhibition and activation among genes and the wider impact this has on cells and on whole organisms through the resulting protein production or lack thereof, aka *gene expression*. In particular, the regulation of genes may involve complex regulatory processes such as auto-regulation and feedback loops, possibly via complex pathways. Studying regulation benefits from using formal tools to give well-founded explanations of the complexities.

Substantial work has focused on stochastic techniques and differential equations (over time) [4]. In this article, we focus on the two best known state-based (aka *logical* and *multivalued*) models, due to Kauffman [5,6] and Thomas [18,19]. The two models are discrete and aim at providing qualitative information about the dynamic aspects of gene regulation [4]. They are underpinned by the definition of a state graph that is intended to represent possible gene-state changes [1,19]. Informally, analyzing the dynamics of the state graph is based on the

---

* Corresponding authors.

C. Priami (Ed.): CMSB 2006, LNBI 4210, pp. 257–270, 2006.

identification of paths having specific topological properties. Notions in dynamical systems are translated into topological properties on the graph. A trajectory is a pathway, a steady state is a sink (i.e., a vertex with no output edge), and periodicity is described by a cycle. Sink cycles (i.e., cycles that has no edges leaving it) and sinks are remarkable topological features because they both embody attractors. Special attention is paid to attractors because they represent robust and steady characteristic modes of a dynamic system. Hence they can be considered as functional features (capabilities) of the system at a more integrated level because, over time, evolution will make the system reach one of them as influenced by external conditions.

Sink cycles, sinks and more generally attractors, can be computationally unified in a homogeneous notion of *sink strongly connected components* (SSCC), Basically, it represents a sub graph where any two vertices are connected together by a circuit that has no edges leaving it in the graph. The topology of the attractor is often interpreted as a characteristic feature of the regulatory dynamics. For instance, cycles and sinks correspond respectively to homeostasis [2] and multi-stationarity. However to embrace the complexity of these dynamics, one must understand and be able to work with them as mathematical objects in a general and ideally algebraic manner to smoothly and coherently address all the known and desirable features of gene regulation and to accommodate future discoveries. In other words, a foundation is called for that, on the one hand, is flexible and general and, on the other hand, employs a conceptual and technical framework that sheds direct light on the issues at hand, i.e., that can bridge the gap between topological features in a state graph and regulatory effects of inter-dependent but autonomous genes.

The cornerstone of our contribution is to show that the steady states of gene regulatory networks, as they are commonly understood, are a recently established kind of Nash-style equilibria, called *change-of-mind equilibria* [14]. The game-theoretic perspective we provide is technically and conceptually beneficial because non-cooperative game theory is the embodiment of the compete-and-coexist reality of genes and because it allows us to leverage the independently developed theory of dynamic equilibria in rewriting game theory. In particular, we show that Kauffman's and Thomas' models can be defined as specific instances of a particular game-skeleton. Technically, this recasts steady states (attractors) and gene regulation to the fixed-point construction underlying our discrete Nash equilibria. In particular, we show that steady states are the least non-empty fixed points (in a lattice of fixed points) of the *update* functions already considered by Kauffman and Thomas.

In Section 2, we briefly account for (rewriting) game theory, computation of discrete Nash equilibria, and the very general game formalism involved, called conversion/preference (C/P) games. In section 3, we review the discrete models for gene regulation introduced by Kauffman and Thomas. In section 4, we show that the two models can be viewed as instances of a C/P game.

In [15], we apply rewriting game theory to protein signalling in mitogen-activated protein kinase (MAPK) cascades, which govern biological responses

Fig. 1. Example of sequential(extensive form) and strategic game(normal form)

such as cell growth. The aim there is more practical than in this article and involves establishing an analytic model for protein signalling in the first place and to develop tool support for it.

2   Rewriting Game Theory

In this section, we first provide a gentle reminder of the relevant ideas in game theory (Section 2.1) and then introduce the principles of a framework for discrete game theory (Section 2.2), before going into more technical details in the remainder of the section. The new framework generalises the notions of strategic games and Nash equilibria without involving probability theory and continuous notions. Good accounts of traditional game theory are [9,13].

2.1   Non-cooperative Game Theory

Non-cooperative game theory is game theory based around the notion of Nash equilibria. Nash equilibria are defined over *strategies* that account for the intended behaviour of all agents/players in a game. We say that an agent is *happy* if he cannot change his contribution to a (combined) strategy and generate a better overall outcome for himself. A (combined) strategy is a Nash equilibrium if all agents are happy with it. Game theory involves a wide spectrum of games and theories. However two kinds of games are usually considered for modelling : *sequential games* and *strategic games*. An example using a sequential game in *extensive form* is in Figure 1, left. An example of a strategic game in *normal-form* is in the figure, right.

   A play of the game on the left is a path from the root to a leaf, where the first (second) number indicates the payoff to agent a_1 (a_2). A strategy over the game, by contrast, is a situation where a choice has been made in all internal nodes, not just in the nodes on a considered path. While it might look like the strategy of a_1 going to the right and a_2 going to the left for payoffs $7,5$ is good, it is not a Nash equilibrium because a_2 can go to the right, for a better payoff. At that point, also a_1 can benefit from changing his choice and, in fact, the only Nash equilibrium in the game is a_1 (a_2) going to the left (to the right), for payoffs $1,0$. Nash equilibria can be guaranteed to exist for all sequential games, a result known as Kuhn's Theorem [7,20].

   In strategic games, players act simultaneously. In contrast to sequential games, Nash equilibria do not always exist in a pure form in strategic games. An example is above on the right. In the example, there are two players: vertical, who chooses a row and gets the first payoff, and horizontal, who chooses a column and gets the second payoff. As can be seen, in no outcome are both players happy, i.e., one player always can and wants to move away. Instead, Nash's Theorem says that a *probabilistic* combination of strategies exists, where the agents are happy with their *expected* payoffs [10,12]. In the example, the only probabilistic Nash equilibrium arises if both agents choose between their two options with equal probability for expected payoffs of a half to each. Addressing the hows and whys of this in general quickly turns in to pure probability theory, with justifications that need not necessarily be meaningful in the application area.

2.2   Conversion/Preference Games

Conversion/preference (C/P) games have been designed as an abstraction over strategic-form games and as a game formalism that introduces as few concepts as possible. This aim leads us to distinguish two relations on strategies, **C**onversion and **P**reference. The key concept of C/P games is the *synopsis*, which abstracts the notion of (combined) strategies. Roughly speaking, conversion says how an agent can move from a synopsis to another; in other words, it says which changes are allowed on synopses for a given agent. An agent makes choices among synopses according to which he prefers over others. It should be noted that conversions and preferences depend on agents. In what follows, conversion is denoted $\succ$ and preference is denoted $\lhd$. Clearly strategic-form games are instances of C/P games, conversions are one dimension move (for instance along a line or a column), while preferences are given by comparisons over payoffs: a synopsis is preferred by an agent over another if his payoff is larger in the former.

Definition 1 (C/P Games [14]). G^{cp} *are 4-tuples* $\langle \mathcal{A}, \mathcal{S}, (\underset{a}{\succ})_{a \in \mathcal{A}}, (\lhd_a)_{a \in \mathcal{A}} \rangle$:

- $\mathcal{A}$ *is a non-empty set of* agents.
- $\mathcal{S}$ *is a non-empty set of* synopses *(read: outcomes of the game).*
- *For* $a \in \mathcal{A}$, $\underset{a}{\succ}$ *is a binary relation over* $\mathcal{S}$, *associating two synopses if agent* a *can* convert *the first to the second.*
- *For* $a \in \mathcal{A}$, $\lhd_a$ *is a binary relation over* $\mathcal{S}$, *associating two synopsis if agent* a *prefers the second to the first.*

The concept of *synopsis* is abstracted from that of *(combined) strategy*. One can move from one synopsis to the other by conversion and compare two synopses by preference.

The idea of the definition is to make explicit the parts of strategic-form games that are relevant to the definition of Nash equilibria and to dispense with any other structural constraints, such as the uniform restriction that 'vertical' can only move up and down. To illustrate, we note that the example we considered earlier amounts to the C/P game in Figure 2.

2.3   Abstract Nash Equilibrium

The following definition says that a synopsis s, i.e., our abstraction over (combined) strategies, is an abstract Nash equilibrium if and only if all agents are

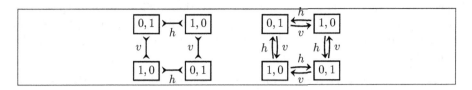

Fig. 2. Conversions and Preferences example

$$\frac{s \succ_a s' \qquad s \lhd_a s'}{s \to_a s'}$$

Fig. 3. The *(free) change-of-mind* relation for agent a in G^{cp}

happy, meaning that whenever an agent can convert s to s' then it is not the case that he prefers s' to s. The notion of abstract Nash equilibrium specialises to Nash's concrete form in the presence of the discussed structural constraints on strategic-form games.

Definition 2 (Abstract Nash Equilibrium [14]). *Given* G^{cp}.

$$Eq^{aN}_{G^{cp}}(s) \quad \triangleq \quad \forall a \in \mathcal{A}, s' \in \mathcal{S} . s \succ_a s' \Rightarrow \neg(s \lhd_a s')$$

We no more use the word abstract for the reason discussed above: in strategic-form games, the notions coincide [14]. Said differently, Definition 2 is merely a more general (and simpler) way of writing what Nash wrote [10,12]. Technically, the form of our definition is intended to facilitate the following definition, thus giving rise to the name *rewriting game theory*.

Definition 3 ([14]). *Given* G^{cp}, *the* change-of-mind *relation,* $\to_a$, *for agent* a *is given in Figure 3. Let* $\to \triangleq \bigcup_{a \in \mathcal{A}} \to_a$.

In other words, a Nash equilibrium is a synopsis for which there is no outgoing change-of-mind step, i.e., an $\to$-irreducible (aka a $\to$-normal form). The set of $\to$-irreducibles is $IrR_{\to}$.

Proposition 4 ([14]). $Eq^{aN}_{G^{cp}}(s) \Leftrightarrow s \in IrR_{\to}$.

The benefits of the changed perspective on game theory are partly conceptual, in the first instance for people who like rewriting, but they are also technical in that Proposition 4 highlights the positive notion, i.e., change-of-mind, that is behind Nash's original definition and through which we get easy access to a range of formal(ist) tools, not least of which is definition and proof by induction.

2.4 A Graph-Theoretic Construction

Returning to our rewriting/graph-theoretic view on game theory, we note that for arbitrary finite graphs only cycles can prevent the existence of sinks. We show in this section how that simple observation suffices for underpinning a discrete version of Nash's Theorem for arbitrary finite C/P games. The relevant graph-theoretic notion we need for capturing all cycles is *strongly connected components*.

- A *graph* is a binary relation on a carrier set, called vertices: $\rightarrow\ \subseteq \mathcal{V} \times \mathcal{V}$.
- The *reflexive, transitive (or pre-order) closure*, $\rightarrow^*$, of a graph, $\rightarrow$, is

$$\frac{v_1 \rightarrow v_2}{v_1 \rightarrow^* v_2} \qquad \overline{v \rightarrow^* v} \qquad \frac{v_1 \rightarrow^* v \quad v \rightarrow^* v_2}{v_1 \rightarrow^* v_2}$$

- The strongly connected component (SCC) of a vertex, v, in a graph is

$$\lfloor v \rfloor \quad \triangleq \quad \{v' \mid v \rightarrow^* v' \wedge v' \rightarrow^* v\}$$

- The set of SCCs of a graph is

$$\lfloor \mathcal{V} \rfloor \quad \triangleq \quad \{\lfloor v \rfloor \mid v \in \mathcal{V}\}$$

- The *shrunken graph* of $\rightarrow\ \subseteq \mathcal{V} \times \mathcal{V}$ is $\curvearrowright\ \subseteq \lfloor \mathcal{V} \rfloor \times \lfloor \mathcal{V} \rfloor$, defined by

$$V_a \curvearrowright V_b \quad \triangleq \quad V_a \neq V_b \wedge (\exists v_a \in V_a, v_b \in V_b \,.\, v_a \rightarrow v_b)$$

The set $\lfloor \mathcal{V} \rfloor$ and the relation $\curvearrowright$ allows us to define a C/P game with the same set of agents, $\lfloor \mathcal{V} \rfloor$ as set of synopses and $\curvearrowright$ as both conversion and preference. We call that game the *"shrunken game"*. The following result says that a Nash equilibrium exists in "shrunken" games.

Theorem 5 ([14]). *For any finite C/P game, $\langle \mathcal{A}, \mathcal{S}, (\succ_a)_{a \in \mathcal{A}}, (\lhd_a)_{a \in \mathcal{A}} \rangle$,*

- *$\langle \mathcal{A}, \lfloor \mathcal{S} \rfloor, (\curvearrowright_a)_{a \in \mathcal{A}}, (\curvearrowright_a)_{a \in \mathcal{A}} \rangle$ have Nash equilibria, $Eq^{aN}_{\lfloor G^{cp} \rfloor}$,*
- *all of which can be found in linear time in the size of $\mathcal{S}$ and $\rightarrow$.*

Nash's Theorem says that probabilistic Nash equilibria exist for all finite strategic-form games. By comparison, the result above says that "shrunken" Nash equilibria always exist for finite members of the much larger class of C/P games. We clarify what the "shrunken" qualifier means next.

2.5   Change-of-Mind Equilibria

The topic of this section is to directly characterise the Nash equilibria prescribed by Theorem 5. Naively speaking, our notion of *change-of-mind* equilibrium is simply the graph underlying the considered compromises between synopses. However, the technical form we use is different for reasons of game-theoretic interpretation [14].

Definition 6 (Change-of-Mind Equilibrium [14]). *Write $\xrightarrow{S}$ for $\rightarrow \cap (S \times S)$, i.e., the graph of a set of synopses. For non-empty S, $\xrightarrow{S}$ is a change-of-mind equilibrium, $Eq^{com}_{G^{cp}}(\xrightarrow{S})$, for G^{cp} if*

$$\forall s \in S, s' \in \mathcal{S} \;\;.\;\; s \rightarrow^* s' \Leftrightarrow s' \in S$$

$$
\begin{array}{ccc}
0,\,1 & \twoheadleftarrow & 1,\,0 \\
\downarrow & & \uparrow \\
1,\,0 & \twoheadrightarrow & 0,\,1
\end{array}
$$

Fig. 4. Change-of-mind equilibrium for our running strategic-form example

As implied, our two notions of Nash equilibria coincide.

Lemma 7 ([14]). $\mathrm{Eq}^{\mathrm{com}}_{\mathrm{G^{cp}}}(\overset{\mathrm{S}}{\rightarrow}) \quad \Leftrightarrow \quad \mathrm{Eq}^{\mathrm{aN}}_{\lfloor \mathrm{G^{cp}} \rfloor}(\mathrm{S})$

The lemma implies that the Nash equilibria prescribed by Theorem 5 have the property that no agent can escape from them (and that only S of the form $\lfloor s \rfloor$ can be change-of-mind equilibria). Agents are allowed to move within the equilibria but they will have to stay within the set perimeter. We will return to the issue of size of the perimeter in Section 5. For now, we note that our running example, see Figure 1, left, and Figure 2, has the change-of-mind equilibrium in Figure 4. We note that both the probabilistic Nash and the change-of-mind equilibria of the example involve all four outcomes. The probabilistic version prescribes an exact *expected* payoff, while the discrete change-of-mind version makes the dynamics behind the equilibrium clear. The two notions may differ quite substantially in general but neither is uniformly smaller, has higher (implied) payoff values, or is better in any similar sense.

3   Basic State-Based Analysis of Gene Regulation

We now leave game for a while and analyse models of gene regulation.

Kauffman's and Thomas' models differ on a number of minor and on one major issue, relative to our presentation. Among the minor ones, we count Kauffman's assumption that i) genes are boolean, i.e., that they can be in exactly two states: active (expressing protein) or inactive and that ii) when one gene regulates another it is always either repressing or activating it. Assumption ii) is reflected in the signs (polarities) annotated to the regulatory-network example at the beginning of Section 1. The major difference between the two approaches concerns the way states are updated. In Thomas' model only one gene is updated at each step (asynchronous update) while in Kauffman's all genes are updated (synchronous update), albeit possibly reflexively. We return to this issue in Section 3.2.

3.1   Regulatory Networks

On the minor issues, we essentially follow Thomas' more general perspective of allowing for a gene to assume a fixed but unbounded number of states (albeit typically 2 or 3) and of using a more detailed way of specifying regulation.

Definition 8 (Regulatory Networks) *are 3-tuples* $\langle \mathrm{G}, \curvearrowright, K \rangle$:

$$G = \{cI, cro\}, \text{ with } cI_0 < cI_1 \text{ and } cro_0 < cro_1 < cro_2$$

$$K_{cI}(\langle \_, cro_0 \rangle) = cI_1$$
$$K_{cI}(\langle \_, cro_1 \rangle) = cI_0$$
$$K_{cI}(\langle \_, cro_2 \rangle) = cI_0$$

$$K_{cro}(\langle cI_0, cro_0 \rangle) = cro_2$$
$$K_{cro}(\langle cI_0, cro_1 \rangle) = cro_2$$
$$K_{cro}(\langle cI_0, cro_2 \rangle) = cro_0$$
$$K_{cro}(\langle cI_1, \_ \rangle) = cro_0$$

Fig. 5. Two-variable regulatory network for phage λ (_ is wildcard)

- G *is a non-empty set of* genes, *ranged over by* g, g_i *and each associated with a non-empty, linearly ordered set of* states, $\langle S_g, <^g \rangle$, *ranged over by* s^g, s_i^g;
- $\curvearrowright \subseteq$ G × G, *a relation, with* $g_1 \curvearrowright g_2$ *saying that* g_1 *may* regulate g_2 — *let* $\mathcal{I}_g \triangleq \{g_i \mid g_i \curvearrowright g\}$ *be the regulatory* inputs *to* g;
- $K_G \triangleq \bigotimes_{g \in G} K_g$, *are* comfort *functions,* $K_g : \bigotimes_{g_i \in \mathcal{I}_g} S_{g_i} \to S_g$, *for each gene saying when* g *is being regulated and what state it is pushed towards.*

Let us say a few words about the concepts defined above. When one considers gene regulation networks from Kauffman's and Thomas' points of view, one encounters three entities (hence a 3-uple) namely a set G of genes, a relation among genes and a family of comfort functions. Each gene g possesses a set S_g of states, which is hierarchic. This hierarchy is a linear order. The relation $\curvearrowright$ is the regulation, i.e., the ability for a gene to regulate another gene. The genes g_i's that regulate the gene g form the set $\mathcal{I}_g$. The comfort functions have a slightly more complex structure. For each gene g there is a function from tuples (g_i) (for $g_i \in \mathcal{I}_g$) of states of genes that regulate g. It says what state g is pushed toward when it is regulated by those g_i's.

We note that G is not restricted to genes, per se, but could also contain, e.g., proteins, or something completely different. We also note that our comfort functions are seeming slightly more general than Thomas' corresponding notion of *logical parameters* for the simple reason that, as given, Definition 8 is more in line with our other definitions; for the examples we consider, we shall not need the extra expressive power. Finally, we will sometimes use the comfort functions, K_g, as if they had type $\bigotimes_{g \in G} S_{g_i} \to S_g$, with the obvious implicit coercion.

As an example, Figure 5 displays a regulatory network similar to the Kauffman-style one at the beginning of Section 1, namely the standard example of *bacteriophage lambda* (*phage* λ), with two genes: *cI* and *cro*.

3.2   Gene-State Updates

Both Kauffman's and Thomas' analyses proceed by considering the state space of a given regulatory network, $\bigotimes_{g \in G} S_g$, and both prevent updates across a state, e.g., cI_0 to cI_2. The rationale for the latter is that moving from a state to another involves a phase transition, which is costly in terms of energy, and two

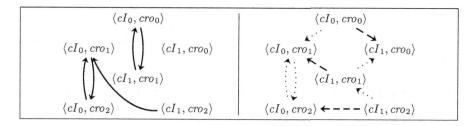

Fig. 6. Kauffman (left) and Thomas (right) analysis of two-variable phage λ

phase transitions should therefore not be considered atomically. They differ in what state they predict the system will move to from a given state.

In Kauffman's case, above left, each gene is prompted for its comfort state *relative to the states of all genes in the given point in the state space* and a *synchronous* move is made towards the combined comfort state, while allowing for at most one phase transition for each gene. If a gene is not regulated upon, i.e., if no comfort state is specified, it retains its state. Reflexive state-space transitions are not considered. The details for the example in Figure 5 are in Figure 6, left.

In Thomas' case, Figure 6 right, each gene is prompted as before but moves are made *asynchronously*, i.e., each state may have several moves out of it, one for each gene being considered. In the figure, we indicate cI-updates with dashed arrows and cro-updates with dotted arrows, although the two are not distinguished in the actual analysis.

3.3 Steady States

In the two state graphs above, $\langle cI_1, cro_0\rangle$ clearly plays a special role: it is a static steady state, i.e., it is the only state that does not have arrows out of it. From this, we can seemingly conclude that if the two genes end up in that configuration, they stay that way. The relevance of state-based analysis comes from the fact that the state in question has been observed to be *(self-)sustainable*: it is phage λ's *lysogenic* state that "involves integration of the phage DNA into the bacterial chromosome [of its host] where it is passively replicated at each cell division — just as though it were a legitimate part of the bacterial genome" [21].

Similarly, there is an inescapable cycle, i.e., a dynamic steady state, involving $\langle cI_0, cro_1\rangle$ and $\langle cI_0, cro_2\rangle$ in both graphs. The implied regulatory flip-flopping between $\langle cI_0, cro_1\rangle$ and $\langle cI_0, cro_2\rangle$ is, in fact, biologically characteristic of phage λ's *lytic* state in which it actively uses its host's transcription mechanism to replicate itself [21].[1]

In our formalism, and despite their obvious topological differences, both the static and the dynamic steady states described are simply change-of-mind-equilibria,

---

[1] The cycle between $\langle cI_0, cro_0\rangle$ and $\langle cI_1, cro_1\rangle$ is a known false positive of Kauffman's model.

which means that they can be uniformly accommodated as far as our general theory goes. More, the biological justification for why the states are special, i.e., that they are inescapable, is the exact the justification for why they are both change-of-mind equilibria.

4 C/P Games-Based Modeling of Gene Regulatory Networks

Our modeling of Kauffman/Thomas-style gene regulation via C/P games will have the update graphs exemplified in Figure 6 as change-of-mind relations. Several ways of specifying the C/P-game 4-tuple, $\langle \mathcal{A}, \mathcal{S}, (\succ_a)_{a \in \mathcal{A}}, (\lhd_a)_{a \in \mathcal{A}} \rangle$, will lead to the desired result. The approach we take interprets the distinction between conversion and preference as chemical reality vs observation of the same. Specifically, we distinguish chemical reactions that genes and proteins are involved in and how closely we choose/are able to observe changes.

Given a regulatory network, $\langle G, \curvearrowright, K_G \rangle$, with associated gene states, $(S_g)_{g \in G}$, we take the gene state space, $S_G \triangleq \bigotimes_{g \in G} S_g$, as our set of synopses, $\mathcal{S}$. Reflecting the (perceived) universality of the considered chemical situation, we insist that the conversion relations of all agents, whatever we specify them to be, are the same. By default, we allow all state changes and leave it to the specific applications to put in place any necessary ad hoc restrictions.[2] Following Thomas, however, we are particularly interested in the at-most-one-phase-transition-at-a-time restriction.

Definition 9. *For linear order* $g_0 < \ldots < g_n$, *let* $g_i \ominus g_j = i - j$, *and let*

$$s \overset{\pm 1}{\succ} s' \quad \triangleq \quad \forall g \in G . \mid \pi_g(s) \ominus \pi_g(s') \mid \leq 1$$

A C/P game whose conversion fulfills the previous definition is called *1-restrained*. Similarly straightforwardly, our preference relation is dictated by the comfort functions, K_g, of a regulation network.

Definition 10. *We say that* s' *is a* comfort approximation *for* g *in* s *if*

$$\text{K-Approx}_g(s, s') \triangleq (\pi_g(s) \leq \pi_g(s') \leq K_g(s)) \vee (\pi_g(s) \geq \pi_g(s') \geq K_g(s))$$

Definition 11. *Given* $\langle G, \curvearrowright, K_G \rangle$ *and for any* $s, s' \in S_G$ *and* $g \in G$, *let*

- $s \lhd_G s' \triangleq \forall g . \text{K-Approx}_g(s, s')$
- $s \lhd_g s' \triangleq \text{K-Approx}_g(s, s') \wedge (\forall g' . g' \neq g \Rightarrow \pi_{g'}(s) = \pi_{g'}(s'))$

be the synchronous *respectively* g-asynchronous *preference relations.*

With this, we see that Kauffman-style regulation analysis is a 1-player regulation game, while Thomas-style regulation analysis is a multi-player game, played by the considered genes. In Kauffman-style the 1-player is the whole set of genes G, whereas in Thomas-style, players are the elements of G.

---

[2] For example, for eliminating "false cycles" arising due to vastly differing kinetics for two or more reactions, e.g., in the standard 4-variable model of phage λ [17].

Theorem 12 (Regulation Games). *Given* $\langle G, \curvearrowright, K_G \rangle$,

- *The Kauffman update function, cf. Figure 6, left, is the change-of-mind relation,* $\rightarrow$, *of* $\langle G, S_G, \overset{\pm 1}{\succ}, \lhd_G \rangle$, *the* 1-restrained synchronous regulation game, *and the steady states are the change-of-mind equilibria,* Eq^{com}.
- *The Thomas update function, cf. Figure 6, right, is the change-of-mind relation,* $\rightarrow$, *of* $\langle G, S_G, (\overset{\pm 1}{\succ})_{g \in G}, (\lhd_g)_{g \in G} \rangle$, *the* 1-restrained asynchronous regulation game, *and the steady states are the change-of-mind equilibria,* Eq^{com}.

Moreover, and in both cases, the static *(dynamic) steady states are the change-of-mind equilibria that are also (not) Nash equilibria,* Eq^{aN}.

Proof. The statements about the update functions follow by construction. The statements about static vs dynamic equilibria are questions of terminology, according to Proposition 4: "singleton change-of-mind equilibria are Nash equilibria". That steady states and change-of-mind equilibria coincide follow from Lemma 7 and (the proof of) Theorem 5, further to the characterisation of steady states as sink strongly connected components in [3].

In Figure 7 we depict the full regulation game of 2-variable λ-phage.

5   A Fixed-Point Construction

The original Thomas characterisation of steady states is in terms of fixed points of the considered update function [17]. As noted earlier, that function is a specific instance of the following function.

Definition 13 (Upgrade [14]). $\mathcal{U}(S) \triangleq \bigcup_{s \in \mathcal{S}} \{ s' \in S \mid s \rightarrow^* s' \}$

We first note that $\mathcal{U}$ always has fixed points.

Lemma 14 ([14]). *The fixed points of* $\mathcal{U}$ *is a non-empty, complete lattice.*

Proof. $\mathcal{U}$ is monotonic on the complete lattice $\mathcal{P}(\mathcal{S})$ because $\rightarrow^*$ is reflexive, and we are done by Tarski's Fixed-Point Theorem [16].

Example fixed-points are the empty set, $\emptyset$, and the whole set, $\mathcal{S}$. The interesting point is that the change-of-mind equilibria are exactly the least non-empty (pre-)fixed-points of the upgrade function.

Lemma 15 ([14]). *Consider some* $\langle \mathcal{A}, \mathcal{S}, (\succ_a)_{a \in \mathcal{A}}, (\lhd_a)_{a \in \mathcal{A}} \rangle$.

$$Eq^{com}_{G^{cp}}(\overset{S}{\rightarrow})$$
$$\Updownarrow$$
$$\mathcal{U}(S) = S \ \wedge \ (\forall S'. \emptyset \subsetneq S' \subsetneq S \Rightarrow \mathcal{U}(S') \nsubseteq S')$$

The characterisations of steady states in [17] and in [3] therefore coincide, with the proviso that the fixed points are least non-empty, and both are instances of our more general theory of dynamic equilibria in rewriting game theory.

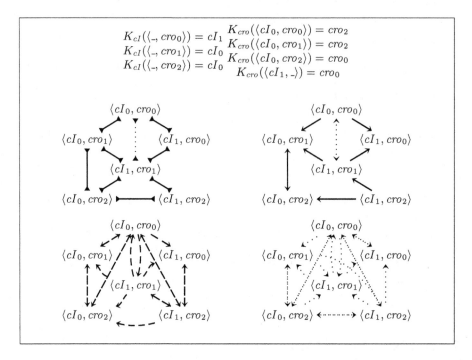

$$K_{cI}(\langle \_, cro_0\rangle) = cI_1 \qquad K_{cro}(\langle cI_0, cro_0\rangle) = cro_2$$
$$K_{cI}(\langle \_, cro_1\rangle) = cI_0 \qquad K_{cro}(\langle cI_0, cro_1\rangle) = cro_2$$
$$K_{cI}(\langle \_, cro_2\rangle) = cI_0 \qquad K_{cro}(\langle cI_0, cro_2\rangle) = cro_0$$
$$K_{cro}(\langle cI_1, \_\rangle) = cro_0$$

Fig. 7. λ-phage C/P game

- The upper left hand-side diagram describes the convertibility relation.
- The upper right hand-side diagram describes the resulting C/P game.
- The lower left hand-side diagram is the preference relation of cI.
- The lower right hand-side diagram is the preference relation of cro.

6 Conclusion

In this article we introduce a game-theory based framework to model gene regulatory networks. We show that a discrete Nash equilibrium can be viewed as a generalization of steady states in discrete models (SSCC). More, we show that Thomas' and Kaufman's models are particular instances of a more general game construction (that conceivably could have other interesting instances). Game theory aims at describing equilibria coming from interactions between agents. One way of viewing Nash-style equilibria is that they are logical expressions capturing the functional units at the level of abstraction above the one at which the considered game exists. In this paper, for example, we have shown that change-of-mind equilibria can be used to predict what gene expression will take place. In other words, we have moved from the chemical abstraction level of protein binding and catalysis captured in expression games, up to the biochemical abstraction level of, e.g., phage λ's lysogenic and lytic states. At the other end of the spectrum, Maynard Smith has shown that a game-theoretic analysis of the ecological concept of fitness leads to the formal substantiation of Darwinian evolution, i.e., "survival of the fittest". Our future work concerns similar treatments of the various abstraction levels in between, namely chemical, biochemical, cellular, multi-cellular and environmental level. Game theory may provide a unified framework to encompass theory occuring at different levels and to provide a suitable framework to deal with interactions between levels in order to get an integrative theory of biology.

References

1. G. Bernot, Cassez, J.P. G. Comet, F. Delaplace, C. Mller, O. Roux, and Roux O.(H.). Semantics of biological regulatory networks. In *Workshop on Concurrent Models in Molecular Biology (BioConcur 2003),*, 2003.
2. Walter Bradford Cannon. *The Wisdom of the Body*. W. W. Norton, New York, 1932.
3. Claudine Chaouiya, Elizabeth Remy, and Denis Thieffry. Petri net modelling of biological regulatory networks. *Proceedings of CMSB-3*, 2005.
4. Hidde de Jong. Modeling and simulation of genetic regulatory systems: A literature review. *Journal of Computational Biology*, 9(1):67–103, 2002.
5. Stuart A. Kauffman. Metabolic stability and epigenisis in randomly constructed genetic nets. *Journal of Theoretical Biology*, 22:437–467, 1969.
6. Stuart A. Kauffman. *The Origins of Order: Self-Organization and Selection in Evolution*. Oxford University Press, 1993.
7. Harold W. Kuhn. Extensive games and the problem of information. *Contributions to the Theory of Games II*, 1953. Reprinted in [8].
8. Harold W. Kuhn, editor. *Classics in Game Theory*. Princeton Uni. Press, 1997.
9. R. B. Myerson. *Game Theory : Analysis of Conflict*. Harvard University Press, 1991.
10. John F. Nash. Equilibrium points in n-person games. *Proceedings of the National Academy of Sciences*, 36, 1950. Reprinted in [8].
11. John F. Nash. *Non-Cooperative Games*. PhD thesis, Princeton University, 1950.
12. John F. Nash. Non-cooperative games. *Annals of Mathematics*, 54, 1951. Reprinted in [8]; published version of [11].
13. M. J. Osborne. *An introduction to game theory*. Oxford University Press, 2003.
14. Stéphane Le Roux, Pierre Lescanne, and René Vestergaard. A discrete Nash theorem with quadratic complexity and dynamic equilibria. Research Report IS-RR-2006-006, JAIST, May 2006.
15. J. Senachak, M. Vestergaard, and R. Vestergaard. Rewriting game theory applied to protein signalling in MAPK cascades. Research Report IS-RR-2006-007, JAIST, May 2006.
16. Alfred Tarski. A lattice-theoretical fixpoint theorem and its applications. *Pacific Journal of Mathematics*, 5:285–309, 1955.
17. R. Thomas and M. Kaufman. Multistationarity, the basis of cell differentiation and memory II: Logical analysis of regulatory networks in terms of feedback circuits. *Chaos*, 11(1):180–195, 2001.
18. René Thomas. Boolean formalization of genetic control circuits. *Journal of Theoretical Biology*, 42(3):563–585, 1973.

19. René Thomas, Denis Thieffry, and Marcelle Kaufman. Dynamical behaviour of biological regulatory networks I: Biological role of feedback loops and practical use of the concept of the loop-characteristic state. *Bulletin of Mathematical Biology*, 57:247–276, 1995.

20. René Vestergaard. A constructive approach to sequential Nash equilibria. *Information Processing Letters*, 97:46–51, 2006.

21. James D. Watson, Tania A. Baker, Stephen P. Bell, Alexander Gann, Michael Levine, and Richard Losick. *Molecular Biology of the Gene, 5th edition*. The Benjamin/Cummings Publishing Company, 2004.

Condition Transition Analysis Reveals TF Activity Related to Nutrient-Limitation-Specific Effects of Oxygen Presence in Yeast

T.A. Knijnenburg[1,2], L.F.A. Wessels[1,3], and M.J.T. Reinders[1]

[1] Information and Communication Theory Group, Faculty of Electrical Engineering, Mathematics and Computer Science, Delft University of Technology, Mekelweg 4, 2628 CD Delft, The Netherlands
t.a.knijnenburg@ewi.tudelft.nl
[2] Kluyver Centre for Genomics of Industrial Fermentation, Julianalaan 67, 2628 BC Delft, The Netherlands
[3] Department of Molecular Biology, The Netherlands Cancer Institute, Plesmanlaan 121, 1066 CX Amsterdam, The Netherlands

Abstract. Regulatory networks are usually presented as graph structures showing the (combinatorial) regulatory effect of transcription factors (TF's) on modules of similarly expressed or otherwise related genes. However, from these networks it is not clear when and how TF's are activated. The actual conditions or perturbations that trigger a change in the activity of TF's should be a crucial part of the generated regulatory network.

Here, we demonstrate the power to uncover TF activity by focusing on a small, homogeneous, yet well defined set of chemostat cultivation experiments, where the transcriptional response of yeast grown under four different nutrient limitations, both aerobically as well as anaerobically was measured. We define a condition transition as an instant change in yeast's extracellular environment by comparing two cultivation conditions, where either the limited nutrient or the oxygen availability is different. Differential gene expression as a consequence of such a condition transition is represented in a tertiary matrix, where zero indicates no change in expression; 1 and -1 respectively indicate an increase and decrease in expression as a consequence of a condition transition. We uncover TF activity by assessing significant TF binding in the promotor region of genes that behave accordingly at a condition transition. The interrelatedness of the conditions in the combinatorial setup is exploited by performing specific hypergeometric tests that allow for the discovery of both individual and combined effects of the cultivation parameters on TF activity. Additionally, we create a weight-matrix indicating the involvement of each TF in each of the condition transitions by posing our problem as an orthogonal Procrustes problem. We show that the Procrustes analysis strengthens and broadens the uncovered relationships.

The resulting regulatory network reveals nutrient-limitation-specific effects of oxygen presence on expression behavior and TF activity. Our analysis identifies many TF's that seem to play a very specific regulatory role at the nutrient and oxygen availability transitions.

C. Priami (Ed.): CMSB 2006, LNBI 4210, pp. 271–284, 2006.
© Springer-Verlag Berlin Heidelberg 2006

1   Introduction

The systems biology view of an organism as an interacting network of genes, proteins and biochemical reactions seems very promising for revealing the underlying networks of transcriptional regulation in *Saccharomyces cerevisiae*. For this yeast enormous amounts of different intracellular data have been measured, enabling the integration of multiple data sources [1]. In inferring regulatory networks common approaches focus on integration of microarray gene expression data, ChIP-chip TF binding data and sequence data (to detect cis regulatory elements) [2]. The resulting networks are usually presented as graph structures showing the (combinatorial) regulatory effect of TF's on modules of similarly expressed or otherwise related genes (e.g [3,4,5]). However, from these networks it not clear when and how TF's are activated. This is quite strange, since the actual conditions or perturbations that trigger a change in the activity of TF's should be a crucial part of the generated regulatory network. Three main reasons for this exclusion can be identified: Firstly, the present inability to directly measure protein levels in vivo prevents direct assessment of the presence of a TF in a particular condition. Secondly, in most cases post-transcriptional and/or post-translational regulation prevent deriving TF activity from gene expression, although an attempt was made based on this assumption [6]. Thirdly, the trend of employing increasingly large compendia of heterogeneous microarray data, where yeast is grown under a wide variety of very different and unrelated conditions, makes it impossible to incorporate all these conditions in a regulatory program. Hence, the functionality of modules and TF's is assigned based on enrichment in annotation categories (e.g. Gene Ontology [7]). This means that the functionality purely depends on the result of clustering, i.e. the grouping of genes, and not specifically on the cultivation conditions under which the expression behavior is characteristic for a module. This approach can only provide a global overview of TF activity and obstructs novel knowledge discovery, since an existing body of knowledge, i.e. the ontologies, is taken as a golden standard.

Here, we demonstrate the power in uncovering TF activity by focusing on a small, homogeneous, yet well defined set of chemostat cultivation experiments, where the transcriptional response of yeast grown under four different nutrient limitations, both aerobically as well as anaerobically was measured (See Table 1 and Figure 1) [8]. In this research we focus on condition *transitions* by comparing gene expression profiles of two cultivation conditions and evaluate whether genes are differentially expressed between these two conditions. TF activity is inferred by assessing significant TF binding in the promotor region of genes that behave accordingly at the transitions. For this, we use the largest available TF binding dataset [9]. The interrelatedness of the cultivation conditions within the systematic combinatorial setup is exploited by performing specific hypergeometric tests. This enabled us to reveal nutrient-limitation-specific effects of oxygen presence on expression behavior and TF activity. Additionally, we create a weight-matrix indicating the involvement of TF's in each of the condition transitions by posing our problem as an orthogonal Procrustes problem. Analysis of this weight matrix broadens the significant relations found by the hypergeometric test. The uncovered regulatory mechanisms offer valuable clues of how yeast changes its metabolism and respiration as a result of specific changes in nutrient and oxygen availability.

2   Methods

2.1   Data and Preprocessing

The employed microarray gene expression data consists of the measured transcriptional response of the yeast *Saccharomyces cerevisiae* to growth limitation by four different macronutrients in both aerobic and anaerobic media. See Table 1. Three independently cultured replicates were performed per experimental condition. For more information see [8]. SAM [10] was employed (with median false discovery rate of 0.01%) to select genes that are differentially expressed amongst the eight conditions. Next, we remove the observed global effect that the presence of oxygen has on the expression level of each gene under all nutrient limitations by a linear regression strategy as described in [11]. Then, for each gene individually the expression levels are discretized by employing a k-means clustering algorithm on the eight mean expression levels (corresponding to the eight conditions) in a one-dimensional space [12]. Here, the Davies-Bouldin index [13] was employed to select between $k = 2$ and $k = 3$. The conditions that comprise the largest cluster are said to have common expression level, while conditions that form a cluster with a higher or lower expression level when compared to the largest cluster are called up- or downregulated, respectively. (In the case that $k = 2$ one cluster has common expression level and the other is either upregulated or downregulated.) Hence, the expression behavior of a gene is defined in terms of up- and/or down regulation under the eight cultivation conditions. Discretized expression patterns of all genes are captured in $\mathbf{G}$, a tertiary matrix of $6383 \times 2 \times 4$. In $\mathbf{G}_{g,o,n}$, $g = \{1 \ldots 6383\}$ are the different genes in the genome, $o = \{1, 2\}$ represents oxygen supply (aerobic and anaerobic respectively) and $n = \{1 \ldots 4\}$ are the four nutrient limitations (carbon, nitrogen, phosphorus and sulfur respectively). Zero indicates common expression level; 1 and -1 indicate upregulation and downregulation respectively. An example:

$$\mathbf{G}_{453,:,:} = \begin{pmatrix} 0 & 0 & 1 & 0 \\ 0 & -1 & 1 & 0 \end{pmatrix}$$

This gene (*MTD1*, indexed as no. 453) is thus upregulated under the phosphorus limitation (both aerobically and anaerobically) and downregulated under the nitrogen limitation in anaerobic growth. Note that genes that are not differentially expressed are assigned zeros in all cultivation conditions.

The TF binding data indicates the number of motifs in the promoter region of each gene for 102 TF's [9]. In this study we have employed motifs that are bound at high confidence ($P \leq 0.001$); not taking into account conservation among other *sensu stricto Saccharomyces* species. The 6383×102 matrix, denoted by $\mathbf{F}$, is binarized, such that $\mathbf{F}_{g,f}$ indicates whether the promoter region of gene g can be bound by TF f.

Table 1. Experimental conditions; the black squares indicate the employed nutrient limitation and oxygen supply

| Experimental | Nutrient limitation | | | | Oxygen supply | |
condition	Carbon	Nitrogen	Phosphorus	Sulfur	Aerobic	Anaerobic
1. ClimAer	■				■	
2. NlimAer		■			■	
3. PlimAer			■		■	
4. SlimAer				■	■	
5. ClimAna	■					■
6. NlimAna		■				■
7. PlimAna			■			■
8. SlimAna				■		■

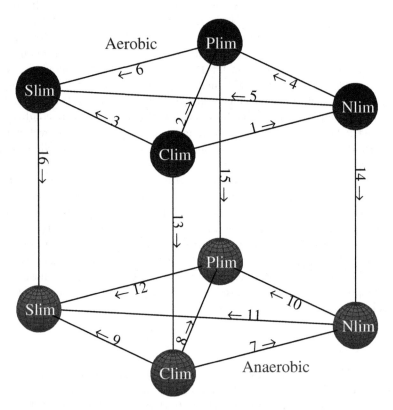

Fig. 1. Cube representing the eight cultivation conditions. Edges indicate defined condition transitions.

2.2   Condition Transition Analysis

From expression matrix $\mathbf{G}$ we derive the condition transition matrix $\mathbf{T}$. We define a condition transition as an instant change in yeast's extracellular environment by comparing two cultivation conditions and assess whether genes exhibit change in expression level when "going" from one cultivation condition to the other. In total we define sixteen condition transitions. These are only the transitions, where either the nutrient limitation or the oxygen availability changes; not both. The transitions are indicated by the edges in the cube of Figure 1. The six nutrient limitation transitions, both aerobically and anaerobically, (edges in the upper and lower face of the cube) are computed by:

$$\mathbf{T}_{g,I(n_1,n_2,o)} = sign(\mathbf{G}_{g,o,n_1} - \mathbf{G}_{g,o,n_2}) \quad \forall\{g,o,n_1 > n_2\} \tag{1}$$

The four oxygen availability transitions (vertical edges) are computed by:

$$\mathbf{T}_{g,12+n} = sign(\mathbf{G}_{g,1,n} - \mathbf{G}_{g,2,n}) \quad \forall\{g,n\} \tag{2}$$

Here, $I(n_1,n_2,o) = [6*(o-1) + n_1 + 4\cdot(n_2-1) - \frac{n_2\cdot(n_2+1)}{2}]$, such that the indices of the different transitions in $\mathbf{T}$ correspond to the numbers assigned to the edges in the cube of Figure 1. $\mathbf{T}$ (6383 × 16) is again a tertiary matrix, where zero indicates no change in expression; 1 and -1 respectively indicate an increase and decrease in expression as a consequence of a condition transition. Now, by consulting the TF binding matrix $\mathbf{F}$, a hypergeometric test can be employed to assess if genes that are up- and/or downregulated at a condition transition are bound (upstream) by a TF much more frequently than would be expected by chance. In more general terms, by employing the hypergeometric distribution we compute the probability of the observed (or more extreme) overlap between two sets of genes under the assumption that these sets of genes were randomly chosen from all genes [14]. Small probabilities (P-values) indicate that the hypothesis that these sets are randomly drawn must be dismissed, thereby acknowledging a significant relation between the two sets. In our setting, one set is constituted of all genes that can be bound by a particular TF, while the other set consists of e.g. all genes upregulated at a particular condition transition.

However, the systematic setup of the cultivation conditions in this dataset, allows for selection of more interesting groups of genes to input into the hypergeometric test. For example, genes that are upregulated at an aerobic nutrient limitation transition, yet not upregulated at the same nutrient limitation transition without the presence of oxygen. More specifically, for each of the six nutrient limitation transitions we define nine different groups of genes allowing us to focus on upregulation (1), downregulation (-1) and differential expression (-1 or 1), both specifically for aerobic or anaerobic growth as well as regardless of the oxygen supply. See Table 2. The 54 groups, augmented with groups of genes up-, downregulated or differentially expressed under the four oxygen availability transitions (Transitions 13-16), are tested for significant association with TF's by employing the hypergeometric test. (To adjust for multiple testing, the P-value cutoff was set, such that we expect one false positive (per-comparison error rate (PCER) of one [15]), corresponding with $P \leq 1.5 \cdot 10^{-4}$.) Figure 2 displays the significant relations in (for reasons of visibility) a part of the cube. We now have a regulatory network, which associates a TF with a cluster of genes that shows

Table 2. Conditions on $\mathbf{T}$ which define nine groups for each nutrient limitation transition ($t = \{1 \ldots 6\}$). The last two columns indicate the vertical placement (vp) and color of TF's that are significantly related to these groups as visualized in Figure 2.

no.	$\mathbf{T}_{g,t}$	$\mathbf{T}_{g,t+6}$	Description	vp	color
I	1	0,-1	Only up under aerobic growth	top	orange
II	0,-1	1	Only up under anaerobic growth	bottom	orange
III	1	1	Up under both aerobic and anaerobic growth	center	orange
IV	-1	0,1	Only down under aerobic growth	top	green
V	0,1	-1	Only down under anaerobic growth	bottom	green
VI	-1	-1	Down under both aerobic and anaerobic growth	center	green
VII	1,-1	0	Only diff. expressed under aerobic growth	top	black
VIII	0	1,-1	Only diff. expressed under anaerobic growth	bottom	black
IX	1,-1	1,-1	Diff. expressed under both aerobic and anaerobic growth	center	black

specific gene expression changes when a transition is made from one condition to the next.

In an attempt to gain more insight into the dynamics and combinatorial effects within the complete generated regulatory network, in stead of performing stringent tests of individual hypotheses, we add an additional step to our analysis. Here, we aim at modeling the expression behavior at all condition transitions $\mathbf{T}$ by employing binding matrix $\mathbf{F}$ and assess the activity of each TF at a condition transition. This approach is based on the simple biological model that ascribes the change of gene expression levels as observed at a condition transition to changes in TF activity; the means by which the organism adapts to the changed extracellular environment. In contrast to the landmark article by Bussemaker *et al.* [16], where expression was explained using cis-regulatory elements, we thus explain expression behavior at transitions by using TF binding data. In a more recent article from Bussemaker's group [17] also TF binding data was used to explain expression. However, they used a continuous score (the logarithm of the binding P-value) to represent the degree of TF binding, while we use the binary one, which indicates simply whether there is the ability to bind or not. Furthermore, we do not employ continuous expression levels, which are a measure of absolute mRNA quantities. We use the discrete elements of $\mathbf{T}$ that represent relative up- and downregulation, since we find this more robust and informative compared to (the difference between) absolute mRNA levels. Another big difference is that we do not use an iterative procedure to solve the problem, but aim at explaining all the transitions using all TF's in one time. Our problem finds its mathematical formulation in the orthogonal Procrustes problem, where we explore the possibility that $\mathbf{F}$ can be rotated into $\mathbf{T}$ by solving:

$$\min \|\mathbf{T}' - \mathbf{W}\mathbf{F}'\|_{Fro} \qquad \text{subject to } \mathbf{W}^T\mathbf{W} = \mathbf{I} \qquad (3)$$

In principle, this is a linear transformation of the points in $\mathbf{F}$ to best conform them to the points in $\mathbf{T}$. In our setting, the change in expression of a gene at a condition transition (as given in $\mathbf{T}$) is approximated by a weighted sum of ones. These ones correspond to the TF's that can bind the upstream region of that

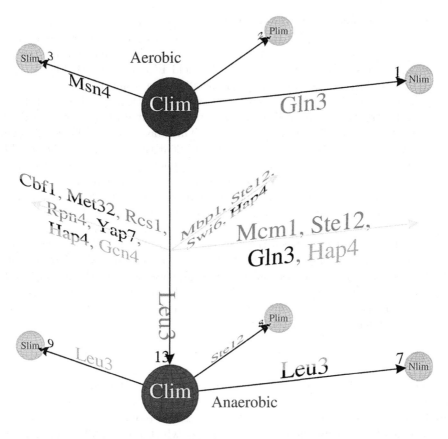

Fig. 2. TF activity for part of the transitions. Green, orange and/or black TF's are significantly related to genes that are downregulated, upregulated or differentially expressed respectively when going from one cultivation condition to the other (in the direction of the arrows). TF's on the top and bottom edges are activated only under aerobic or anaerobic growth respectively; TF's in the center of a surface indicate activation independent of the presence of oxygen. For example, Mcm1, Ste12, Gln3 and Hap4 are associated with transitions from carbon limitation to nitrogen limitation, independent of the presence of oxygen.

particular gene (as given in **F**). Thus, the elements in **W** represent a measure of the activity of a TF at a condition transition. Properties of the Procrustes rotation are the closed solution (via a SVD decomposition [18]) in minimizing the *Frobenius norm* (sum of squared errors) and the orthonormality of weight matrix **W**. A prerequisite for this rotation is that the number of columns (TF's) in **F′** should match the number of columns (condition transitions) in **T′**. Since our main focus is on TF's that regulate differently at nutrient limitation transitions as a consequence of oxygen supply, we select only the first twelve columns from **T**. The twelve selected TF's are those that (according to the hypergeometric test) are most significantly related to up- or downregulation, specifically under aerobic or anaerobic growth (i.e. related to groups I, II, IV and V in Table 2). Furthermore, we only employ those genes which exhibit different expression between aerobic and anaerobic growth for at least one of the six nutrient limitation transitions. These adjustments on $\mathbf{F}$ and $\mathbf{T}$ yield $\mathbf{F}'$ and $\mathbf{T}'$ (both 1493×12), which are employed in Eq. (3). Figure 3 visualizes the resulting $\mathbf{W}$.

Permutation tests were performed to assess the statistical significance of these weights. The rows (genes) of $\mathbf{T}'$ were randomly permuted after which the Procrustes rotation (Eq. (3)) was recomputed. This was done 10,000 times. The Wilcoxon signed rank test was applied to check if the original weights could be the medians of the distributions of weights generated by the permutations. The extremely low P-values for almost all weights indicated that this hypothesis should be dismissed. (Results not shown.) This attaches, at least, a statistical meaning to the derived weight matrix. More interestingly, for each of the twelve TF's and each of this six nutrient limitation transitions we assessed the significance of the difference between the assigned weight under aerobic growth and the weight under anaerobic growth. A P-value was computed by determining the fraction of permutations in which the difference between the aerobic and anaerobic weight was larger than for the original (non-permuted) data. Significant differences ($P \leq 0.05$) point towards oxygen specific regulation of a TF at a specific nutrient limitation transition.

3   Results

The network of TF activity, as partly presented in Figure 2, provides many very specific clues towards the transcriptional regulation of yeast's metabolism and respiration. Some of these can be linked to existing biological knowledge quite easily. One obvious example is the TF Hap4, of which the mRNA abundance is decreased by the presence of glucose [19]. This explains downregulation of the regulon of Hap4 in the three nutrient transitions moving away from the carbon limitation. Furthermore, in the carbon to sulfur limitation transition, we find Met32, a known transcriptional regulator of methionine metabolism [20], as well as Cbf1, which is part of the transcription activation complex Cbf1-Met4-Met28 [21]. To find TF Gln3 at the transition from carbon limited growth to growth where nitrogen becomes the limiting nutrient is also not surprising. Ammonium, the nitrogen source used in these experiments and generally considered to be the preferred nitrogen source for *S. cerevisiae*, is in excess under carbon-limited growth, while absent under nitrogen-limited growth. It is well known that high concentrations of ammonium lead to nitrogen catabolite repression (NCR), a transcriptional regulation mechanism that represses pathways for the use of alternative nitrogen sources [22]. Gln3 is one of the four so-called GATA factors active in NCR to adapt to the change in need of alternative nitrogen sources at this transition. It is however surprising that Gln3 is significantly related to genes upregulated, especially under aerobic conditions. Also unexpectedly, Leu3, a regulator for genes of branched-chain amino acid biosynthesis pathways, is significantly related to genes downregulated, especially at the anaerobic transition from carbon to nitrogen limitation.

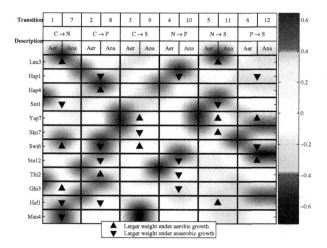

Fig. 3. Visualization of **W**, representing the TF activity of twelve TF's under the six nutrient limitation transitions, both aerobically and anaerobically. Large positive weights (red) indicate involvement in upregulation, negative weights (blue) refer to downregulation. Triangles indicate a significance difference in weights ($P \leq 0.05$) for a nutrient limitation transition between the aerobic and anaerobic case.

Here, we come to the crux of our work. Our approach is able to infer TF activity related to very specific changes in combinatorial cultivation parameters. The algorithm that is especially designed for the combinatorial setup of nutrient limitations and oxygen supply in the employed microarray dataset, not only provides unprecedented detailed insight into the behavior of yeast's metabolism and respiration at the transcriptional level, but also in terms of TF activity. Thus, we do not find many TF's that are globally related to particular nutrients. (These have already been identified in previous studies, e.g. [4,12]). More specifically, we identify lots of TF's that are not primarily related to the metabolism of a particular nutrient, yet seem to play a more specific and subtle (and as of yet unknown) regulatory role at these transitions between nutrient limitations. The involvement of these TF's demonstrate the complex and multiple regulatory roles that they exhibit in transcriptional regulation in different processes.

The involvement of a particular TF in different processes has of course been established by many independent studies. For example, Mcm1 is a known multifunctional protein which plays a role both in the initiation of DNA replication (cell-cycle) and in the transcriptional regulation of diverse genes [23]. A more recent study also suggests that in response to changes in their nutritional states, yeast cells modulate the activity of global regulators like Mcm1 via posttranscriptional regulation induced by the flux of glycolysis [24]. The identification of Mcm1 as a regulator in the carbon to nitrogen limitation transition, where the glycolysis flux changes dramatically, thus strengthens and even broadens this hypothesis.

In general, the results provide new regulatory roles for many TF's in metabolism and respiration. Additionally, the results underline the complexity of transcriptional regulation in the cell, especially when taking into account the fact that changes in nutrient and oxygen availability can not be seen in isolation from (or even modulate) cell-cycle (e.g. [25]) and energy processes (e.g. [26]) and is even known to evoke stress responses (e.g [27,28]). To strengthen this notion, enrichment in MIPS [29] and GO [7] functional categories was computed. Table 3 displays the results for the transition from carbon to nitrogen limitation. These results also indicate that many non-metabolic processes play a role in the nutrient and oxygen availability transitions.

Table 3. Significantly enriched ($P \leq 5 \cdot 10^{-5}$) MIPS and GO functional categories for the nine groups defined at the carbon to nitrogen limitation transition. Processes other than metabolism, energy and cellular transport are underlined.

no.	MIPS	GO
I	metabolism	
II	energy, oxydative stress response	response to stress
III	metabolism, complex cofactor/cosubstrate binding	
IV	tetracyclic and pentacyclic triterpenes biosynthesis	lipid metabolism, steroid metabolism and biosynthesis
V	metabolism of the pyruvate family and D-alanine, mitochondrion	cellular biosynthesis, nitrogen compound biosynthesis, a.o.
VI		lipid metabolism, steroid metabolism and biosynthesis
VII	mitotic cell cycle and cell cycle control, cellular transport, a.o.	
VIII	energy, respiration, a.o.	aerobic respiration, generation of precursor metabolites, a.o.
IX	energy, respiration, transported compounds, a.o.	oxidative phosphorylation, transport, a.o.

In the remainder of this section we focus on three identified TF's and hypothesize about their putative role in regulation at specific transitions. Here, we also demonstrate the power of the Procrustes approach in clarifying more subtle patterns of regulation.

Leu3

Leu3 is the main transcriptional regulator of branched-chain amino acid metabolism and has been extensively studied [30,31]. To exactly meet the demands of protein synthesis, the activity of Leu3 is modulated by α-isopropylmalate (α-IPM), an intermediate of the branched-chain amino acid pathway. As a result Leu3 can act as both an activator and a repressor. Our findings indicate an oxygen-specific role of Leu3 in several nutrient limitation transitions. Figure 4 displays the expression behavior at transitions for the regulon of Leu3. Many genes are downregulated at the $C \rightarrow N$ and $C \rightarrow S$ transitions under anaerobic conditions in comparison to the same transitions grown under aerobic conditions. (This can be seen by the much larger number of green boxes in transition 7 w.r.t. transition 1 and similarly for transitions 9 and 3). Furthermore, when going from aerobic to anaerobic carbon-limited growth many genes are upregulated. (All above mentioned relations were found significant in the hypergeometric tests, as can be seen in Figure 2.) The involvement of Leu3 as a repressor and activator at these transitions has not been established before.

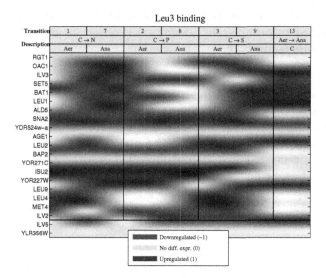

Fig. 4. Part of the transition matrix **T**, indicating the expression behavior of all genes to which TF Leu3 can bind (upstream) for the transitions that are displayed in Figure 2.

Personal communication with the first author of [31] lead to the observation that the expression pattern of Leu3's regulon under anaerobic growth is quite remarkable. If it were the case that also at the anaerobic $C \rightarrow P$ transition many genes were downregulated, one could associate this to mitochondrial capacity [8], since the synthesis route of α-IPM is mainly located in the mitochondrion. However, this is not the case. Possibly, regulation of Leu3 under anaerobic growth can be linked to different concentrations of α-IPM, caused by different concentrations of Acetyl-CoA and ATP/ADP that change at the transitions. However, this is not more than speculation at this point.

Yap7

The TF Yap7 was only significantly associated with upregulation of genes when going from nitrogen to sulfur limited ($N \rightarrow S$) aerobic growth. (This result is not visible in Figure 2.) The Procrustes analysis, however, shows a more interesting pattern of regulation. In Procrustes, the TF binding data set is employed to explain the different expression behavior between *all* the aerobic and anaerobic nutrient limitation transitions simultaneously. (This in contrast to employing the hypergeometric distribution, where hypotheses can only be tested individually.) Furthermore, the orthonormality constraint emphasizes the difference in activity of a TF at different transitions. When investigating the weights assigned to Yap7, we see that not only in $N \rightarrow S$ the weight is significantly larger under aerobic growth, but also in the case of the other transitions moving towards sulfur limited growth; $C \rightarrow S$ and $P \rightarrow S$. (See Figure 3.) For the other transitions the weights are near zero. Thus, we can hypothesize that Yap7 (a member of the yeast bZip family of proteins, of which two other members can only be linked indirectly to sulfur metabolism [32]) is involved in regulation under aerobic sulfur-limited growth, thereby assigning a very specific putative regulatory role for this poorly studied TF.

Ste12

Also in the case of Ste12, the Procrustus rotation confirms and broadens the relationships as established by the hypergeometric tests. From the literature it follows that Ste12 is a transcription factor that binds to the pheromone response element (PRE) to regulate genes required for mating and also functions with Tec1 to regulate genes required for pseudohyphal growth [20]. Additionally to these functionalities, we find it to upregulate genes when entering a phosphorus-limited state, especially when no oxygen is present. See the condition transition weights for Ste12 in Figure 3. (Note that the $S \rightarrow P$ is not in the table, but it is justified to expect that these weights will be the complement of the $P \rightarrow S$ transition.)

4   Discussion

Today's main use and strength of bioinformatics tools is generating hypotheses on all types of relationships and functionalities of and between quantifiable parameters inside and outside the cell. Specific biological experiments are, however, still required to validate the automatically generated hypotheses before accepting them as newly discovered knowledge. The common trend of focusing on large compendia of intracellular measurement datasets is often in contrast with the biologist's very specific field of research. These broad approaches are able to recognize global patterns in the data, but miss specific and subtle effects that characterize the complex reality of the cell.

In this research we applied a tailor-made informatics approach on a small, well defined dataset. This enabled us to provide the biologist with very detailed hypotheses about the specific biological processes of interest. The basis for this work is the systematic combinatorial setup of the cultivation conditions under which yeast was grown in highly controllable chemostats. Incorporation of TF binding data through stringent statistical tests as well as a Procrustes rotation, led us to infer the activity of TF's at transitions between the different cultivation conditions. In contrast to common approaches the generated regulatory network thus shows the actual changes in conditions that lead to the activation of TF's. Incorporation of (changes in) conditions is a crucial part of regulatory networks and in the quest for simulation of the complete regulatory mechanisms within the cell, will be part of more elaborate future analysis. Additionally, future work will aim at interpreting the uncovered results, not only by literature, but also by performing specific follow-up experiments. Furthermore, the uncovered results have proved to be very interesting, and therefore encourage application of similar techniques to other systematically setup datasets.

References

1. N. Banerjee and M.Q. Zhang. Functional genomics as applied to mapping transcription regulatory networks. *Curr. Opin. Microbiol.*, 5:313–317, 2002.
2. G. Chua, M.D. Robinson, Q. Morris, and T.R. Hughes. Transcriptional networks: reverse-engineering gene regulation on a global scale. *Curr. Opin. Microbiol.*, 7(6):638–646, 2004.
3. Y. Pilpel, P. Sudarsanam, and G.M. Church. Identifying regulatory networks by combinatorial analysis of promoter elements. *Nat. Genet.*, 29(2):153–159, 2001.
4. Z. Bar-Joseph, G.K. Gerber, T.I. Lee, N.J. Rinaldi, J.Y. Yoo, F. Robert, D.B. Gorden, E. Fraenkel, T.S. Jaakkola, R.A. Young, and D.K. Gifford. Computational discovery of gene modules and regulatory networks. *Nat. Biotechnol.*, 21(11):1337–1342, 2003.
5. W. Wang, J.M. Cherry, Y. Nochomovitz, E. Jolly, D. Botstein, and H.Li. Inference of combinatorial regulation in yeast transcriptional networks: a case study of sporulation. *Proc. Natl. Acad. Sci. U.S.A.*, 102(6):1998–2003, 2005.
6. E. Segal, M. Shapira, A. Regev, D. Pe'er, D. Botstein, D. Koller, and N. Friedman. Module networks: identifying regulatory modules and their condition-specific regulators from gene expression data. *Nat. Genet.*, 34(2):166–176, 2003.
7. M. Ashburner, C.A. Ball, G.M. Rubin, G. Sherlock, et al. Gene ontology: tool for the unification of biology. the gene ontology consortium. *Nat. Genet.*, 25(1):25–29, 2000.
8. S.L. Tai, V.M. Boer, P. Daran-Lapujade, M.C. Walsh, J.H. de Winde, J.M. Daran, and J.T. Pronk. Two-dimensional transcriptome analysis in chemostat cultures. *J. Biol. Chem.*, 280(1):437–447, 2005.
9. C.T. Harbison, D.B. Gordon, T.I. Lee, N.J. Rinaldi, E. Fraenkel, R.A. Young, et al. Transcriptional regulatory code of a eukaryotic genome. *Nature*, 431(7004):99–104, 2004.
10. V.G. Tusher, R. Tibshirani, and G. Chu. Significance analysis of microarrays applied to the ionizing radiation response. *Proc. Natl. Acad. Sci. USA*, 98(9):5116–5121, 2001.
11. T.A. Knijnenburg et al. unpublished results.
12. T.A. Knijnenburg, J.M. Daran, P. Daran-Lapujade, M.J.T. Reinders, and L.F.A. Wessels. Relating transcription factors, modules of genes and cultivation conditions in saccharomyces cerevisiae. *IEEE CSBW'05*, pages 71–72, 2005.
13. D.L. Davies and D.W. Bouldin. A cluster separation measure. *IEEE Trans. Patt. Anal. Machine Intell.*, PAMI-1:224–227, 1979.
14. Y. Barash, G. Bejerano, and N. Friedman. A simple hypergeometric approach for discovering putative transcription factor binding sites. *Algorithms in Bioinformatics: Proc. First InternationalWorkshop, number 2149 in LNCS*, pages 278–293, 2001.
15. Y. Ge, S. Dudoit, and T. P. Speed. Resampling-based multiple testing for microarray data analysis. *TEST*, 12(1):1–77, 2003.
16. H.J. Bussemaker, H. Li, and E.D. Siggia. Regulatory element detection using correlation with expression. *Nat. Genet.*, 27(2):167–174, 2001.
17. F. Gao, B.C. Foat, and H.J. Bussemaker. Defining transcriptional networks through integrative modeling of mrna expression and transcription factor binding data. *BMC Bioinformatics*, 5(31), 2004.
18. G.H. Golub and C.F. Van Loan. *Matrix Computations*. The John Hopkins University Press, Maryland, U.S.A., 1996.

19. S.L. Folsburg and L. Gaurente. Identification and characterization of hap4: a third component of the ccaat-bound hap2/hap3 heteromer. *Genes Dev.*, 3(8):1166–1178, 1989.

20. Yeast protein database http://www.proteome.com.

21. P.L. Blaiseau and D. Thomas. Multiple transcriptional activation complexes tether the yeast activator met4 to dna. *EMBO J.*, 17(21):6327–6336, 1998.

22. B. Magasanik and C. A. Kaiser. Nitrogen regulation in saccharomyces cerevisiae. *Gene*, 290:1–18, 2002.

23. S.Passmore, R. Elbe, and B.K.Tye. A protein involved in minichromosome maintenance in yeast binds a transcriptional enhancer conserved in eukaryotes. *Genes Dev.*, 3:921–935, 1989.

24. Y. Chen and B.K.Tye. The yeast mcm1 protein is regulated posttranscriptionally by the flux of glycolysis. *Mol. Cell. Biol.*, 15(8):4631–4639, 1995.

25. L.L. Newcomb, J.A. Diderich, M.G. Slattery, and W. Heideman. Glucose regulation of saccharomyces cerevisiae cell cycle genes. *Eukaryot. Cell.*, 2(1):143–149, 2003.

26. J. Wu, N. Zhang, A. Hayes, K. Panoutsopoulou, and S.G. Oliver. Global analysis of nutrient control of gene expression in saccharomyces cerevisiae during growth and starvation. *Proc. Natl. Acad. Sci. U.S.A.*, 101(9):3148–3153, 2004.

27. D.J. Jamieson. Oxidative stress responses of the yeast saccharomyces cerevisiae. *Yeast*, 14(16):1511–1527, 1998.

28. A.P. Gasch and M. Werner-Washburne. The genomics of yeast responses to environmental stress and starvation. *Funct. Integr. Genomics*, (4-5):181–192, 2002.

29. H.W. Mewes, K. Albermann, K. Heumann, S. Liebl, and F. Pfeiffer. Mips: a database for protein sequences, homology data and yeast genome information. *Nucleic Acids Research*, 25(1):28–30, 1997.

30. G.B. Kohlhaw. Leucine biosynthesis in fungi: entering metabolism through the back door. *Microbiol. Mol. Biol. Rev.*, 67(1):1–15, 2003.

31. V.M. Boer, J.M. Daran, M.J. Almering, J.H. de Winde, and J.T. Pronk. Contribution of the saccharomyces cerevisiae transcriptional regulator leu3p to physiology and gene expression in nitrogen- and carbon-limited chemostat cultures. *FEMS Yeast Res.*, 5(10):885–897, 2005.

32. D. Mendoza-Cozatl, H. Loza-Tavera, A. Hernandez-Navarro, and R. Moreno-Sanchez. Sulfur assimilation and glutathione metabolism under cadmium stress in yeast, protists and plants. *FEMS Microbiol. Rev.*, 29(4):653–671, 2005.

An In Silico Analogue of In Vitro Systems Used to Study Epithelial Cell Morphogenesis

Mark R. Grant and C. Anthony Hunt

Joint UCSF/UCB Bioengineering Graduate Group and The Biosystems Group, Department of Biopharmaceutical Sciences, The University of California, San Francisco, CA 94143, USA
a.hunt@ucsf.edu, mgrant@calmail.berkeley.edu

Abstract. In vitro model systems are used to study epithelial cell growth, morphogenesis, differentiation, and transition to cancer-like forms. MDCK cell lines (from immortalized kidney epithelial cells) are widely used examples. Prominent in vitro phenotypic attributes include stable cyst formation in embedded culture, inverted cyst formation in suspension culture, and lumen formation in overlay culture. We present a low-resolution system analogue in which space, events, and time are discretized; object interaction uses a two-dimensional grid similar to a cellular automaton. The framework enables "cell" agents to act independent using an embedded logic based on axioms. In silico growth and morphology can mimic in vitro observations in four different simulated environments. Matched behaviors include stable "cyst" formation. The in silico system is designed to facilitate experimental exploration of outcomes from changing components and features, including the embedded logic (the in silico analogue of a mutation or epigenetic change). Some simulated behaviors are sensitive to changes in logic. In two cases, the change caused cancer-like growth patterns to emerge.

Keywords: Agent-based, cystogenesis, epithelial, model, morphogenesis, simulation, synthetic, systems biology, complex systems.

1 Introduction

To better understand mammalian physiology we often study simpler systems: in vitro model systems, such as 3D cultures of epithelial cells. To understand how molecular level events are causally linked in vitro to emergent, system level, phenotypic attributes, we need synthetic, in silico analogues of those in vitro systems that are suitable for experimentation: they need to exhibit properties and characteristics (PCs) that overlap in useful ways with those of the in vitro referent. We report significant, early progress toward that goal for cultured MDCK [1] and related cells.

The in vitro morphological phenotype of a MDCK cell system depends on the cells' environment. Examples of common properties and characteristics (PCs) are illustrated in Fig. 1. In 3D embedded cultures, formation of stable, self-enclosed monolayers (cysts) is the dominant morphological characteristic. O'Brien et al. have observed that cyst formation, along with structures formed in other in vitro environments,

C. Priami (Ed.): CMSB 2006, LNBI 4210, pp. 285–297, 2006.
© Springer-Verlag Berlin Heidelberg 2006

seems to be driven by each cell's pursuit of three types of membrane surfaces, called free, lateral, and basal [1]. They suggest that this drive is intrinsic to epithelial cell differentiation and morphogenesis. Epithelial cells respond to signals resulting from interactions with particular components and properties of their environment, with other cells via cadherins and other cell-cell adhesion molecules, with matrix via integrins and other molecules, and with free surface via apically-located integrins, cilia, and flow sensing. Our plan has been to bring these mechanisms into focus iteratively by building and validating detailed analogues beginning with the one presented here: a four-component, discrete event, discrete time, and discrete space analogue. It is simple yet exhibits a rich in silico "phenotype." It is capable of recapitulating in silico the key outcomes in the growth of MDCK cells [2-4] shown in Fig. 1. These successful simulations stand as hypotheses of the mechanism(s) that causes those PCs. Models that are more detailed, built by extending the approach used here, are expected to follow.

2 Methods

To avoid confusion and clearly distinguish in vitro components from corresponding simulation components, such as "cells," "matrix," "cyst," and "lumen," we use small caps when referring to simulation components and properties.

We use the synthetic modeling method [5,6]. Steels and Brooks contrast the synthetic and inductive methods [7]. In inductive modeling, one usually creates a mapping between the envisioned system structure and components of the analyzed data, and then represents those data components with mathematical equations. The synthetic method, in contrast, works forward from domain (components) to range (data). It is more concerned with *how* properties are generated. It is especially suitable for representing spatial and discrete event phenomena. That makes it ideally suited for assembling analogues from components and exploring the resulting behaviors. Synthetic modeling of cell systems requires knowledge of their function and behavior; for the in silico model

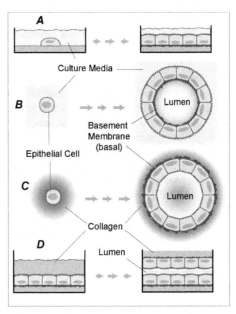

Fig. 1. Targeted in vitro phenotypic attributes: **A**: A single epithelial cell plated on a layer of collagen (surface culture) generates a uniform monolayer; **B**: Cyst formation in suspension culture; cell polarity is inverted relative to that in C; **C**: Representation of a cross-section of an epithelial cyst in vitro formed in embedded culture. Starting with a single cell embedded in a collagen gel: cell division, apoptosis, and shape change over a period of several days leads to a lumen-containing cyst; **D**: Lumen formation in collagen overlay experiments.

described herein, considerable cell biology and molecular knowledge is available. Ideas about plausible mechanisms and of relevant observables are available. Strategies are needed for how the analogue and referent observables will be measured and compared.

2.1 Framework, Analogue, and Specifications

We used MASON (cs.gmu.edu/~eclab/projects/mason/) to construct the simulation framework. MASON is based on the Swarm simulation package. It is an optimized, and extensible Java-based simulation toolkit. We used the JFreeChart library for plotting purposes and import and export of XML-formatted data. In order to take full advantage of the flexibility of object-oriented programming we discretize space and time and assume that all events can be represented as being discrete. The software used along with executable applets are available at

<http://128.218.188.102:8080/growthmodel/index.php>.

We accumulated relevant wet-lab observations by identifying experiments and observations that could be directly compared with simulated output. We also looked for and characterized in silico phenotypic attributes; we then searched the literature for wet-lab evidence that supported or invalidated these attributes. Our conceptual model is that epithelial cells in vitro behave as if they are following innate mandates that result indirectly from the biological counterpart of axioms of development and differentiation. Meeting the mandates is necessary for "success" in vitro. In total, they are sufficient. Each cell responds to stimuli based on its need to comply with all of the mandates. Further, each cell has internal systems contained by an operational interface (not identical to the cell membrane) that mediates interaction between the external environment and internal systems. The observed repertoire of behaviors is hypothesized to be a consequence of stimulus-response mappings between the inputs and outputs of that interface. Consequently, that interface is the logical starting place for building an analogue. What are appropriate initial spatial and temporal resolutions?

We speculated that by treating each cell and similar sized units of environment as single (atomic) objects driven by an axiomatic mechanism, we could capture important PCs. We assumed that the cell's operational interface accepts stimuli at intervals, processes them, and later, when appropriate, the cell responds resulting in a change in the cell or its immediate environment. We assumed that a repertoire of stimulus-response sequences precedes each evident behavior, and that response times for different behaviors can differ. Different cells in culture will be undergoing different state behaviors in parallel. For the model discussed herein, we elected to use an average response time to represent all behaviors. That interval is represented by one simulation cycle. Because events in two different in vitro systems can proceed at different paces, one simulation cycle can map to different intervals of wet-lab time.

Events such as a macromolecular binding and signaling are below the analogue's level of temporal resolution, but that can be changed. Processes such as changes in cell shape or in local matrix organization are not ignored; they are below the level of spatial resolution. Their contribution is conflated into axioms[1] (illustrated in Fig. 2) and components. A specific stimulus-response mapping between inputs from the environment and outputs, such as initiation of apoptosis, are below the levels of temporal and event resolution. We assume that they can be represented collectively by the axioms and decisional process that governs behaviors. CELL behaviors are independently scheduled; they are executed in sequence, not in parallel as is typical for a cellular automaton [8]. This enables exploration of outcomes from interactions of CELLS undergoing different behaviors. Additionally, the logic governing CELL behavior is embedded within that CELL (the CELL is an agent).

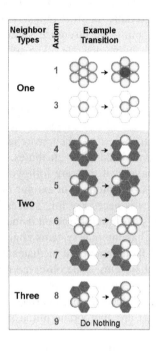

Fig. 2. Illustrations (2D hexagonal grid) of the axioms that govern a CELL'S action during any simulation cycle. Dark hexagons: FREE SPACE; circles with gradient shading: CELLS; and white hexagons: MATRIX. Axioms are organized by the number of object types in the local neighborhood: one or more of CELL, MATRIX, and FREE SPACE. Axiom 2 (die if all neighbors are FREE SPACE) is not shown because it was not used. Each axiom defines a mapping from a precondition to a new condition for the center CELL. The in silico decisional process (sketched in Fig. 3) is based on the arrangement of three object types adjacent to the CELL. Multiple locations can meet the axiom's requirement daughter placement. When that occurs, one is selected at random. When none of the conditions of the first eight axioms is met, Axiom 9 applies: the CELL does nothing. The wet-lab observations and data that motivated the first six axioms are as follows (axiom : [reference(s)]): 1 : [15,20,21]; 2 : [20,21]; 3 : [2-4]; 4 : [3,15]; 5 : [21]; 6 : [2,3,22], 7&8 : [1].

Four simulated environments represent four referent culture conditions. Each consists of a 2D (or 3D) grid, hexagonal or square, typically 100 x 100. Just one of three different types of components is assigned to each grid point: MATRIX, FREE SPACE, or CELL. MATRIX represents a cell-sized region of culture medium that contains matrix. FREE SPACE represents a similar sized region of cell-free or matrix-free medium, such as a portion of a luminal space. A CELL represents an MDCK II epithelial cell. When a higher level of intracellular resolution is needed, it is straightforward to replace a CELL with a component that is a composite object representing some combination of internal systems. When doing so, we can elect to keep the resolution of the environment as it is, or, with equal ease replace selected units with composite objects that

---

[1] *Axiom* emphasizes that computer programs are mathematical, formal systems and the initial mechanistic premises in our simulations are analogous to axioms in formal systems. Here, an axiom is an assumption about what conclusion can be drawn (action taken) from what precondition for the purposes of further analysis or deduction, and for developing the analogue system.

represent more environmental detail. The temporal resolution too can be increased (or decreased) when that is needed. All of these changes can be carried out within, and accommodated by the framework.

The only components in the simulation that schedule events are CELLS. Each CELL is an agent. Each CELL schedules a new event for itself at the next simulation cycle, and it then executes a program, an in silico decisional process (Fig. 3). For each neighborhood arrangement, there is only one action option. All behaviors are assumed to be influenced by only the immediate environment. The ordering of CELL event execution during each simulation cycle is randomized. For the simulations discussed, one simulation cycle corresponds to one unit of in vitro time.

The axioms specify how a CELL will behave when in contact with only one type of local environment (only MATRIX, CELLS, or FREE SPACE, where the latter can represent a luminal environment), two types (MATRIX and CELLS, MATRIX and FREE SPACE, and CELLS and FREE SPACE), or all three types.

2.2 Axiom Development

Matrix attachment is required for long-term survival of epithelial cells. Therefore, one might consider implementing a function that allows for survival only in the presence of matrix. However, it is possible for an epithelial cell to generate matrix de novo in the absence of existing matrix, thus "rescuing" itself. In order to allow for unanticipated behaviors in all possible environments, we assumed that cell action in a particular environment is independent of its actions in others and from any past action. For a hexagonal grid containing arrangements of three different components (CELL, MATRIX, and FREE SPACE), there are 729 possible neighborhood arrangements. The number is reduced when mirror images are assumed equivalent. The number is reduced further by noting that several arrangements are identical after axial rotations. We classified the remaining arrangements based on their perceived similarities with respect to overall cell behavior. We arrived at a set of arrangements in which cell behavior could be functionally distinct. Guided by specific literature observations of in vitro systems, we assigned a prediction as to how an epithelial cell would behave in each. We found no relevant experimental observations for several classes. For those we assumed that epithelial cells desire to maintain existing surfaces, and generate additional environment types, in a pursuit of three surface types [1]. Unrealistic alternative outcomes were not considered. Merging similar be-

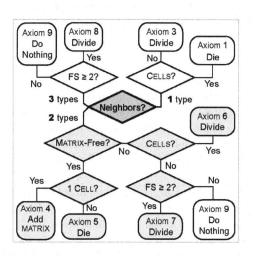

Fig. 3. A diagram of the in silico decisional protocol by which each cell matches its precondition (neighboring components and their arrangement (see Fig. 2) with just one axiom. FS: FREE SPACE.

havior predictions resulted in the simplified collection of axioms listed below (Fig. 2) and the decisional protocol that determines which axiom applies (Fig. 3).

During each simulation cycle (time step), each CELL uses a decision tree to select one of five actions: do nothing, die, add MATRIX, and divide (two axioms manage daughter placement differently). First, each CELL determines if it has one, two, or three types of neighbors. If there is only one type, then follow either Axiom 1 or 3. Two types: follow either Axioms 4, 5, 6, or 7. Three: follow Axiom 8. If 1-8 do not apply, apply Axiom 9. The axioms are as follows.

Axiom 1: only CELL neighbors: die
Axiom 2: only FREE SPACE neighbors: die (does not apply for these simulations)
Axiom 3: only MATRIX neighbors: divide
Axiom 4: neighbors are FREE SPACE and just one CELL: add MATRIX between self and that CELL
Axiom 5: neighbors are FREE SPACE and ≥ two CELLS: die
Axiom 6: neighbors are ≥ one CELL and MATRIX: divide and replace a MATRIX with daughter to maximize her CELL neighbors
Axiom 7: neighbors are MATRIX and ≥ two FREE SPACE: divide and place daughter in a FREE SPACE to maximize her MATRIX neighbors
Axiom 8: three types of neighbors with MATRIX neighbored by two adjacent FREE SPACES: divide and place daughter in a FREE SPACE to maximize her MATRIX neighbors

2.3 Simulation Experiments

To insure that the scheduling of CELL events was nondeterministic, random number generator seeds were changed for each simulation. Typically, one-to-two hundred simulation cycles were run for each simulation, with up to two hundred simulations per experimental condition. Images of each completed simulation were collected for subsequent comparison. CELL numbers were saved at each step of a simulation, and for each simulation run.

We studied growth rates and structures formed in multiple simulated conditions. For simulated surface culture, we ran 50 simulations of 20 simulation cycles each. Each used 3D square grids of dimensions 2 x 100 x 100: first, the bottom 1 x 100 x 100 section was filled with MATRIX. Next, a CELL was placed at the center of the top 1 x 100 x 100 grid. Similar in vitro surface culture data (days 0–5) was available [9]. For graphical representation of these studies, we assumed each simulation cycle corresponds to 20.4 hours in vitro.

We also studied growth properties of CELLS in simulated embedded culture. For simulated embedded culture we ran 200 simulations of 100 simulation cycles each. An in vitro embedded culture is represented by a 2D grid having MATRIX in all locations. To initiate a simulation, one or more CELLS replace MATRIX. Data from in vitro embedded culture data was available for comparison with simulation outcomes [3]. CELL number and position information was saved for each simulation run. For graphical representation of embedded conditions, we assumed a simulation cycle correspondence of 12.0 hours in vitro.

In vitro suspension culture was represented by a grid in which FREE SPACE is assigned initially to all locations. A simulation is initiated by having one or more CELLS replace FREE SPACE. Results from simulated suspension culture of normal CELLS are identical between simulation runs, therefore only image captures were taken. CELL number data was not saved.

Last, simulation of an in vitro overlay or sandwich culture starts with a single horizontal monolayer of CELLS bordered above and below by a region of MATRIX, placed in a 2D grid. Image captures from simulation runs were taken, but no corresponding CELL number data was available for comparison so this data was not acquired from the simulation.

As a strategy to get a variety of CYST sizes in simulated suspension culture and to simulate the effect of altered matrix production, Axioms 4 and 5 were replaced by a new one. It specified the placement of MATRIX at a neighboring FREE SPACE position with a maximal number of CELL neighbors, as determined by the scheduled CELL. Model outcomes were studied in simulated embedded conditions, except that a CELL with the altered axioms was used.

To examine the importance of the orientation of daughter placement on simulated morphology, Axiom 8 was changed to allow for placement of the daughter into any neighboring FREE SPACE location. Image captures from simulations in a range of simulated environments (surface, suspension, embedded, and overlay) were taken for study. These simulated environments were constructed as for normal CELLS, except CELLS with this altered axiom were used.

3 Results

3.1 Targeted Phenotypic Attributes

MDCK cells grown on collagen I generate a simple monolayer that covers the entire surface [10]. Most cell division takes place on the outer edge of an expanding colony [11]. Analogue simulations successfully represent these phenotypic attributes. As one would expect, given the simulation axioms summarized in Fig. 2, MONOLAYER formation always occurred on flat MATRIX surfaces (Fig. 4C). The pattern of growth in simulated surface culture be-

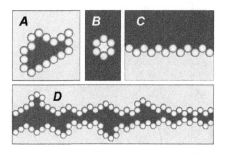

Fig. 4. Example output from the four simulated environments. Spheres represent CELLS; black space represents FREE SPACE or simulated internal lumen space; white space represents MATRIX. **A**: stable CYST formation in simulated embedded culture; **B**: stable CYST formation in simulated suspension culture; **C**: MONOLAYER formation in simulated surface culture stable; **D**: LUMEN formation in simulated overlay culture.

gins with an initial exponential growth phase that lasts for three simulated days. It closely matches published observations of MDCK growth rates in surface cultures [9] (Fig. 5A).

Overlay of a MDCK cell monolayer with Collagen I induces a morphological response [12] that results in lumen spaces completely surrounded by cells [2]. These events include migration, division, apoptosis, and shape changing. Simulation of an overlay culture resulted in the formation of structures similar to those just described (Fig 4D): regions of FREE SPACE are formed surrounded by single layers of CELLS, each bordering MATRIX.

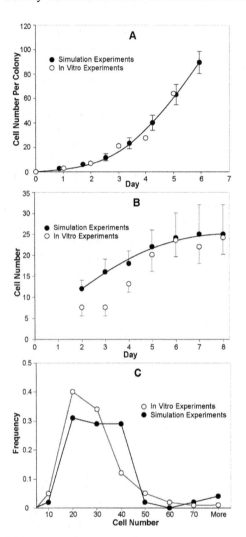

MDCK cells grown in type I collagen (embedded culture) exhibit clonal growth in three stages [13,14]: 1) repeated rounds of cell division during the first two days of culture, followed by 2) the formation of a central lumen accompanied by an increase in cell numbers and cyst diameter between day two and seven. 3) Thereafter, cyst size plateaus with little apparent change thereafter. An example CYST is provided in Fig. 4A.

In silico experiments demonstrated five important similarities with in vitro cyst morphogenesis. 1) LUMEN formation occurred after 2–2.5 simulated days. Lumen formation was observed to occur in vitro by as early as day two [3]. 2) CYST expansion was arrested (Fig. 5B) [3]. 3) Stable CYSTS contained a central region of FREE SPACE lined by a single layer of CELLS. 4) CELL division and CELL death continued after the initiation of LUMEN formation [15]. 5) There was variation in CELL numbers per mature CYST, similar to in vitro observations (Fig. 5C).

The fourth targeted attribute is formation of cysts in suspension cultures. Cystogenesis in vitro begins with the formation of aggregates of two to ten cells. Thereafter, over approximately ten days, an "inverted" cyst (Fig. 4B) forms, expands, and then stabilizes. During the process, basement membrane components, including collagen IV and laminin I, accumulate in the lumen and line the inner surface of the cyst [3]. Some matrix is apparently produced *de*

Fig. 5. Comparisons of data from in silico (n = 50) and in vitro experiments, assuming a mean CELL division time of 20.4 hours: **A**: CELL numbers per COLONY in simulated surface culture; **B**: CELL numbers per CYST in simulated embedded culture, where two simulation cycles corresponds to 1 simulated day; **C**: comparison of in vitro cell numbers per cyst cross-section after ten days in embedded culture with CELL numbers per CYST cross-section after twelve simulation steps (n = 50).

novo [16]. We limited MATRIX production in the simulation to one situation: when a CELL has only one other CELL neighbor, and no other MATRIX neighbors (Axiom 4). This abstract behavioral specification is sufficient to enable formation of small, stable CYSTS in simulated suspension culture (Fig. 4B), but fails to represent any of the other characteristics.

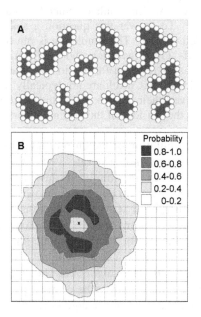

Fig. 6. CYST properties: *A*: examples of CYSTS formed in simulated embedded culture. The selection illustrates the range in CYST size and shape. *B*: the contour plots show the probability of CELL location, relative to CYST center (x), for a stable CYST formed in simulated embedded culture. The location of each CYST'S "center" was defined as the average grid location for the CELLS forming that CYST. Relative to each CYST'S center, the frequency of finding a cell at each grid location was determined for two hundred embedded culture simulations that each ran for 100 steps (all CYSTS stabilized). The contour plot shows that averaged over many simulations, CYSTS tend to be circular with one LUMEN.

There are three noteworthy differences between in vitro and in silico behaviors in simulated embedded culture. In vitro, there is occasional expansion of cysts [3] as consequence of division by cells that have already made contact with three environments. There is no corresponding behavior during simulations, because the operative mechanisms are below the level of resolution of this analogue. The second difference is the variability in CYST shapes. Some stable CYSTS are quite aspherical (wrinkled and puckered). Figure 6A contains a representative set. Such shapes are rarely observed in vitro. Positive pressure differentials are thought to exist between the inside and the outside of cysts in vitro. Consequently, any homogeneous luminal space would tend to become spherical. We can simulate the consequences of such pressure differences. A consequence would be that all of the simulated CYST cross-sections in Fig. 6A would be more circular. Because appropriate in vitro data is lacking, however, we elected not to include such effects. Nevertheless, as the contour plot in Fig. 6B shows, on average, the current in silico cysts tend to be circular. A third difference is that not all cells placed in embedded culture in vitro form lumen-filled cysts. A few masses have no lumens at all; plausible explanations have not been offered.

3.2 Experimentation and Exploration

Cells in epithelial tumors appear less polarized [17]. We simulated an aspect of loss of polarity by disrupting the directional CELL placement in Axiom 8 so that daughter CELLS could be placed in any FREE SPACE. The result was continued CELL proliferation resulting in the formation of amorphous combinations of CELLS, MATRIX, and FREE SPACE (Fig. 8), including de novo MATRIX production (a result of Axiom 4). That, combined with significant CELL removal due to Axiom 5, led to a proliferative, amorphous phenotypic attributes in all four environments reminiscent of cancerous growths. A comparison of growth rates in simulated embedded conditions for normal CELLS and the altered CELL behaviors demonstrates the failure of growth arrest in both cases (not shown).

Also of interest are observed behaviors that were not part of the original set of targeted attributes. Does the placement of an inverted CYST in simulated embedded culture result in the formation of a normal CYST? To answer this question a stable inverted CYST was placed in simulated embedded culture. The simulation is run for 50 simulation cycles and the outcomes observed. CYST inversion and stabilization does occur. However, proliferation and CELL death is required for this to take place. The in vitro evidence indicates that proliferation and CELL death are not typically required. The cells of the cyst apparently simply invert their polarity and in the process activate enzymes to digest the matrix that was on the "inside" of the original inverted suspension cyst [18,19].

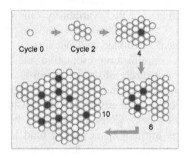

Fig. 7. Screen shots taken during exploratory simulation of the consequences of model changes: The consequences of liberalized matrix production: the sequence shows the effect of altered matrix production rules on cyst formation in simulated embedded culture. Cycle 0 is addition of one normal cell to a matrix-filled the grid. Alteration of Axioms 4 and 5 to allow for de novo matrix production in any environment consisting of free space and cell neighbors leads to a distinct pattern of unstable growth with matrix and free space production within clusters of cells. Cyst formation and growth arrest are not observed.

A limitation of the current model is that it does not generate a variety of CYST sizes in simulated suspension culture. As a strategy to get a variety, we liberalized MATRIX production: we replaced Axioms 4 and 5 with a axiom that allowed for MATRIX production in any environment consisting of FREE SPACE and CELL neighbors. The axiom change did not solve the original problem. There was no change in results in simulated suspension culture. However, it did result in the interesting and unexpected simulated behavior illustrated in Fig. 7: cellularization and MATRIX production in the center of growing clusters. Normal CYST formation and growth arrest failed to occur. This growth characteristic is distinct form the division-direction altered CELLS in Fig. 8. This axiom change, which can represent a mechanistic alteration caused by epigenetic phenomena, produces a hyper-proliferative, cancer-like response solely due to aberrant MATRIX production. No in vitro experiments have explored such behavior.

4   Discussion

We achieved our principal goal: to instantiate, in silico, an analogue of cell morphogenesis in vitro based on the sensing and production of three primary environment components. The analogues can generate structures that mimic those formed in in vitro conditions: stable cyst formation in embedded culture, inverted cyst formation in suspension culture, and lumen formation in overlay culture. Experimentation (e.g., Figs. 7 and 8) shows that if some axioms are relaxed, the targeted PCs are not achieved and instead cancer-like growth is observed. The results of one of our experiments presents an interesting hypothesis: that relaxation of conditions for de novo matrix production could lead to dysregulated growth of epithelial cells. Together these observations help build the case that simulation analogues such as this one can be useful in identifying new plausible epigenetic or genetic mechanisms that could contribute to cancer-like growth.

The relationship between an in silico analogue and its in vitro referent can become similar to the relationship between an in vitro model and its referent, typically a feature of a tissue, often within a patient. The in vitro model is obviously a simplified abstraction of the in vivo target. Similarly, our in silico model is a simplified abstraction of its in vitro referent. The in vitro and in vivo systems have their own unique PCs. Their two sets of experimentally measured properties are intended to overlap in specific, scientifically useful ways. The region of overlap is the validity of the in vitro analogue. The extent to which that overlap is complete is the accuracy of the analogues. There are also significant, possibly larger, non-overlapping regions. It is much easier to do experiments on the in vitro model rather than the in vivo referent. A similar relationship is feasible between in silico analogues and their in vitro referents. The experimental usefulness of such analogues will likely be bolstered by practical as well as ethical considerations.

Although not represented explicitly in the current analogue, polarization is implicit in Axi-

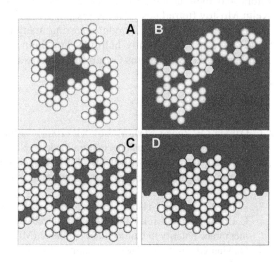

Fig. 8. Effects of changes in axioms on CELL growth properties. Shown are early cycle examples of the consequence of weakening the direction of CELL division in Axiom 8, as described in the text: simulated embedded culture (**A**), suspension culture (**B**), overlay culture (**C**), and surface culture (**D**). FREE SPACE is shaded dark. The light shaded hexagons are units of MATRIX.

oms 6–9. CELL actions are based not only on whether a particular environment component is present, but also on the relative locations of adjacent environment components. Cell polarization in vitro is presumed to be a feature of the larger mechanism governing cell action. Polarization in vitro likely represents a different cell class, enabling a different response to a given environmental perturbation. A new analogue that includes a separately validated class representing polarized cells can be validated against the current, acceptable analogue, as well as against an expanded set of targeted attributes.

The ability to produce and observe many different PCs from a relatively simple model make it clear that more needs to be done to systematically expand the *variety* of system level observations made on the experimental in vitro systems. Currently available experimental data provides sparse and spotty coverage of the available in vitro behavior space and thus the in vitro phenotype, in part because attention is often focused on specific molecular details. A way of selecting new wetlab experiments may be to give a degree of priority to those that may invalidate, or not, the current best *in silico* analogues.

Acknowledgments. This work was abstracted in part from material assembled by MRG for his Ph.D. dissertation. We are grateful for the funding provided by the CDH Research Foundation, and for the support provided by our cell biology collaborators and their groups: Profs. Thea Tlsty (tetrad.ucsf.edu/faculty.php?ID=78) and Keith Mostov (tetrad.ucsf.edu/faculty.php?ID=45). Our dialogues with Glen Ropella, Hal Berman, and Nancy Dumont covering many technical, biological, and theoretical issues have been especially helpful. For their support helpful advice, we thank the members of the BioSystems Group (biosystems.ucsf.edu).

References

1. O'brien LE, Zegers MM, Mostov KE (2002) Opinion: Building epithelial architecture: insights from three-dimensional culture models. Nat Rev Mol Cell Biol 3: 531-537.
2. Hall HG, Farson DA, Bissel MJ (1982) Lumen formation by epithelial cell lines in response to collagen overlay: a morphogenetic model in culture. Proc Natl Acad Sci 79: 4672-4676.
3. Wang AZ, K. OG, Nelson WJ (1990) Steps in the morphogenesis of a polarized epithelium I. Uncoupling the roles of cell-cell and cell-substratum contact in establishing plasma membrane polarity in multicellular epithelial (MDCK) cysts. J Cell Sci 95: 137-151.
4. Ojakian GK, Ratcliffe DR, Schwimmer R (2001) Integrin regulation of cell-cell adhesion during epithelial tubule formation. J Cell Sci 114: 941-952.
5. Ropella GEP, Hunt CA, Nag DA (2005) Using heuristic models to bridge the gap between analytic and experimental models in biology. Spring Simulation Multiconference, The Society for Modeling and Simulation International, April 2-8, San Diego, CA.
6. Ropella GEP, Hunt CA, Sheikh-Bahaei S (2005b) Methodological considerations of heuristic modeling of biological systems. The 9th World Multi-Conference on Systemics, Cybernetics and Informatics, July 10-13, Orlando Fl.
7. Steels L, Brooks R (1995) The artificial life route to artificial intelligence: building embodied, situated agents. Hillsdale, NJ: Lawrence Earlbaum Associates.
8. Ermentrout GB, Edelstein-Keshet L (1993) Cellular automata approaches to biological modeling. J theor Biol 160: 97-133.
9. Taub M, Chuman L, Saier MHJ, Sato G (1979) Growth of madin-darby canine kdiney epithelial cell (MDCK) line in hormone-supplemented, serum-free medium. Proc Natl Acad Sci 76: 3338-3342.
10. Madin SH, Darby NB (1975) As catalogued (1958) in American Type Culture Collection Catalogue of Strains. 2.
11. Nelson CM, Jean RP, Tan JL, Liu WF, Sniadecki NJ, et al. (2005) Emergent Patterns of growth controlled by multicellular form and mechanics. Proc Natl Acad Sci 102: 11594-11599.

12. Schwimmer R, Ojakian GK (1995) The a2b1 integrin regulates collagen-mediated MDCK epithelial membrane remodeling and tube formation. Journal of Cell Science 108: 2487-2498.

13. Warren SL, Nelson WJ (1987) Nonmitogenic morphoregulatory action of pp60^{v-src} on multicellular epithelial structures. Molec Cell Biol 7: 1326-1337.

14. Wang AZ, Ojakian GK, Nelson WJ (1990) Steps in the morphogenesis of a polarized epithelium. J Cell Sci 95: 153-165.

15. Lin H-H, Yang T-P, Jian S-T, Yang H-Y, Tang M-J (1999) Bcl-2 overexpression prevents apoptosis-induced Madin-Darby canine kidney simple epithelial cyst formation. Kidney International 55: 168-178.

16. Nelson WJ (2003) Epithelial cell polarity from the outside looking in. News Physiol Sci 18: 143-146.

17. Thiery JP (2002) Epithelial-mesenchymal transitions in tumor progression. Nat Rev Cancer 10: 442-454.

18. Wodarz A (2000) Tumor suppressors: linking cell polarity and growth control. Curr Biol 10: R624-R626.

19. Chambard M, Verrier B, Gabrion J, Mauchamp J (1984) Polarity Reversal of inside-out Thyroid-Follicles Cultured within Collagen Gel - Reexpression of Specific Functions. Biology of the Cell 51: 315-326.

20. Tang M-J, Hu J-J, Lin H-H, Chiu W-T, Jiang S-T (1998) Collagen gel overlay induces apoptosis of polarized cells in culture: disoriented cell death? Am J Physiol 275: C921-31.

21. Meredith JEJ, Fazeli B, Schwartz MA (1993) The extracellular matrix as a cell survival factor. Mol Biol Cell 4: 953-961.

22. Hayashi T, Carthew RW (2004) Surface mechanics mediate pattern formation in the developing retina. Nature 431: 647-652.

A Numerical Aggregation Algorithm for the Enzyme-Catalyzed Substrate Conversion

Hauke Busch[1], Werner Sandmann[2], and Verena Wolf[3]

[1] German Cancer Research Center, D-69120 Heidelberg, Germany
h.busch@dkfz.de
[2] University of Bamberg, D-96045 Bamberg, Germany
werner.sandmann@wiai.uni-bamberg.de
[3] University of Mannheim, D-68131 Mannheim, Germany
wolf@informatik.uni-mannheim.de

Abstract. Computational models of biochemical systems are usually very large, and moreover, if reaction frequencies of different reaction types differ in orders of magnitude, models possess the mathematical property of stiffness, which renders system analysis difficult and often even impossible with traditional methods. Recently, an accelerated stochastic simulation technique based on a system partitioning, the slow-scale stochastic simulation algorithm, has been applied to the enzyme-catalyzed substrate conversion to circumvent the inefficiency of standard stochastic simulation in the presence of stiffness. We propose a numerical algorithm based on a similar partitioning but without resorting to simulation. The algorithm exploits the connection to continuous-time Markov chains and decomposes the overall problem to significantly smaller sub-problems that become tractable. Numerical results show enormous efficiency improvements relative to accelerated stochastic simulation.

Keywords: Biochemical Reactions, Stochastic Model, Markov Chain, Aggregation.

1 Introduction

The complexity of living systems has led to a rapidly increasing interest in modeling and analysis of biochemically reacting systems. Different types of computational mathematical models exist, where quantitative and temporal relationships are often given in terms of *rates* and the specific meaning of these rates depends on the chosen model type. Of course, the different model types are intimately related since they represent the same type of system. A comprehensive treatment of computational models can be found in [2].

Models, not only in the context of biochemical systems, are distinguished in terms of their states and state changes (*transitions*) where a state consists of a collection of variables that sufficiently well represents the relevant[1] parameters

---

[1] Any model is a simplified abstraction of the real system and both suitability of a model and the relevant parameters depend on the scope of the study.

C. Priami (Ed.): CMSB 2006, LNBI 4210, pp. 298–311, 2006.
© Springer-Verlag Berlin Heidelberg 2006

of the original system at any time. The set of all states, also referred to as the *state space*, may be either discrete, meaning only a countable number of states that can be mapped to a subset of the natural numbers $\mathbb{N}$, or the state space may be continuous. In both discrete and continuous state space models the state transitions may occur deterministically or stochastically.

For a long time the model type of choice for biochemically reacting systems was a deterministic model with continuous state space, based on the law of mass action and expressed in terms of the *chemical rate equations* leading to a system of nonlinear ordinary differential equations (ODE) that often turns out to be difficult to solve. The stochastic approach, motivated by the observation that biochemical reactions occur randomly, leads to a system of partial differential equations, the *chemical master equation* (CME).

Since direct solution of the CME is often analytically intractable, stochastic simulation is in widespread use to analyze biochemically reacting systems. In particular, Gillespie's stochastic simulation algorithm [12,13] and its enhancements [11,8] that are slightly modified implementations are very popular. The algorithm basically consists of generating exponentially distributed times between successive reactions and drawing uniformly distributed numbers from the unit interval, the latter to decide which type of reaction occurs next. In that way the temporal evolution of the system is imitated by simulating an associated discrete-state Markov process or – in other words – an associated continuous-time Markov chain [3,10,14]. Stochastic simulation of continuous-time Markov chains is well known at the latest since the early 1960s as indicated by [10,17] and the references therein.

Although the CME arises from a stochastic model there is no need to apply stochastic solution methods. In particular there is a significant difference between a stochastic model and a stochastic simulation, although in the systems biology literature "the stochastic approach" and "the stochastic simulation algorithm" are often taken as the same thing. To open access to a wider range of analysis methodologies, it is important and useful to realize and exploit the link between biochemical reactions and Markov processes. Computational probability [16] has spent much effort to solve Markov processes analytically/numerically without resorting to stochastic simulation. In particular in computer systems performance analysis Markov chains with extremely large state spaces arise very often, and "numerical solution of Markov chains" is a vital research area [23,24].

A major drawback of stochastic simulation is the random nature of simulation results. Despite the fact that Gillespie's algorithm is termed exact, a stochastic simulation can never be exact. Mathematically, it constitutes a statistical estimation procedure implying that the results are subject to statistical uncertainty and in order to draw meaningful conclusions it is necessary to make statistically valid statements on the results. The exactness of Gillespie's algorithm is only "in the sense that it takes full account of the fluctuations and correlations" [13] of reactions within a single simulation run. It is common sense in stochastic simulation theory and practice [20] that one should never rely on a single simulation run and Gillespie mentioned that it is "necessary to make several simulation runs from time 0 to the chosen time t, all identical with each other except for the initialization of the random number generator". In fact the reliability of simulation results strongly depends on a sufficiently large number of simulation runs, and a proper determination of that number has to be carefully done in terms of mathematical statistics (cf. [17,20]).

Furthermore, stochastic simulation is inherently costly. In many cases even a single simulation run is extremely computer time demanding and thus reducing the space complexity compared to numerical methods has to be paid by a significant increase of time complexity. Therefore, often approximations, as for example the explicit τ-leaping method [15], are required to achieve simulation speed up. As an immediate consequence even the exactness in the sense stated above gets lost. Serious difficulties arise, both for deterministic and stochastic models, in the presence of multiple time scales or stiffness. Several approximate stochastic simulation algorithms such as the implicit τ-leap method [22], the slow-scale stochastic simulation algorithm [5] and the multiscale stochastic simulation algorithm [6] have been proposed to deal with these specific problems. As a representative stiff reaction set, we consider the enzyme-catalyzed substrate conversion

$$S_1 + S_2 \underset{c_2}{\overset{c_1}{\rightleftharpoons}} S_3 \xrightarrow{c_3} S_1 + S_4 \tag{1}$$

of a *substrate* S_2 into a *product* S_4 via an *enzyme-substrate complex* S_3, *catalyzed* (accelerated) by an enzyme S_1. Stiffness and different time scales arise, if the reversible reaction is much faster than the irreversible one. This is expressed by the condition $c_2 \gg c_3$ on the *stochastic reaction rate constants* (see 2.1 for details). Approximate stochastic simulation algorithms for (1) have been recently proposed in [21] and [7]. Both approaches are closely related in that they are based on the idea of partitioning the system and solving subproblems by different simulation techniques. A similar idea also appeared in [18]. Techniques based on partitioning the system are often also referred to as *aggregation techniques*.

As outlined above a clear disadvantage of stochastic simulation compared to numerical analysis, provided that such an analysis would be possible, is the random nature of simulation results. Thus, we argue that if a problem may be tackled both by stochastic simulation and by numerical analysis, the latter should be preferred. We propose an aggregation technique based on a partitioning similar to that in [21] and [7] but without resorting to simulation. Instead, in our method all resulting subproblems become tractable and are solved numerically. Since the simulation methods mentioned above are approximations, they obviously have two sources of inaccuracy, the approximation error due to the partitioning and the inherent statistical uncertainty of stochastic simulation, whereas our method only has the approximation error.

The basic ingredients of our method are the continuous-time Markov chain interpretation as an abstraction from the original system under consideration and the specific aggregation of states and transitions. We thereby revisit ideas from the analysis of fault-tolerant computer systems [1] and we appropriately modify these ideas according to our requirements. The remainder of this paper is organized as follows. In section 2 we formally describe our model and discuss solution approaches. The numerical aggregation algorithm (NAA) is derived in section 3, and its accuracy and efficiency are demonstrated in section 4. Finally, section 5 concludes the paper and gives directions of further research.

2   Mathematical Model

The general stochastic framework for biochemical systems leading to the CME has been well known for a long time (cf. [12,13]). Here, we first establish and elucidate the intimate connection to continuous-time Markov chains (CTMC) and we introduce our notations thereby focusing on system (1). Then a brief exposition of numerical solution methods and the arising problems is given with a particular emphasis on large and stiff systems.

2.1   Biochemical Reactions and Markov Chains

Let $X(t) = \big(X_1(t), X_2(t), X_3(t), X_4(t)\big)$ be a vector such that $X_i(t), i \in \{1, 2, 3, 4\}$ is a discrete random variable describing the number of molecules of species S_i at time instant t. If $X(t) = x := (x_1, x_2, x_3, x_4) \in \mathbb{N}^4$, the system is in state x at time t, meaning that for each S_i the current number of molecules is x_i. Assume, that initially the number of enzyme molecules is $x_1^{(0)}$ and for the substrate it is $x_2^{(0)}$ whereas no molecules of the enzyme-substrate complex or the product are present. For all possible states of the system: $x_1 + x_3 = x_1^{(0)}$ and $x_2 + x_4 = x_2^{(0)}$. Hence, the maximum numbers of molecules of S_1 and S_3 are $x_1^{(0)}$ and for S_2 and S_4 they are $x_2^{(0)}$. This implies a state space of size $n = (x_1^{(0)} + 1) \cdot (x_2^{(0)} + 1)$. Note that n is usually very large. For example, if $x_1^{(0)} = 200$ and $x_2^{(0)} = 3000$, it follows $n = 201 \cdot 3001 \approx 6 \cdot 10^5$. Here, the state space grows exponentially in the number of involved species. Moreover, the number of molecules of a species can be very large. Let $\mathcal{S} := \{(x_1, \ldots, x_4) : x_1 + x_3 = x_1^{(0)} \wedge x_2 + x_4 = x_2^{(0)}\}$ be the state space of $X(t)$.

System (1) consists of the three biochemical reactions

$$R_1 : \; S_1 + S_2 \xrightarrow{c_1} S_3, \qquad R_2 : \; S_3 \xrightarrow{c_2} S_1 + S_2, \qquad R_3 : \; S_3 \xrightarrow{c_3} S_1 + S_4,$$

where the stochastic interpretation is that the *reaction rates* (also called *transition rates*) are proportional to the number of participating molecules and to the stochastic reaction rate constants c_j, $j \in \{1, 2, 3\}$. For details and a rigorous formal justification see [12,14]. The *propensity function* that gives the transition rates of the reactions R_j is defined by $\lambda_1(x) = c_1 x_1 x_2, \lambda_2(x) = c_2 x_3, \lambda_3(x) = c_3 x_3$. Note that the c_j do not depend on the specific time t. The next state of the system only depends on x and the reaction type. If there exists a transition from state x to state x' with transition rate $q(x, x') \in \{\lambda_1(x), \lambda_2(x), \lambda_3(x)\}$ then we write $x \xrightarrow{q(x,x')} x'$. More precisely, for $x = (x_1, x_2, x_3, x_4)$

$R_1 : x \xrightarrow{c_1 x_1 x_2} (x_1 - 1, x_2 - 1, x_3 + 1, x_4)$, if $x_1, x_2 > 0$ and $x_3 < x_1^{(0)}$,

$R_2 : x \xrightarrow{c_2 x_3} (x_1 + 1, x_2 + 1, x_3 - 1, x_4)$,   if $x_1 < x_1^{(0)}, x_2 < x_2^{(0)}$ and $x_3 > 0$,

$R_3 : x \xrightarrow{c_3 x_3} (x_1 + 1, x_2, x_3 - 1, x_4 + 1)$,   if $x_1 < x_1^{(0)}, x_4 < x_2^{(0)}$ and $x_3 > 0$.

The probability of leaving x within a small time interval of length Δt via a reaction of type R_j is given by $\lambda_j(x)\Delta t$. Correspondingly, the probability of staying in x within this interval is given by $1 - \Lambda(x)\Delta t$ where $\Lambda(x) := \lambda_1(x) + \lambda_2(x) + \lambda_3(x)$ equals the sum of all outgoing rates of x and is called the *exit rate*. If $\Lambda(x)$ is small, state x is *slow* whereas otherwise x is a *fast* state which is due to the fact that $1/\Lambda(x)$ is the mean sojourn time in x. Let $p_t(x)$ be the probability that $X(t) = x$. Then

$$p_{t+\Delta t}(x) = (1 - \Lambda(x)\Delta t) \cdot p_t(x) + \sum_{x' : x \neq x', x' \xrightarrow{q(x',x)} x} q(x', x)\Delta t \cdot p_t(x').$$

This leads to the differential equations

$$\dot{p}_t = \frac{d}{dt} p_t = \lim_{\Delta t \to 0} \frac{p_{t+\Delta t} - p_t}{\Delta t} = Q p_t$$

where $p_t \in \mathbb{R}_{\geq 0}^n$ is the vector with entries $p_t(x)$ and $Q \in \mathbb{R}^{n \times n}$ is defined by[2]

$$Q(x, x') := \begin{cases} -\Lambda(x), & \text{if } x = x', \\ q(x, x'), & \text{if } x \xrightarrow{q(x,x')} x', \\ 0, & \text{otherwise.} \end{cases}$$

Process $X(t)$ is called a *(homogeneous) continuous-time Markov chain* (or CTMC for short). A CTMC is uniquely described by the *(infinitesimal) generator matrix* Q and an initial distribution (cf. [3,10]). In general the stochastic interpretation of chemical equations in the style of (1) always yields a CTMC as indicated in [12,13].

2.2   Numerical Solution of Markov Chains

Stochastic systems in general, and in particular Markov chains, are analyzed with respect to their temporal evolution where one distinguishes *transient* and *steady-state* analysis. The latter refers to systems in equilibrium whereas the former refers to the phase where an equilibrium has not yet been reached. A large amount of work exists on the numerical solution of Markov chains [24], where numerical solution means to compute probability distributions, either time-dependent transient distributions or steady-state distributions.

---

[2] We assume that the state space is mapped to $\mathbb{N}$.

As already stated, numerical analysis has lots of advantages over stochastic simulation. Unfortunately, the complexity of most real-life systems, such as biological or chemical systems and many more, leads to models with very large state spaces, a problem known as *state space explosion*. Direct numerical solution of such large models requires enormous computational effort and may be even impossible due to high space complexity.

Combating the problem of state space explosion has received much attention in computational probability and it turned out that sophisticated *abstraction techniques* can achieve dramatic computational speed up with accurate results. In this context, abstraction means to construct a less detailed model from the original one. One popular abstraction method is *aggregation* which is based on a partitioning of the state space. The result is a smaller aggregated model where each state corresponds to exactly one *aggregate*. The aggregated model is then analyzed yielding an approximation for the original one.

Numerical solution of large Markov chains is often inspired by models (such as queueing or Petri nets) arising in operations research or performance analysis of computer systems. Since in these areas, analysis most often aims at steady-state solutions, there are significantly fewer approaches to transient analysis [23], which moreover is more complicated. On the contrary, in systems biology transient solutions are usually more important than steady-state solutions since in many cases the latter do not give useful results. For example, if one considers the system given by (1) the expected number of molecules of the product S_4 at time t is of interest. In the steady-state, all molecules of the substrate S_2 are converted. Hence, steady-state analysis does not give new insights. The question of interest is how fast all molecules of the substrate are converted. But this also pertains to transient analysis.

Additional difficulties arise in case of stiffness, as already explained in the introduction. Many systems can be decomposed into several subsystems. For instance, (1) consists of the reactions R_1, R_2, R_3 and therefore of three subsystems. If the time a subsystem needs to reach equilibrium differs in orders of magnitude from the time until other subsystems reach equilibrium, meaning different time scales, i.e. $c_2 \gg c_3$, the system is called *stiff*. Both numerical analysis and stochastic simulation perform poorly in case of stiff systems.

3 Numerical Aggregation Algorithm

In the following we describe an efficient and extensible technique for the transient analysis of the stiff system given by (1). We obtain an aggregated model where numerical analysis is far less costly than for the original model. The basic idea of our numerical aggregation algorithm (NAA) is to consider a relatively small aggregated model for reaction R_3 based on an analysis of the submodels for R_1 and R_2. The decomposition into two different parts yields a significant speed up for the numerical computation of transient measures.

3.1   Aggregation

For aggregation techniques one has to define an appropriate partitioning of the state space of the CTMC. Here we focus on techniques where the aggregated model induced by the partitioning is again a CTMC. States of the original model are called *micro states* whereas states of the aggregated model are called *macro states* or *aggregates*. The aggregated model has to be considerably smaller than the original model to achieve computational advantages. On the other hand, the accuracy of the aggregation method strongly depends on the choice of the aggregates since each macro state approximates the behavior of its micro states.

Previous approaches of aggregation methods have in common that states belonging to one aggregate have "similar" or even "equal" performance properties. Many aggregation methods are based on *lumpability* [19], a structural property of Markov chains, where all states belonging to the same aggregate change with (nearly) the same transition rate to another aggregate. For example, in case of exact lumpability [4] the aggregated model is again a Markov chain and transition rates between macro states equal the uniquely determined transition rates between the micro states of the respective aggregates. The probability of being in a certain macro state in the aggregated model then equals the sum of the probabilities to be in any of the micro states that constitute this aggregate. For biochemical reactions this approach is in most cases not appropriate since transition rates depend linearly on the numbers of molecules of certain species but these numbers are state parameters. Therefore, aggregated models resulting from techniques based on lumpability may not be sufficiently small or may not yield accurate approximations.

Another approach is to partition the state space with respect to different speeds of states and to define the transition rates between macro states on the basis of an analysis of each aggregate considered in isolation. The aggregation methods in [9,1] decompose the state space into fast and slow subsets. The steady-state solution is computed for each fast subset and the transition rates between the macro states are then given by the rates of the original model weighted with the steady-state probabilities of the corresponding micro states. The stochastic simulations in [7,18,21] are based on a similar idea. The accuracy of these approximation techniques relies on the fact that within subsets of fast states equilibrium is reached much earlier than in subsets of slow states. This approach is also well suited in the context of (non-simulative) numerical solution of biochemically reacting systems.

Let us now draw our attention to the system given by (1). The exit rate $\Lambda(x) = x_1 \cdot x_2 \cdot c_1 + x_3 \cdot c_2 + x_3 \cdot c_3$ determines the sojourn time in state x and therefore the speed of x. Assume that $c_2 \gg c_3$ and $c_1 \gg c_3$. Then nearly all states in $\mathcal{S}$ are fast since they have either at least one fast transition via R_2 because $x_3 > 0$ or via R_1 because $x_1 \cdot x_2 \gg 1$. Hence there may be slow transitions between states, e.g. there can be a fast reaction and a corresponding slow reverse reaction as long as the states' speeds are of the same order of magnitude. Since $x_3 = 0$ implies $x_1 = x_1^{(0)}$ the only slow state in the system is $x = (x_1^{(0)}, 0, 0, x_2^{(0)})$, i.e. the *absorbing* state (with exit rate zero) where no substrate molecules are left and no further reactions are possible.

We decompose $\mathcal{S}$ such that the slow transitions of type R_3 occur only between different macro states and fast transitions of type R_1 and R_2 only exist within a macro state. This yields exactly one aggregate for each value $x_4 = k$, $0 \leq k \leq x_2^{(0)}$, and all micro states within an aggregate are fast except for the aggregate consisting of only one single element, i.e. the absorbing state. Hence, if the aggregates are considered as isolated submodels, steady-state is reached very fast. The main idea of the NAA is to define the slow transition rates of the aggregated CTMC by multiplying for each submodel c_3 with the expected number of enzyme-substrate complex molecules in equilibrium. This ensures that the aggregated model is slow compared to the fast submodels. Note that in general for an aggregated model it holds that transition rates of aggregates consisting of only one single element are always equal to the corresponding rates of the original model. Hence, the assumption that steady-state is reached fast within an aggregate has to be checked only for aggregates with more than one element for which the transition rates in the aggregated CTMC are approximations.

If we drop the assumption on c_1, i.e. if we allow $c_1 \approx c_3$, the states where no molecules of S_3 are in the system, are slow but belong to an aggregate with fast micro states. Although the NAA and the underlying partitioning are actually not designed for such systems, numerical results indicate that the algorithm is accurate even for $c_1 \approx c_3$.

For an extension of the NAA only fast micro states should be aggregated whereas each slow micro state forms a single macro state. Note that for this partitioning the aggregated Markov chain might be small but does not necessarily possess the simple structure as it is the case for the partitioning proposed above for system (1).

3.2   Analysis of the Fast Subsystems

Assume that the state space is partitioned as explained above for the case $c_2 \gg c_3$ and $c_1 \gg c_3$. Aggregate $\mathcal{A}_k$, $0 \leq k \leq x_2^{(0)}$ is defined as $\mathcal{A}_k = \{(x_1, x_2, x_3, x_4) \in \mathcal{S} : x_4 = k\}$. It is easy to see that $|\mathcal{A}_k| = \min\{x_1^{(0)}, x_2^{(0)} - k\} + 1$, because if initially $x_1^{(0)}$ enzyme molecules are present, each reaction of type R_1 moves the system to the next state within the aggregate where the number of molecules of S_3 is increased by one, and back to the previous state via R_2. If already most of the substrate molecules are converted, only $x_2^{(0)} - x_4$ substrate molecules are left for binding. The CTMC induced by $\mathcal{A}_k$ is derived by considering $\mathcal{A}_k$ as state space and transitions according to R_1 and R_2.

Example: Assume that $x_1^{(0)} = 20$ and $x_2^{(0)} = 300$. The Markov chain that corresponds to $\mathcal{A}_k$ ($0 \leq k \leq 280$) is illustrated in Figure 1. The nodes represent states with entries x_1, x_2, x_3, x_4 and the edges are labeled with transition rates.

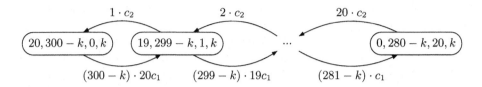

Fig. 1. Fast submodel $\mathcal{A}_k$

After the partitioning step we compute the average number of enzyme-substrate complex molecules under the assumption that $\mathcal{A}_k$ has reached equilibrium[3]. For any fixed k let p_i be the steady-state probability to be in state (x_1, x_2, x_3, x_4) of $\mathcal{A}_k$ with $x_3 = i$. The simple structure of $\mathcal{A}_k$ leads to the direct solution

$$p_i = \prod_{j=1}^{i} \frac{(x_2^{(0)} + 1 - k - j)(x_1^{(0)} + 1 - j)c_1}{jc_2} \cdot p_0 \quad \text{and} \quad \sum_{i=0}^{m_k} p_i = 1 \quad (2)$$

where $0 \leq k \leq x_2^{(0)}$ and $0 < i \leq \min\{x_1^{(0)}, x_2^{(0)} - k\} =: m_k$. More precisely, $\mathcal{A}_k$ is a *bounded birth-death-process* and the derivation of the steady-state probabilities that leads to (2) can be found in standard literature [3,10]. The expected number of enzyme-substrate complex molecules in $\mathcal{A}_k$ is then given by

$$\bar{x}_3^{(k)} = \sum_{i=0}^{m_k} i \cdot p_i. \quad (3)$$

For each aggregate $\mathcal{A}_k$, the value $\bar{x}_3^{(k)}$ can be calculated directly from (2) and (3) and determine the transition rates of the aggregated CTMC as described in the next section.

3.3   Analysis of the Slow Aggregated Model

The aggregated CTMC consists of $x_2^{(0)} + 1$ macro states (i.e. the aggregates $\mathcal{A}_k$). As illustrated in Figure 2, each macro state $\mathcal{A}_k$ is connected with its successor state $\mathcal{A}_{k+1}$ via a transition of type R_3. Hence, the aggregated Markov chain describes a sequence of exponentially distributed phases. For fixed k the expected sojourn time in $\mathcal{A}_k$ is $1/\bar{x}_3^{(k)}c_3$, the reciprocal of the exit rate. Therefore, the expected number of S_4 molecules at time instant t in the original model can be approximated by $E[X_4(t)] \approx \bar{x}_4(t)$ where $\bar{x}_4(t)$ is such that

$$\sum_{k=0}^{\bar{x}_4(t)-1} 1/(\bar{x}_3^{(k)}c_3) < t \leq \sum_{k=0}^{\bar{x}_4(t)} 1/(\bar{x}_3^{(k)}c_3)$$

---

[3] We use the notation $\mathcal{A}_k$ also as shorthand term for the CTMC associated with aggregate $\mathcal{A}_k$.

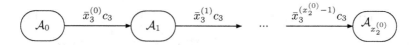

Fig. 2. The aggregated model

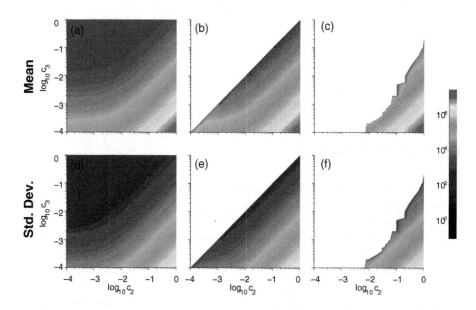

Fig. 3. Mean and standard deviation of the conversion time obtained by exact solution (left), NAA (middle), and ssSSA (right)

if $t > 1/(\bar{x}_3^{(0)}c_3)$ and $\bar{x}_4(t) = 0$ otherwise. In a similar way, an approximation for the variance of $X_4(t)$ is given by

$$\text{VAR}[X_4(t)] \approx \sum_{k=0}^{\bar{x}_4(t)} 1/(\bar{x}_3^{(k)}c_3)^2.$$

The approximations for $E[X_4(t)]$ and $\text{VAR}[X_4(t)]$ are the better the greater t. For small t, the assumption that the subsystems are already in equilibrium is not correct as opposed to the case where t is much greater than the average sojourn times in the micro states. As already explained, nearly all micro states are fast which means that their average sojourn times are small, maximum of the same order of magnitude as $1/\min\{c_1, c_2\}$. Hence, the approximation yields accurate results for $t \geq 1/\min\{c_1, c_2\}$.

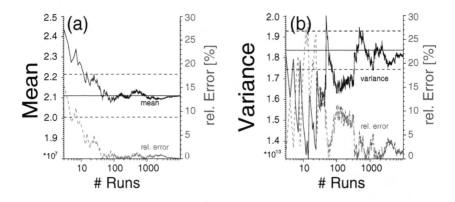

Fig. 4. Relative error of the ssSSA algorithm

4 Numerical Results

We extensively applied our numerical aggregation algorithm (NAA) to a variety of different parameter sets and system sizes. Here, we present some representative data to demonstrate the accuracy and efficiency of the algorithm.

Accuracy is best demonstrated by comparison to exact results. Since these can be only obtained for relatively small systems, for this purpose we choose the initial parameters $x_1^{(0)} = 30$ and $x_2^{(0)} = 300$, which means that we start with 300 substrate and 30 enzyme molecules, and the reaction is finished when all substrate molecules are converted. We also compare our algorithm with the slow-scale stochastic simulation algorithm (ssSSA) that has been recently applied to system (1) in [7]. Exact results and those yielded by NAA and ssSSA, resp., for the mean and the standard deviation of the time until all substrate molecules are converted are shown in Figure 3. Note that both the NAA and the ssSSA deal with stiff systems, i.e. $c_2 \gg c_3$, and thus for fixed $c_1 = 0.0001$ we vary c_2, c_3 in the appropriate range.

The exact results are computed by standard direct solution methods (cf. [3,10]) that are able to deal with such small systems. The ssSSA results for each data point are averaged over 500 simulation runs with different seeds for the random number generator. As illustrated in Figure 3 our NAA yields very accurate results even in parameter regions where $c_2 > c_3$ holds and the stiffness condition $c_2 \gg c_3$ does not.

That it is really necessary to make several simulation runs of the ssSSA and average them to an overall result can be seen by Figure 4 where the mean and the variance obtained by the ssSSA together with the corresponding relative errors are plotted against the number of simulation runs. It is illustrated that the relative errors are not strictly monotone decreasing with an increasing number of simulation runs, which is due to the random nature of stochastic simulation and the thereby implied statistical uncertainty. In fact it is well known that to

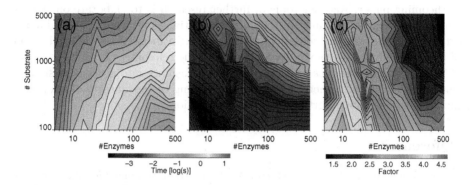

Fig. 5. Computer times for ssSSA (left), NAA (middle), and acceleration factor (right); scales in powers of 10

reduce the relative error of a stochastic simulation by a factor of ℓ, approximately ℓ^2 times as many simulation runs are required. For details see e.g. [20].

The efficiency of the NAA is demonstrated by means of run time comparisons. Here, we can include larger systems. Figure 5 shows the computer times needed by the ssSSA and the NAA for different numbers of substrate and enzyme molecules and the *acceleration factor*, i.e. the factor of computer time savings provided by the NAA compared to the ssSSA. The colored scales in Figure 5 are in terms of powers of 10 meaning that $-3, \ldots, 1$ and $1.5, \ldots, 4.5$ are the corresponding logarithms of the computer times and the acceleration factor, respectively. As can be seen, the NAA runs at least more than 10 times faster than the ssSSA and even up to more than 10^4 times faster in parameter regions where only a small number of enzyme molecules is present. Thus, the NAA provides significant efficiency improvements compared to the ssSSA.

5   Conclusion

We have presented the numerical aggregation algorithm (NAA), a novel approximate analysis method for very large stiff biochemically reacting systems that are neither efficiently tractable by standard numerical analysis techniques nor by direct stochastic simulation. The algorithm is based on the Markov chain interpretation and state space partitioning (aggregation) of the system. Compared to currently available accelerated approximate stochastic simulation algorithms the results obtained by the NAA are at least as accurate and besides do not possess any statistical uncertainty. In addition to eliminating statistical uncertainty while preserving accuracy, striking efficiency improvements by computer time savings up to orders of magnitudes are achieved by the NAA.

Accuracy and efficiency have been illustrated by numerical results for the stiff enzyme-catalyzed substrate conversion but the NAA is extensible to more general systems, which is part of ongoing work that will be dealt with in a forthcoming paper. Another topic of further research is to elaborate on the inherent statistical uncertainty of stochastic simulation that has not received much attention in the system biology literature so far. In fact, formal determination of the required number of simulation runs and the reliability of results in terms of mathematical statistics will give important insights into stochastic simulation and its drawbacks thereby further emphasizing the advantages of numerical methods that do not resort to stochastic simulation.

References

1. A. Bobbio and K. S. Trivedi (1986). An Aggregation Technique for the Transient Analysis of Stiff Markov Chains. *IEEE Trans. Comp.* C-35 (9), 803–814.
2. J. M. Bower and H. Bolouri, eds., (2001). *Computational Modeling of Genetic and Biochemical Networks.* The MIT Press.
3. P. Bremaud (1998). *Markov Chains.* Springer.
4. P. Buchholz (1994). Exact and Ordinary Lumpability in Finite Markov Chains. *Journal of Applied Probability* 31, 59–74.
5. Y. Cao, D. T. Gillespie, and L. R. Petzold (2005a). The Slow-Scale Stochastic Simulation Algorithm. *J. Chem. Phys.* 122, 014116.
6. Y. Cao, D. T. Gillespie, and L. R. Petzold (2005b). Multiscale Stochastic Simulation Algorithm with Stochastic Partial Equilibrium Assumption for Chemically Reacting Systems. *J. Comp. Phys.* 206 (2), 395–411.
7. Y. Cao, D. T. Gillespie, and L. R. Petzold (2005c). Accelerated Stochastic Simulation of the Stiff Enzyme-Substrate Reaction. *J. Chem. Phys.* 123 (14), 144917.
8. Y. Cao, H. Li, and L. R. Petzold (2004). Efficient Formulation of the Stochastic Simulation Algorithm for Chemically Reacting Systems. *J. Chem. Phys.* 121 (9), 4059–4067.
9. P. J. Courtois (1977). *Decomposability: Queueing and Computer System Applications.* Academic Press.
10. D. R. Cox and H. D. Miller (1965). *Theory of Stochastic Processes.* Chapman & Hall.
11. M. A. Gibson and J. Bruck (2000). Efficient Exact Stochastic Simulation of Chemical Systems with Many Species and Many Channels. *J. Phys. Chem.* A, 104, 1876–1889.
12. D. T. Gillespie (1976). A General Method for Numerically Simulating the Time Evolution of Coupled Chemical Reactions. *J. Comp. Phys.* 22, 403–434.
13. D. T. Gillespie (1977). Exact Stochastic Simulation of Coupled Chemical Reactions. *J. Phys. Chem.*, 81 (25), 2340–2361.
14. D. T. Gillespie (1992). *Markov Processes.* Academic Press.
15. D. T. Gillespie (2001). Approximate Accelerated Stochastic Simulation of Chemically Reacting Systems. *J. Chem. Phys.* 115 (4), 1716–1732.
16. W. Grassmann, editor (2000). Computational Probability. *Kluwer Academic Publishers*
17. J. M. Hammersley and D. C. Handscomb (1964). *Monte Carlo Methods.* Methuen.
18. E. L. Haseltine and J. B. Rawlings (2002). Approximate Simulation of Coupled Fast and Slow Reactions for Chemical Kinetics. *J. Chem. Phys.* 117, 6959–6969.
19. J. G. Kemeny and J. L. Snell (1960). *Finite Markov Chains.* Van Nostrand.
20. A. M. Law and W. D. Kelton (2000). *Simulation Modeling and Analysis.* 3rd ed., McGraw Hill.

21. C. V. Rao and A. P. Arkin (2003). Stochastic Chemical Kinetics and the Quasi-Steady-State Assumption: Application to the Gillespie Algorithm. *J. Chem. Phys.* 118, 4999–5010.

22. M. Rathinam, L. R. Petzold, Y. Cao, and D. T. Gillespie (2003). Stiffness in Stochastic Chemically Reacting Systems: The Implicit Tau-Leaping Method. *J. Chem. Phys.* 119, 12784–12794.

23. E. de Souza e Silva and H. R. Gail (2000). Transient Solutions for Markov Chains. Chapter 3 in W. K. Grassmann (ed.), *Computational Probability*, pp. 43–81. Kluwer Academic Publishers.

24. W. J. Stewart (1994). *Introduction to the Numerical Solution of Markov Chains.* Princeton University Press.

Possibilistic Approach to Biclustering: An Application to Oligonucleotide Microarray Data Analysis

Maurizio Filippone[1], Francesco Masulli[1], Stefano Rovetta[1],
Sushmita Mitra[2,3], and Haider Banka[2]

[1] DISI, Dept. Computer and Information Sciences, University of Genova and CNISM,
16146 Genova, Italy
{filippone, masulli, rovetta}@disi.unige.it
[2] Center for Soft Computing: A National Facility, Indian Statistical Institute,
Kolkata 700108, India
{hbanka_r, sushmita}@isical.ac.in
[3] Machine Intelligence Unit, Indian Statistical Institute,
Kolkata 700108, India

Abstract. The important research objective of identifying genes with similar behavior with respect to different conditions has recently been tackled with biclustering techniques. In this paper we introduce a new approach to the biclustering problem using the Possibilistic Clustering paradigm. The proposed Possibilistic Biclustering algorithm finds one bicluster at a time, assigning a membership to the bicluster for each gene and for each condition. The biclustering problem, in which one would maximize the size of the bicluster and minimizing the residual, is faced as the optimization of a proper functional. We applied the algorithm to the Yeast database, obtaining fast convergence and good quality solutions. We discuss the effects of parameter tuning and the sensitivity of the method to parameter values. Comparisons with other methods from the literature are also presented.

1 Introduction

1.1 The Biclustering Problem

In the last few years the analysis of genomic data from DNA microarray has attracted the attention of many researchers since the results can give a valuable information on the biological relevance of genes and correlations between them [1].

An important research objective consists in identifying genes with similar behavior with respect to different conditions. Recently this problem has been tackled with a class of techniques called *biclustering* [2,3,4,5].

Let x_{ij} be the expression level of the i-th gene in the j-th condition. A *bicluster* is defined as a subset of the $m \times n$ data matrix X. A bicluster [2,3,4,5] is a pair $(\mathbf{g}, \mathbf{c})$, where $\mathbf{g} \subset \{1, \ldots, m\}$ is a subset of genes and $\mathbf{c} \subset \{1, \ldots, n\}$ is a subset of conditions. We are interested in largest biclusters from DNA microarray data that do not exceed an assigned homogeneity constraint [2] as they can supply relevant biological information.

The size (or volume) n of a bicluster is usually defined as the number of cells in the gene expression matrix X belonging to it, that is the product of the cardinalities $n_g = |\mathbf{g}|$ and $n_c = |\mathbf{c}|$:

$$n = n_g \cdot n_c \tag{1}$$

C. Priami (Ed.): CMSB 2006, LNBI 4210, pp. 312–322, 2006.

Let

$$d_{ij}^2 = \frac{(x_{ij} + x_{IJ} - x_{iJ} - x_{Ij})^2}{n} \tag{2}$$

where the elements x_{IJ}, x_{iJ} and x_{Ij} are respectively the bicluster mean, the row mean and the column mean of X for the selected genes and conditions:

$$x_{IJ} = \frac{1}{n} \sum_{i \in \mathbf{g}} \sum_{j \in \mathbf{c}} x_{ij} \tag{3}$$

$$x_{iJ} = \frac{1}{n_c} \sum_{j \in \mathbf{c}} x_{ij} \tag{4}$$

$$x_{Ij} = \frac{1}{n_g} \sum_{i \in \mathbf{g}} x_{ij} \tag{5}$$

We can define now G as the mean square residual, a quantity that measures the bicluster homogeneity [2]:

$$G = \sum_{i \in \mathbf{g}} \sum_{j \in \mathbf{c}} d_{ij}^2 \tag{6}$$

The residual quantifies the difference between the actual value of an element x_{ij} and its expected value as predicted from the corresponding row mean, column mean, and bicluster mean.

To the aim of finding large biclusters we must perform an optimization that maximizes the bicluster cardinality n and at the same time minimizes the residual G, that is reported to be an NP-complete task [6]. The high complexity of this problem has motivated researchers to apply various approximation techniques to generate near optimal solutions. In the present work we take the approach to combine the criteria in a single objective function.

1.2   Overview of Previous Works

A survey on biclustering is given in [1] where a categorization of the different heuristic approaches is shown, such as iterative row and column clustering, divide and conquer strategy, greedy search, exhaustive biclustering enumeration, distribution parameter identification and others.

In the microarray analysis framework, the pioneering work by Cheng and Church [2] employs a set of greedy algorithms to find one or more biclusters in gene expression data, based on a mean squared residue as a measure of similarity. One bicluster is identified at a time iteratively. The masking of null values of the discovered biclusters are replaced by large random numbers that helps to find new biclusters at each iteration. Nodes are deleted and added and also the inclusion of inverted data is taken into consideration when finding biclusters. The masking procedure [7] results in a phenomenon of *random interference*, affecting the subsequent discovery of large-sized biclusters. A two-phase probabilistic algorithm termed Flexible Overlapped Clusters (FLOC) has been proposed by Yang et al. [7] to simultaneously discover a set of possibly overlapping biclusters. Initial biclusters are chosen randomly from the original data matrix.

Iteratively genes and/or conditions are added and/or deleted in order to achieve the best potential residue reduction. Bipartite graphs are also employed in [8], with a bicluster being defined as a subset of genes that jointly respond across a subset of conditions. The objective is to identify the maximum-weighted subgraph. Here a gene is considered to be responding under a condition if its expression level changes significantly, under that condition over the connecting edge, with respect to its normal level. This involves an exhaustive enumeration, with a restriction on the number of genes that can appear in the bicluster.

Other methods have been successfully employed in the Deterministic Biclustering with Frequent pattern mining algorithm (DBF) [9] to generate a set of good quality biclusters. Here concepts from the Data Mining practice are exploited. The changing tendency between two conditions is modeled as an item, with the genes corresponding to transactions. A frequent item-set with the supporting genes forms a bicluster. In the second phase, these are iteratively refined by adding more genes and/or conditions.

Genetic algorithms (GAs) have been employed by Mitra et al. [10] with local search strategy for identifying overlapped biclusters in gene expression data. In [11], a simulated annealing based biclustering algorithm has been proposed to provide improved performance over that of [2], escaping from local minima by means of a probabilistic acceptance of temporary worsening in fitness scores.

1.3   Outline of the Paper

In this paper we introduce a new approach to the biclustering problem using the possibilistic clustering paradigm [12]. The proposed Possibilistic Biclustering algorithm (PBC) finds one bicluster at a time, assigning a membership to the bicluster for each gene and for each condition. The membership model is of the fuzzy possibilistic type [12].

The paper is organized as follows: in section 2 the possibilistic paradigm is illustrated; section 3 presents the possibilistic approach to biclustering, and section 4 reports on experimental results. Section 5 is devoted to conclusions.

2   Possibilistic Clustering Paradigm

The central clustering paradigm is implemented in several algorithms including C-Means [13], Self Organizing Map [14] Fuzzy C-Means [15], Deterministic Annealing [16], Alternating Cluster Estimation [17], and many others. Often, central clustering algorithms impose a *probabilistic constraint*, according to which the sum of the membership values of a point in all the clusters must be equal to one. This competitive constraint allows the unsupervised learning algorithms to find the barycenter of fuzzy clusters, but the obtained evaluations of membership to clusters are not interpretable as a *degree of typicality*, and moreover can give sensibility to outliers, as isolated outliers can hold high membership values to some clusters, thus distorting the position of centroids.

The possibilistic approach to clustering proposed by Keller and Krishnapuram [12], [18] assumes that the membership function of a data point in a *fuzzy* set (or cluster) is absolute, i.e. it is an evaluation of a degree of typicality not depending on the membership values of the same point in other clusters.

Let $X = \{\mathbf{x}_1, \ldots, \mathbf{x}_r\}$ be a set of unlabeled data points, $Y = \{\mathbf{y}_1, \ldots, \mathbf{y}_s\}$ a set of cluster centers (or prototypes) and $U = [u_{pq}]$ the *fuzzy membership matrix*. In the Possibilistic C-Means (PCM) Algorithms the constraints on the elements of U are relaxed to:

$$u_{pq} \in [0,1] \quad \forall p, q; \tag{7}$$

$$0 < \sum_{q=1}^{r} u_{pq} < r \quad \forall p; \tag{8}$$

$$\bigvee_{p} u_{pq} > 0 \quad \forall q. \tag{9}$$

Roughly speaking, these requirements simply imply that cluster cannot be empty and each pattern must be assigned to at least one cluster. This turns a standard fuzzy clustering procedure into a mode seeking algorithm [12].

In [18], the objective function contains two terms, the first one is the objective function of the CM [13], while the second is a penalty (regularization) term considering the entropy of clusters as well as their overall membership values:

$$J_m(U, Y) = \sum_{p=1}^{s} \sum_{q=1}^{r} u_{pq} E_{pq} + \sum_{p=1}^{s} \frac{1}{\beta_p} \sum_{q=1}^{r} (u_{pq} \log u_{pq} - u_{pq}), \tag{10}$$

where $E_{pq} = \|\mathbf{x}_q - \mathbf{y}_p\|^2$ is the squared Euclidean distance, and the parameter β_p (that we can term *scale*) depends on the average size of the p-th cluster, and must be assigned before the clustering procedure. Thanks to the regularizing term, points with a high degree of typicality have high u_{pq} values, and points not very representative have low u_{pq} values in all the clusters. Note that if we take $\beta_p \to \infty \quad \forall p$ (i.e., the second term of $J_m(U, Y)$ is omitted), we obtain a trivial solution of the minimization of the remaining cost function (i.e., $u_{pq} = 0 \quad \forall p, q$), as no probabilistic constraint is assumed.

The pair (U, Y) minimizes J_m, under the constraints 7-9 only if [18]:

$$u_{pq} = e^{-E_{pq}/\beta_p} \quad \forall p, q, \tag{11}$$

and

$$\mathbf{y}_p = \frac{\sum_{q=1}^{r} \mathbf{x}_q u_{pq}}{\sum_{q=1}^{r} u_{pq}} \quad \forall p. \tag{12}$$

Those conditions for minimizing the cost function $J_m(U, Y)$. Eq.s 11 and 12 can be interpreted as formulas for recalculating the membership functions and the cluster centers (Picard iteration technique), as shown, e.g., in [19].

A good initialization of centroids must be performed before applying PCM (using, e.g., Fuzzy C-Means [12], [18], or Capture Effect Neural Network [19]). The PCM works as a refinement algorithm, allowing us to interpret the membership to clusters as cluster typicality degree, moreover PCM shows a high outliers rejection capability as it makes their membership very low.

Note that the lack of probabilistic constraints makes the PCM approach equivalent to a set of s independent estimation problems [20]:

$$(u_{pq}, \mathbf{y}) = \arg \bigwedge_{u_{pq}, \mathbf{y}} \left[\sum_{q=1}^{r} u_{pq} E_{pq} + \frac{1}{\beta_p} \sum_{q=1}^{r} (u_{pq} \log u_{pq} - u_{pq}) \right] \quad \forall p, \quad (13)$$

that can be solved independently one at a time through a Picard iteration of eq. 11 and eq. 12.

3   The Possibilistic Approach to Biclustering

In this section we generalize the concept of biclustering in a fuzzy set theoretical approach. For each bicluster we assign two vectors of membership, one for the rows and one other for the columns, denoting them respectively $\mathbf{a}$ and $\mathbf{b}$. In a crisp set framework row i and column j can either belong to the bicluster ($a_i = 1$ and $b_j = 1$) or not ($a_i = 0$ or $b_j = 0$). An element x_{ij} of X belongs to the bicluster if both $a_i = 1$ and $b_j = 1$, i.e., its membership u_{ij} to the bicluster is:

$$u_{ij} = \text{and}(a_i, b_j) \quad (14)$$

The cardinality of the bicluster is then defined as:

$$n = \sum_i \sum_j u_{ij} \quad (15)$$

A fuzzy formulation of the problem can help to better model the bicluster and also to improve the optimization process. In a fuzzy setting we allow membership u_{ij}, a_i and b_j to belong in the interval $[0, 1]$. The membership u_{ij} of a point to the bicluster can be obtained by an integration of row and column membership, for example by:

$$u_{ij} = a_i b_j \quad \text{(product)} \quad (16)$$

or

$$u_{ij} = \frac{a_i + b_j}{2} \quad \text{(average)} \quad (17)$$

The fuzzy cardinality of the bicluster is defined as the sum of the memberships u_{ij} for all i and j as in eq. 15. We can generalize eqs. 3 to 6 as follows:

$$d_{ij}^2 = \frac{(x_{ij} + x_{IJ} - x_{iJ} - x_{Ij})^2}{n} \quad (18)$$

where:

$$x_{IJ} = \frac{\sum_i \sum_j u_{ij} x_{ij}}{\sum_i \sum_j u_{ij}} \quad (19)$$

$$x_{iJ} = \frac{\sum_j u_{ij} x_{ij}}{\sum_j u_{ij}} \quad (20)$$

$$x_{Ij} = \frac{\sum_i u_{ij} x_{ij}}{\sum_i u_{ij}} \tag{21}$$

$$G = \sum_i \sum_j u_{ij} d_{ij}^2 \tag{22}$$

Then we can tackle the problem of maximizing the bicluster cardinality n and minimizing the residual G using the fuzzy possibilistic paradigm. To this aim we make the following assumptions:

- we treat one bicluster at a time;
- the fuzzy memberships a_i and b_j are interpreted as typicality degrees of gene i and condition j with respect to the bicluster;
- we compute the membership u_{ij} using eq. 17.

All those requirements are fulfilled by minimizing the following functional J_{B} with respect to **a** and **b**:

$$J_{\mathrm{B}} = \sum_i \sum_j \left(\frac{a_i + b_j}{2} \right) d_{ij}^2 + \lambda \sum_i (a_i \ln(a_i) - a_i) + \mu \sum_j (b_j \ln(b_j) - b_j) \tag{23}$$

The parameters λ and μ control the size of the bicluster by penalizing to small values of the memberships. Their value can be estimated by simple statistics over the training set, and then hand-tuned to incorporate possible a-priori knowledge and to obtain the desired results.

Setting the derivatives of J_{B} with respect to the memberships a_i and b_j to zero:

$$\frac{\partial J}{\partial a_i} = \sum_j \frac{d_{ij}^2}{2} + \lambda \ln(a_i) = 0 \tag{24}$$

$$\frac{\partial J}{\partial b_j} = \sum_i \frac{d_{ij}^2}{2} + \mu \ln(b_j) = 0 \tag{25}$$

we obtain these solutions:

$$a_i = \exp \left(-\frac{\sum_j d_{ij}^2}{2\lambda} \right) \tag{26}$$

$$b_j = \exp \left(-\frac{\sum_i d_{ij}^2}{2\mu} \right) \tag{27}$$

As in the case of standard PCM those necessary conditions for the minimization of J_{B} together with the definition of d_{ij}^2 (eq. 18) can be used by an algorithm able to find a numerical solution for the optimization problem (Picard iteration). The algorithm, that we call Possibilistic Biclustering (PBC), is shown in table 1.

The parameter ε is a threshold controlling the convergence of the algorithm. The memberships initialization can be made randomly or using some a priori information about relevant genes and conditions. Moreover, the PBC algorithm can be used as a refinement step for other algorithms using as initialization the results already obtained from them.

After convergence of the algorithm the memberships a and b can be defuzzified by comparing with a threshold (e.g. 0.5). In this way the results obtained with PBC can be compared with those of other techniques.

4   Results

4.1   Experimental Validation

We applied our algorithm to the *Yeast* database which is a genomic database composed by 2884 genes and 17 conditions[1] [21] [22] [23]. We removed from the database all genes having missing expression levels for all the conditions, obtaining a set of 2879 genes.

We performed many runs varying the parameters λ and μ and considering a thresholding for the memberships a and b of 0.5 for the defuzzification. In figure 1 the effect of the choice of these two parameters on the size of the bicluster can be observed. Increasing them results in a larger bicluster.

In figure 1 each result corresponds to the average on 20 runs of the algorithm. Note that, even if the memberships are initialized randomly, starting from the same set of parameters, it is possible to achieve almost the same results. Thus PBC is slightly sensitive to initialization of memberships while strongly sensitive to parameters λ and μ. The parameter ε can be set considering the desired precision on the final memberships. Here it has been set to 10^{-2}.

In table 2 a set of obtained biclusters is shown with the achieved values of G. In particular it is very interesting the ability of PBC to find biclusters of a desired size just tuning the parameters λ and μ. A plot of a small and a large biclusters can be found in fig. 2.

The PBC algorithm has been written in C and R language [24], and run on a Pentium IV 1900 MHz personal computer with $512 Mbytes$ of ram under a Linux operating system. The running time for each set of parameters was $7.5s$, showing that the complexity of the algorithm depends only on the size of the data set.

---

[1] http://arep.med.harvard.edu/biclustering/yeast.matrix

Table 1. Possibilistic Biclustering (PBC) algorithm

---

1. Initialize the memberships a and b
2. Compute d_{ij}^2 $\forall i, j$ using eq. 18
3. Update a_i $\forall i$ using eq. 26
4. Update b_j $\forall j$ using eq. 27
5. **if** $\|a' - a\| < \varepsilon$ and $\|b' - b\| < \varepsilon$ then **stop**
6. **else jump** to step 2

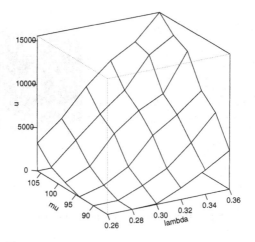

Fig. 1. Size of the biclusters vs. parameters λ and μ

Table 2. Comparison of the biclusters obtained by our algorithms on yeast data. The G value, the number of genes n_g, the number of conditions n_c, the cardinality of the bicluster n are shown with respect to the parameters λ and μ.

λ	μ	n_g	n_c	n	G
0.25	115	448	10	4480	56.07
0.19	200	457	16	7312	67.80
0.30	100	654	8	5232	82.20
0.32	100	840	9	7560	111.63
0.26	150	806	15	12090	130.79
0.31	120	989	13	12857	146.89
0.34	120	1177	13	15301	181.57
0.37	110	1309	13	17017	207.20
0.39	110	1422	13	18486	230.28
0.42	100	1500	13	19500	245.50
0.45	95	1622	12	19464	260.25
0.45	95	1629	13	21177	272.43
0.46	95	1681	13	21853	285.00
0.47	95	1737	13	22581	297.40
0.48	95	1797	13	23361	310.72

4.2   Comparative Study

Table 3 lists a comparison of results on *Yeast* data, involving performance of other, related biclustering algorithms with a $\delta = 300$ (δ is the maximum allowable residual for G). The deterministic DBF [9] discovers 100 biclusters, with half of these lying in the size range 2000 to 3000, and a maximum size of 4000. FLOC [7] uses a probabilistic approach to find biclusters of limited size, that is again dependent on the initial choice of random seeds. FLOC is able to locate large biclusters. However DBF generates a lower mean squared residue, which is indicative of increased similarity between genes in the

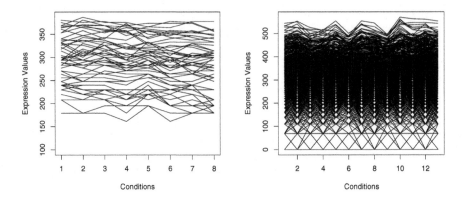

Fig. 2. Plot of a small and a large bicluster

biclusters. Both these methods report an improvement over the pioneering algorithm by Cheng et al. [2], considering mean squared residue as well as bicluster size.

Single-objective GA with local search has also been used [25], to generate considerably overlapped biclusters .

Table 3. Comparative study on *Yeast* data

Method	avg. G	avg. n	avg. n_g	avg. n_c	Largest n
DBF [9]	115	1627	188	11	4000
FLOC [7]	188	1826	195	12.8	2000
Cheng-Church [2]	204	1577	167	12	4485
Single-objective GA [10]	52.9	571	191	5.13	1408
Multi-objective GA [10]	235	10302	1095	9.29	14828
Possibilistic Biclustering	297	22571	1736	13	22607

The average results reported in table 3 concerning the Possibilistic Biclustering algorithm have been obtained involving 20 runs over the same set of parameters λ and μ. The biclusters obtained where very similar, obtaining G close to $\delta = 300$ for all of them and the achieved bicluster size is on average very high. From table 3, we see that the Possibilistic Approach has better performances in finding large biclusters in comparison with others methods.

5   Conclusions

In this paper we proposed the PBC algorithm, a new approach to biclustering based on the possibilistic paradigm. The problem of minimizing the residual G and maximize the size n, has been tackled by optimizing a functional which takes into account these requirements. The proposed method allows to find one bicluster at a time of the desired size.

The results show the ability of the PBC algorithm to find biclusters with low residuals. The quality of the large biclusters obtained is better in comparison with other biclustering methods.

The method will be the subject of further study. In particular, several criteria for automatically selecting the parameters λ and μ can be proposed, and different ways to combine a_i and b_j into u_{ij} can be discussed. Moreover, a biological validation of the obtained results is under study.

Acknowledgment

Work funded by the Italian Ministry of Education, University and Research (2004 "Research Projects of Major National Interest", code 2004062740).

References

1. Madeira, S.C., Oliveira, A.L.: Biclustering algorithms for biological data analysis: A survey. IEEE Transactions on Computational Biology and Bioinformatics **1** (2004) 24–45
2. Cheng, Y., Church, G.M.: Biclustering of expression data. In: Proceedings of the Eighth International Conference on Intelligent Systems for Molecular Biology, AAAI Press (2000) 93–103
3. Hartigan, J.A.: Direct clustering of a data matrix. Journal of American Statistical Association **67(337)** (1972) 123–129
4. Kung, S.Y., Mak, M.W., Tagkopoulos, I.: Multi-metric and multi-substructure biclustering analysis for gene expression data. Proceedings of the 2005 IEEE Computational Systems Bioinformatics Conference (CSB'05) (2005)
5. Turner, H., Bailey, T., Krzanowski, W.: Improved biclustering of microarray data demonstrated through systematic performance tests. Computational Statistics and Data Analysis **48**(2) (2005) 235–254
6. Peeters, R.: The maximum edge biclique problem is NP-Complete. Discrete Applied Mathematics **131** (2003) 651–654
7. Yang, J., Wang, H., Wang, W., Yu, P.: Enhanced biclustering on expression data. In: Proceedings of the Third IEEE Symposium on BioInformatics and Bioengineering (BIBE'03). (2003) 1–7
8. Tanay, A., Sharan, R., Shamir, R.: Discovering statistically significant biclusters in gene expression data. Bioinformatics **18** (2002) S136–S144
9. Zhang, Z., Teo, A., Ooi, B.C.a.: Mining deterministic biclusters in gene expression data. In: Proceedings of the Fourth IEEE Symposium on Bioinformatics and Bioengineering (BIBE'04). (2004) 283–292
10. Mitra, S., Banka, H.: Multi-objective evolutionary biclustering of gene expression data. to appear (2006)
11. Bryan, K., Cunningham, P., Bolshakova, N.: Biclustering of expression data using simulated annealing. In: 18th IEEE Symposium on Computer-Based Medical Systems (CBMS 2005). (2005) 383–388
12. Krishnapuram, R., Keller, J.M.: A possibilistic approach to clustering. Fuzzy Systems, IEEE Transactions on **1**(2) (1993) 98–110
13. Duda, R.O., Hart, P.E.: Pattern Classification and Scene Analysis. Wiley (1973)
14. Kohonen, T.: Self-Organizing Maps. Springer-Verlag New York, Inc., Secaucus, NJ, USA (2001)

15. Bezdek, J.C.: Pattern Recognition with Fuzzy Objective Function Algorithms. Kluwer Academic Publishers, Norwell, MA, USA (1981)

16. Rose, K., Gurewwitz, E., Fox, G.: A deterministic annealing approach to clustering. Pattern Recogn. Lett. **11**(9) (1990) 589–594

17. Runkler, T.A., Bezdek, J.C.: Alternating cluster estimation: a new tool for clustering and function approximation. Fuzzy Systems, IEEE Transactions on **7**(4) (1999) 377–393

18. Krishnapuram, R., Keller, J.M.: The possibilistic c-means algorithm: insights and recommendations. Fuzzy Systems, IEEE Transactions on **4**(3) (1996) 385–393

19. Masulli, F., Schenone, A.: A fuzzy clustering based segmentation system as support to diagnosis in medical imaging. Artificial Intelligence in Medicine **16**(2) (1999) 129–147

20. Nasraoui, O., Krishnapuram, R.: Crisp interpretations of fuzzy and possibilistic clustering algorithms. Volume 3., Aachen, Germany (1995) 1312–1318

21. Tavazoie, S., Hughes, J.D., Campbell, M.J., Cho, R.J., Church, G.M.: Systematic determination of genetic network architecture. Nature Genetics **22**(3) (1999)

22. Ball, C.A., Dolinski, K., Dwight, S.S., Harris, M.A., Tarver, L.I., Kasarskis, A., Scafe, C.R., Sherlock, G., Binkley, G., Jin, H., Kaloper, M., Orr, S.D., Schroeder, M., Weng, S., Zhu, Y., Botstein, D., Cherry, M.J.: Integrating functional genomic information into the saccharomyces genome database. Nucleic Acids Research **28**(1) (2000) 77–80

23. Aach, J., Rindone, W., Church, G.: Systematic management and analysis of yeast gene expression data (2000)

24. R Foundation for Statistical Computing Vienna, Austria: R: A language and environment for statistical computing. (2005)

25. Bleuler, S., Prelić, A., Zitzler, E.: An EA framework for biclustering of gene expression data. In: Congress on Evolutionary Computation (CEC-2004), Piscataway, NJ, IEEE (2004) 166–173

Author Index

Bacci, G.   1
Bagnoli, F.   196
Banka, H.   312
Banks, R.   127
Bassetti, B.   227
Baxter, D. A.   242
Bockmayr, A.   169
Busch, H.   298
Busi, N.   17
Bussolino, F.   184
Byrne, J.H.   242

Calder, M.   63
Cataldo, E.   242
Cavaliere, M.   108
Cavalli, F.   184
Chettaoui, C.   257
Chiarugi, D.   93
Chinellato, M.   93
Coniglio A.   184

de Candia, A.   184
Degano, P.   93
Delaplace, F.   257
Di Talia, S.   184
Duguid, A.   63

Fages, F.   48
Filippone, M.   312

Gamba, A.   184
Gilmore, S.   63
Gössler, G.   212
Grant, M.R.   285

Heath, J.   32
Hillston, J.   63
Hunt, C.A.   285

Jona, P.   227

Knijnenburg, T.A.   271
Kwiatkowska, M.   32

Lagomarsino, M. Cosentino   227
Lawrence, Neil D.   155
Lescanne, P.   257
Liò, P.   196
Lo Brutto, G.   93

Marangoni, R.   93
Masulli, F.   312
Miculan, M.   1
Mitra, S.   312

Norman, G.   32

Parker, D.   32
Prandi, D.   78

Rattray, M.   155
Reinders, M.J.T.   271
Rovetta, S.   312

Sandmann, W.   298
Sanguinetti, G.   155
Sedwards, S.   108
Serini, G.   184
Sguanci, L.   196
Siebert, H.   169
Soliman, S.   48
Steggles, L.J.   127

Tiuryn, J.   142
Tymchyshyn, O.   32

Vestergaard, M.   257
Vestergaard, R.   257

Wessels, L.F.A.   271
Wilczyński, B.   142
Wipat, A.   127
Wolf, V.   298

Lecture Notes in Bioinformatics

Vol. 4216: M.R. Berthold, R. Glen, I. Fischer (Eds.), Computational Life Sciences II. XIII, 269 pages. 2006.

Vol. 4210: C. Priami (Ed.), Computational Methods in Systems Biology. X, 323 pages. 2006.

Vol. 4205: G. Bourque, N. El-Mabrouk (Eds.), Comparative Genomics. X, 231 pages. 2006.

Vol. 4175: P. Bücher, B.M.E. Moret (Eds.), Algorithms in Bioinformatics. XII, 402 pages. 2006.

Vol. 4146: J.C. Rajapakse, L. Wong, R. Acharya (Eds.), Pattern Recognition in Bioinformatics. XIV, 186 pages. 2006.

Vol. 4115: D.-S. Huang, K. Li, G.W. Irwin (Eds.), Computational Intelligence and Bioinformatics, Part III. XXI, 803 pages. 2006.

Vol. 4075: U. Leser, F. Naumann, B. Eckman (Eds.), Data Integration in the Life Sciences. XI, 298 pages. 2006.

Vol. 4070: C. Priami, X. Hu, Y. Pan, T.Y. Lin (Eds.), Transactions on Computational Systems Biology V. IX, 129 pages. 2006.

Vol. 3939: C. Priami, L. Cardelli, S. Emmott (Eds.), Transactions on Computational Systems Biology IV. VII, 141 pages. 2006.

Vol. 3916: J. Li, Q. Yang, A.-H. Tan (Eds.), Data Mining for Biomedical Applications. VIII, 155 pages. 2006.

Vol. 3909: A. Apostolico, C. Guerra, S. Istrail, P. Pevzner, M. Waterman (Eds.), Research in Computational Molecular Biology. XVII, 612 pages. 2006.

Vol. 3886: E.G. Bremer, J. Hakenberg, E.-H.(S.) Han, D. Berrar, W. Dubitzky (Eds.), Knowledge Discovery in Life Science Literature. XIV, 147 pages. 2006.

Vol. 3745: J.L. Oliveira, V. Maojo, F. Martín-Sánchez, A.S. Pereira (Eds.), Biological and Medical Data Analysis. XII, 422 pages. 2005.

Vol. 3737: C. Priami, E. Merelli, P. Gonzalez, A. Omicini (Eds.), Transactions on Computational Systems Biology III. VII, 169 pages. 2005.

Vol. 3695: M.R. Berthold, R. Glen, K. Diederichs, O. Kohlbacher, I. Fischer (Eds.), Computational Life Sciences. XI, 277 pages. 2005.

Vol. 3692: R. Casadio, G. Myers (Eds.), Algorithms in Bioinformatics. X, 436 pages. 2005.

Vol. 3680: C. Priami, A. Zelikovsky (Eds.), Transactions on Computational Systems Biology II. IX, 153 pages. 2005.

Vol. 3678: A. McLysaght, D.H. Huson (Eds.), Comparative Genomics. VIII, 167 pages. 2005.

Vol. 3615: B. Ludäscher, L. Raschid (Eds.), Data Integration in the Life Sciences. XII, 344 pages. 2005.

Vol. 3594: J.C. Setubal, S. Verjovski-Almeida (Eds.), Advances in Bioinformatics and Computational Biology. XIV, 258 pages. 2005.

Vol. 3500: S. Miyano, J. Mesirov, S. Kasif, S. Istrail, P. Pevzner, M. Waterman (Eds.), Research in Computational Molecular Biology. XVII, 632 pages. 2005.

Vol. 3388: J. Lagergren (Ed.), Comparative Genomics. VII, 133 pages. 2005.

Vol. 3380: C. Priami (Ed.), Transactions on Computational Systems Biology I. IX, 111 pages. 2005.

Vol. 3370: A. Konagaya, K. Satou (Eds.), Grid Computing in Life Science. X, 188 pages. 2005.

Vol. 3318: E. Eskin, C. Workman (Eds.), Regulatory Genomics. VII, 115 pages. 2005.

Vol. 3240: I. Jonassen, J. Kim (Eds.), Algorithms in Bioinformatics. IX, 476 pages. 2004.

Vol. 3082: V. Danos, V. Schachter (Eds.), Computational Methods in Systems Biology. IX, 280 pages. 2005.

Vol. 2994: E. Rahm (Ed.), Data Integration in the Life Sciences. X, 221 pages. 2004.

Vol. 2983: S. Istrail, M.S. Waterman, A. Clark (Eds.), Computational Methods for SNPs and Haplotype Inference. IX, 153 pages. 2004.

Vol. 2812: G. Benson, R.D. M. Page (Eds.), Algorithms in Bioinformatics. X, 528 pages. 2003.

Vol. 2666: C. Guerra, S. Istrail (Eds.), Mathematical Methods for Protein Structure Analysis and Design. XI, 157 pages. 2003.